Communications
in Computer and Information Science　　1466

Editorial Board Members

Joaquim Filipe ⓘ
Polytechnic Institute of Setúbal, Setúbal, Portugal

Ashish Ghosh
Indian Statistical Institute, Kolkata, India

Raquel Oliveira Prates ⓘ
Federal University of Minas Gerais (UFMG), Belo Horizonte, Brazil

Lizhu Zhou
Tsinghua University, Beijing, China

More information about this series at http://www.springer.com/series/7899

Bing Qin · Zhi Jin · Haofen Wang · Jeff Pan ·
Yongbin Liu · Bo An (Eds.)

Knowledge Graph and Semantic Computing

Knowledge Graph Empowers New Infrastructure Construction

6th China Conference, CCKS 2021
Guangzhou, China, November 4–7, 2021
Proceedings

 Springer

Editors
Bing Qin
Harbin Institute of Technology
Harbin, China

Haofen Wang
Tongji University
Shanghai, China

Yongbin Liu
University of South China
Hengyang, China

Zhi Jin 🆔
Peking University
Beijing, China

Jeff Pan
University of Edinburgh
Edinburgh, UK

Bo An 🆔
Chinese Academy of Sciences
Beijing, China

ISSN 1865-0929 ISSN 1865-0937 (electronic)
Communications in Computer and Information Science
ISBN 978-981-16-6470-0 ISBN 978-981-16-6471-7 (eBook)
https://doi.org/10.1007/978-981-16-6471-7

This Springer imprint is published by the registered company Springer Nature Singapore Pte Ltd.
The registered company address is: 152 Beach Road, #21-01/04 Gateway East, Singapore 189721, Singapore

Preface

This volume contains the papers presented at CCKS 2021: the China Conference on Knowledge Graph and Semantic Computing held during November 4–7, 2021, in Guangzhou, China.

CCKS is organized by the Technical Committee on Language and Knowledge Computing of the Chinese Information Processing Society, and represents the merger of two previously-held relevant forums, i.e., the Chinese Knowledge Graph Symposium (CKGS) and the Chinese Semantic Web and Web Science Conference (CSWS). CKGS was previously held in Beijing (2013), Nanjing (2014), and Yichang (2015). CSWS was first held in Beijing in 2006 and has been the main forum for research on Semantic (Web) technologies in China for a decade. Since 2016, CCKS has brought together researchers from both forums and covered a wider range of fields, including knowledge graphs, the Semantic Web, linked data, natural language processing, knowledge representation, graph databases, information retrieval, and knowledge aware machine learning. It aims to become the top forum on knowledge graph and semantic technologies for Chinese researchers and practitioners from academia, industry, and government.

The theme of this year's conference was "Knowledge Graph Empowers New Infrastructure Construction". Enclosing this theme, the conference scheduled various activities, including keynotes, academic workshops, industrial forums, evaluation and competition, knowledge graph summit reviews, presentation of academic papers, etc. The conference invited Jiawei Han (Michael Aiken Chair and Professor at the Department of Computer Science, University of Illinois at Urbana-Champaign), Jie Tang (Full Professor and the Associate Chair of the Department of Computer Science and Technology, Tsinghua University), and Ming Zhou (Chief Scientist at Sinovation Ventures and former president of the Association for Computational Linguistics) to present the latest progress and development trends in mining structured knowledge, complex reasoning and graph neural networks, and self-supervised learning, respectively. The conference also invited industrial practitioners to share their experience and promote industry-university-research cooperation.

As for peer-reviewed papers, 170 submissions were received in the following six areas:

– Knowledge Graph Representation and Reasoning
– Knowledge Acquisition and Knowledge Graph Construction
– Linked Data, Knowledge Integration, and Knowledge Graph Storage Management
– Natural Language Understanding and Semantic Computing
– Knowledge Graph Applications (Semantic Search, Question Answering, Dialogue, Decision Support, and Recommendation)
– Knowledge Graph Open Resources

During the reviewing process, each submission was assigned to at least three Program Committee members. The committee decided to accept 56 papers (28 in English, including 19 full papers and 9 short papers). This CCIS volume contains revised versions of the 28 English papers.

The hard work and close collaboration of a number of people have contributed to the success of this conference. We would like to thank the Organizing Committee and Program Committee members for their support, and the authors and participants who are the primary reason for the success of this conference. We also thank Springer for their trust and for publishing the proceedings of CCKS 2021.

Finally, we appreciate the sponsorships from EpiK Tech and Meituan as chief sponsors; Tencent Technology and Haizhi Xingtu Technology as diamond sponsors; Global Tone Communication Technology, Oppo, and PlantData as platinum sponsors; Xiaomi, Baidu, Yidu Cloud, Huawei, IFLYTEK, and Vesoft as gold sponsors; and Ant Group, Zhipu.ai, and Yunfu Technology as silver sponsors.

July 2021

Bing Qin
Zhi Jin
Haofen Wang
Jeff Pan
Yongbin Liu
Bo An

Organization

CCKS 2021 was organized by the Technical Committee on Language and Knowledge Computing of the Chinese Information Processing Society.

General Chairs

Bing Qin	Harbin Institute of Technology, China
Zhi Jin	Peking University, China

Program Committee Chairs

Haofen Wang	Tongji University, China
Jeff Pan	The University of Edinburgh, UK

Local Chairs

Shengyi Jiang	Guangdong University of Foreign Studies, China
Jianfeng Du	Guangdong University of Foreign Studies, China

Publicity Chairs

Zhixu Li	Fudan University, China
Saike He	Institute of Automation, Chinese Academy of Sciences, China

Publication Chairs

Yongbin Liu	University of South China, China
Bo An	Institute of Software, Chinese Academy of Sciences, China

Tutorial Chairs

Changliang Li	Kingsoft Office, China
Shizhu He	Institute of Automation, Chinese Academy of Sciences, China

Evaluation Chairs

Ming Liu	Harbin Institute of Technology, China
Jiangtao Zhang	PLA No. 305 Hospital, China

Top Coference Reviewing Chair

Zhichun Wang   Beijing Normal University, China

Young Scholar Forum Chairs

Bin Xu   Tsinghua University, China
Xiaoling Wang   East China Normal University, China

Poster/Demo Chairs

Xiaolong Jin   Institute of Computing Technology, Chinese Academy of
                        Sciences, China
Tianxing Wu   Southeast University, China

Sponsorship Chairs

Junyu Lin   Institute of Information Engineering, Chinese Academy of
                        Sciences, China
Lei Hou   Tsinghua University, China

Industry Track Chairs

Hao Chao   Vivo, China
Tong Ruan   East China University of Science and Technology, China

Website Chair

Xiao Ding   Harbin Institute of Technology, China

Area Chairs

Knowledge Graph Representation and Reasoning

Xiaowang Zhang   Tianjin University, China
Ningyu Zhang   Zhejiang University, China

Knowledge Acquisition and Knowledge Graph Construction

Qili Zhu   Shanghai Jiao Tong University, China
Yi Cai   South China University of Technology, China

Linked Data, Knowledge Integration, and Knowledge Graph Storage Management

Wei Hu   Nanjing University, China
Shengping Liu   Unisound, China

Natural Language Understanding and Semantic Computing

Xipeng Qiu   Fudan University, China
Baotian Hu   Harbin Institute of Technology, China

Knowledge Graph Applications (Semantic Search, Question Answering, Dialogue, Decision Support, and Recommendation)

Minlie Huang   Tsinghua University, China
Yao Meng   Lenovo, China

Knowledge Graph Open Resources

Meng Wang   Southeast University, China
Ningyu Zhang   Zhejiang University, China

Program Committee

Shuqing Bu   National Library of China, China
Yi Cai   South China University of Technology, China
Yixin Cao   National University of Singapore, Singapore
Hongxu Chen   The University of Queensland, Australia
Mingyang Chen   Zhejiang University, China
Jiaoyan Chen   University of Oxford, UK
Xiang Chen   Zhejiang University, China
Gong Cheng   Nanjing University, China
Shumin Deng   Zhejiang University, China
Jiwei Ding   Nanjing University, China
Bin Dong   Ricoh Software Research Center, China
Cuiyun Gao   The Chinese University of Hong Kong, China
Yuxia Geng   Zhejiang University, China
Shengrong Gong   Changshu Institute of Technology, China
Yuhang Guo   Beijing Institute of Technology, China
Hongqi Han   Institute of Scientific and Technical Information of China,
                            China
Ruifang He   Tianjin University, China
Wei Hu   Nanjing University, China
Baotian Hu   Harbin Institute of Technology, China
Minlie Huang   Tsinghua University, China
Seung-Won Hwang   Seoul National University, South Korea
Shanshan Jiang   Ricoh Software Research Center, China

Guoqiang Li	Shanghai Jiao Tong University, China
Weizhuo Li	Nanjing University of Posts and Telecommunications, China
Dongfang Li	Harbin Institute of Technology, China
Xutao Li	Harbin Institute of Technology, China
Piji Li	Tencent AI Lab, China
Huiying Li	Southeast University, China
Bohan Li	Nanjing University of Aeronautics and Astronautics, China
Luoqiu Li	Zhejiang University, China
Yuan-Fang Li	Monash University, Australia
Yang Li	Alibaba Group, China
Jing Li	The Hong Kong Polytechnic University, China
Yongbin Liu	University of South China, China
Shengping Liu	Unisound, China
Wenqiang Liu	Tencent, China
Xing Liu	Third Xiangya Hospital, Central South University, China
Xusheng Luo	Alibaba Group, China
Xinyu Ma	Southeast University, China
Yinglong Ma	North China Electric Power University, China
Yao Meng	Lenovo Research, China
Qingliang Miao	AISpeech, China
Youcheng Pan	Harbin Institute of Technology, China
Jeff Pan	The University of Edinburgh, UK
Liang Pang	Institute of Computing Technology, Chinese Academy of Sciences, China
Peng Peng	Hunan University, China
Xu Qin	Southeast University, China
Xipeng Qiu	Fudan University, China
Pengjie Ren	Shandong University, China
Minglun Ren	Hefei University of Technology, China
Wei Shen	Nankai University, China
Bi Sheng	Southeast University, China
Chuan Shi	Beijing University of Posts and Telecommunications, China
Kaisong Song	Alibaba Group, China
Zequn Sun	Nanjing University, China
Hai Wan	Sun Yat-sen University, China
Huaiyu Wan	Beijing Jiaotong University, China
Meng Wang	Southeast University, China
Senzhang Wang	Beihang University, China
Beilun Wang	Southeast University, China
Zhigang Wang	Tsinghua University, China
Ruijie Wang	University of Zurich, Switzerland
Haofen Wang	Tongji University, China
Peng Wang	Southeast University, China
Longyue Wang	Tencent AI Lab, China
Xing Wang	Tencent, China
Tao Wang	South China University of Technology, China

Sponsors

Chief Sponsor

Diamond Sponsors

Platinum Sponsors

Gold Sponsors

Silver Sponsors

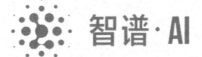

Contents

Knowledge Graph Open Resources

Knowledge Graph Representation and Reasoning

EBSD Grain Knowledge Graph Representation Learning for Material Structure-Property Prediction

Chao Shu, Zhuoran Xin, and Cheng Xie[✉]

School of Software, Yunnan University, Kunming 650504, China
xiecheng@ynu.edu.cn

Abstract. The microstructure is an essential part of materials, storing the genes of materials and having a decisive influence on materials' physical and chemical properties. The material genetic engineering program aims to establish the relationship between material composition/process, organization, and performance to realize the reverse design of materials, thereby accelerating the research and development of new materials. However, tissue analysis methods of materials science, such as metallographic analysis, XRD analysis, and EBSD analysis, cannot directly establish a complete quantitative relationship between tissue structure and performance. Therefore, this paper proposes a novel data-knowledge-driven organization representation and performance prediction method to obtain a quantitative structure-performance relationship. First, a knowledge graph based on EBSD is constructed to describe the material's mesoscopic microstructure. Then a graph representation learning network based on graph attention is constructed, and the EBSD organizational knowledge graph is input into the network to obtain graph-level feature embedding. Finally, the graph-level feature embedding is input to a graph feature mapping network to obtain the material's mechanical properties. The experimental results show that our method is superior to traditional machine learning and machine vision methods.

Keywords: Knowledge graph · EBSD · Graph neural network · Representation learning · Materials genome · Structure-property

1 Introduction

Material science research is a continuous understanding of the organization's evolution, and it is also a process of exploring the quantitative relationship between organizational structure and performance. In the past, the idea of material research was to adjust the composition and process to obtain target materials with ideal microstructure and performance matching. However, this method relies on a lot of experimentation and trial-error experience and is inefficient. Therefore, to speed up the research and development (R&D) of materials, the Material Genome Project [5] has been proposed in various countries.

© Springer Nature Singapore Pte Ltd. 2021
B. Qin et al. (Eds.): CCKS 2021, CCIS 1466, pp. 3–15, 2021.
https://doi.org/10.1007/978-981-16-6471-7_1

The idea of the Material Genome Project is to establish the internal connections between ingredients, processes, microstructures, and properties, and then design microstructures that meet the material performance requirements [5,9]. According to this connection, the composition and process of the material are designed and optimized. Therefore, establishing the quantitative relationship between material composition/process, organizational structure, and performance is the core issue of designing and optimizing materials.

At present, most tissue structure analysis is based on image analysis technology to extract specific geometric forms and optical density data [12]. However, the data obtained by this method is generally limited to the quantitative information about one-dimensional or two-dimensional images, and it is not easy to directly establish a quantitative relationship between tissue structure and material properties. The method has obvious limitations. In addition, current material microstructure analysis (e.g., metallographic analysis, XRD analysis, EBSD analysis) is often qualitative or partially quantitative and relies on manual experience [12]. It is still impossible to directly calculate material properties based on the overall organizational structure.

In response to the above problems, this paper proposes a novel data-driven [18] material performance prediction method based on the EBSD [4]. EBSD is currently one of the most effective material characterization methods. This characterization data not only contains structural information but is also easier for computers to understand. Therefore, we construct a digital knowledge graph [14] representation based on EBSD, then design a representation learning network to embed graph features. Finally, we use neural network [11] to predict material performance with graph embedding. We conducted experiments on magnesium metal and compared our method with traditional machine learning methods and computer vision methods. The results show the scientific validity of our proposed method and the feasibility of property calculation. The contribution of this page include:

1. We design an EBSD grain knowledge graph that can digitally represent the mesoscopic structural organization of materials.
2. We propose an EBSD representation learning method that can predict material's performance based on the EBSD organization representation.
3. We establish a database of structural performance calculations that expand the material gene database.

2   Related Work

2.1   Data-Driven Material Structure-Performance Prediction

Machine learning algorithms can obtain abstract features of data and mine the association rules behind the data. Machine learning algorithms have accelerated the transformation of materials R&D to the fourth paradigm (i.e., Data-driven R&D model). Machine learning is applied to material-aided design.

Ruho Kondo et al. used a lightweight VGG16 networks to predict the ionic conductivity in ceramics based on the microstructure picture [7].

Zhi-Lei Wang et al. developed a new machine learning tool, Material Genome Integrated System Phase and Property Analysis (MIPHA) [16]. They use neural networks to predict the stress-strain curves and mechanical properties based on constructed quantitative structural features. Pokuri et al. used deep convolutional neural networks to map microstructures to photovoltaic performance, and learn structure-attribute relationships of the data [10]. They designed a CNN-based model to extract the active layer morphology feature of thin-film OPVs and predict photovoltaic performance.

Machine learning methods based on numerical and visual features can detect the relationship between organization and performance. However, the microstructure of materials contains essential structural information and connection relationships, and learning methods based on descriptors and images will ignore this information.

2.2 Knowledge Graph Representation Learning

Knowledge Graph is an important data storage form in artificial intelligence technology. It forms a large amount of information into a form of graph structure close to human reasoning habits and provides a way for machines to understand the world better. Graph representation learning [3] gradually shows great potential.

In medicine, knowledge graphs are commonly used for embedding representations of drugs. The knowledge graph embedding method is used to learn the embedding representation of nodes directly and construct the relationship between drug entities. The constructed knowledge graphs can be used for downstream prediction tasks. Lin Xuan et al. propose a graph neural network based on knowledge graphs (KGNN) to solve the problem of predicting interactions in drug knowledge graphs [8].

Similarly, in the molecular field, knowledge graphs are used to characterize the structure of molecules/crystals [2,6,17]. Nodes can describe atoms, and edges can describe chemical bonds between atoms. The molecular or crystal structure is seen as an individual "graph". By constructing a molecular network map and applying graph representation learning methods, the properties of molecules can be predicted.

In the biological field, graphs are used for the structural characterization of proteins. The Partha Talukdar research group of the Indian Institute of Science did work on the quality assessment of protein models [13]. In this work, they used nodes to represent various non-hydrogen atoms in proteins. Edges connect the K nearest neighbors of each node atom. Edge distance, edge coordinates, and edge attributes are used as edge characteristics. After generating the protein map, they used GCN to learn atomic embedding. Finally, the non-linear network is used to predict the quality scores of atomic embedding and protein embedding.

Compared with the representation of descriptors and visual features, knowledge graphs can represent structural information and related information. The EBSD microstructure of the material contains important grain structure information and connection relationships. Therefore, this paper proposes the representation method of the knowledge graph and uses it for the prediction of organizational performance.

3   Representation of the EBSD Grain Knowledge Graph

In this part of the work, we construct a knowledge graph representation of the micro-organization structure. As shown in the Fig. 1, the left image is the scanning crystallographic data onto the sample, and the right is the Inverse Pole Figure map of the microstructure. The small squares in the Fig. 1(b) represents the grains. Based on this grain map data, we construct a grain knowledge graph representation. Because the size, grain boundary, and orientation of the crystal grains affect the macroscopic properties of the material, such as yield strength, tensile strength, melting point, and thermal conductivity [1]. Therefore, in this article, we choose the grain as the primary node in the map, and at the same time, we discretize the main common attributes of the grain size and orientation as the attribute node. Then, according to the grain boundaries of the crystal grains, we divided the two adjacent relationships between the crystal grains, namely, strong correlation and weak correlation. Finally, affiliation with grains and attribute nodes is established.

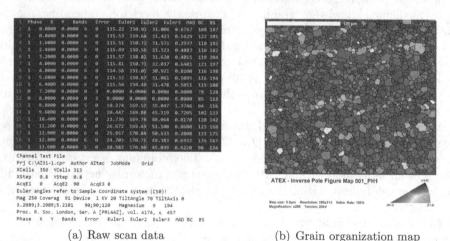

(a) Raw scan data   (b) Grain organization map

Fig. 1. EBSD scan organization information.

3.1   Nodes Representation

Grain Node. We segment each grain in the grain organization map and map it to the knowledge graph as a grain node. First, we use Atex software to count and segment all the grains in a grain organization map. Then We individually number each grain so that all the grains are uniquely identified, and finally, we build the corresponding nodes in the graph. As shown in the Fig. 2, the left side corresponds to the grains of the Fig. 1(b), and the right side are the node we want to build. The original grain corresponds to the grain node one-to-one. The grain node is the main node entity in the graph, reflecting the existence and distribution of the grain.

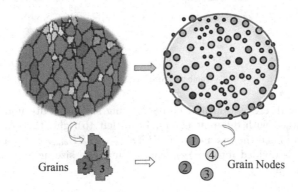

Fig. 2. The grain node corresponds to the original grain.

Grain Size Attribute Node. Next, we construct grain size attribute nodes used to discretize and identify the grain size. First, we discretize the size of the crystal grains. As shown in the Fig. 3, the color represents the difference in the size of the grains, $SIZE_{max}$ represents the largest-scale grain size, and $SIZE_{min}$ represents the smallest-scale grain size. The grain size levels are divided into N_{SIZE}, and the interval size of each level is $(SIZE_{max} - SIZ_{min})/N_{SIZE}$. We regard each interval as a category, as shown in the Eq. 1, for each grain, we divide it into corresponding category according to its size. We use the discretization category to represent the grain size instead of the original value. Then we construct a size attribute node for each size category, as shown in Fig. 3. Finally, we use the one-hot method to encode these N_{SIZE} categories and use the one-hot encoding as the feature of the size attribute node.

$$L\_S_{node} = \lceil Grain.size/\lceil (SIZE_{max} - SIZE_{min})/N_{SIZE}\rceil\rceil \qquad (1)$$

where L_S_{node} represents the size category of the grain. $Grain.size$ is the circle equivalent diameter of the grain, $\lceil\rceil$ means rounding up.

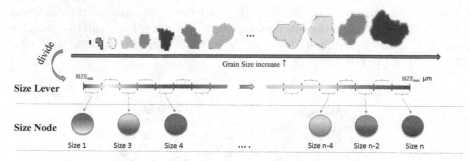

Fig. 3. Grain size discretization and corresponding size attribute nodes. The size of the grains is marked with different colors. From left to right, the grains are getting bigger and bigger. Then the size interval $[SIZE_{min}, SIZE_{max}]$ of the grains is found, and this interval is divided into N_{SIZE} parts. Finally, a size node is constructed for each divided interval.

Grain Orientation Attribute Node. In this work, Euler angles are used to identify the orientation of grains. Similarly, we also discretize the Euler angles, as shown in Fig. 4. The orientation of the grains is determined by the euler angles in three directions, so we discretize the euler angles in the three directions and combine them. The obtained three Euler angle interval combinations are the discretized types of orientation. Specifically as shown in the Eq. 2, we first calculate the maximum and minimum values of the three Euler angles $\phi(\phi1, \phi, \phi2)$ for all grains, namely $\phi_{max} = \{\phi1_{max}, \phi_{max}, \phi2_{max}\}$, $\phi_{min} = \{\phi1_{min}, \phi_{min}, \phi2_{min}\}$. Then each Euler angle $\phi(\phi1, \phi, \phi2)$ is divided into N_ϕ equal parts, the length of each part is $(\phi_{max} - \phi_{min})/N_\phi$. Finally, the N_ϕ equal parts of each Euler angle are cross-combined to obtain N_ϕ^3 combinations. We regard each combination as a kind of orientation, i.e., there are N_ϕ^3 orientation categories. For each grain, we can map it to one of N_ϕ^3 categories according to its three Euler angles $\phi(\phi1, \phi, \phi2)$, As shown in the Eq. 2. In this way, all crystal grains are divided into a certain type of orientation. We construct an orientation attribute node for each type of orientation to represent orientation information. Similarly, we use the one-hot method to encode these N_ϕ^3 categories individually. Each orientation category will be represented by a N_ϕ^3-dimensional one-hot vector used as the feature of the corresponding orientation attribute node.

$$L\_O_{node} = \{\lceil Grain.\phi1/\lceil(\phi1_{max} - \phi1_{min})/N_\phi\rceil\rceil,$$
$$\lceil Grain.\phi/\lceil(\phi_{max} - \phi_{min})/N_\phi\rceil\rceil, \tag{2}$$
$$\lceil Grain.\phi2/\lceil(\phi2_{max} - \phi2_{min})/N_\phi\rceil\rceil\}$$

where $Grain.\phi1$, $Grain.\phi$ and $Grain.\phi2$ are the euler angles in the three directions. L_O_{node} represents the orientation category to which the grains are classified. $\lceil \rceil$ refers to rounding up.

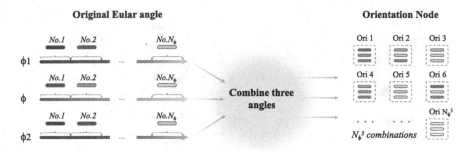

Fig. 4. Grain orientation discretization and corresponding orientation attribute nodes. On the left are three directions Euler angles, whose angles are represented by the RGB color. Each Euler angle is divided into N_ϕ parts, and then each equal part of each Euler angle is combined with one equal part of the remaining Euler angle. Each combination is regarded as an orientation category, and a node is constructed for this.

3.2 Edges Representation

After the construction of the node, the edges between the nodes need to be constructed. The nodes reflect the entities in the graph, and the edges contain the structural information of the graph. We build edges in the grain knowledge graph based on crystallographic knowledge. The constructed edge represents the association between nodes, including position association and property association. The edges between grain nodes reflect position information and grain boundaries; the edges between grain nodes and grain attribute nodes describe the properties of the grains.

Edge Between Grain Nodes. The contact interface between the grains is called the grain boundary, representing the transition of the atomic arrangement from one orientation to another. Generally speaking, grain boundaries have a significant impact on the various properties of the metal. In order to describe the boundary information of grains, we construct edges between grain nodes. First, we obtain the neighboring grains of each grain and construct the connection between the neighbor grain nodes. In order to further restore more complex grain spatial relationships, we set up a knowledge of neighboring rules. As shown in the Eq. 3, we use lp to represent the ratio of the bordering edge length of the grain to the total perimeter of the grain. Then we set a threshold λ, as shown in the Eq. 4, when the lp of grain A and grain B is greater than or equal to λ, we set the relationship between the A node and the B node to be a strong correlation; otherwise, it is set to weak correlation. Figure 5 shows the edge between grain nodes.

$$lp = bound\_length/perimeter \tag{3}$$

$$Rel\_G\_G(lp) = \begin{cases} \text{Strong association,} & lp < \lambda \\ \text{Weak association,} & lp \geq \lambda \end{cases} \tag{4}$$

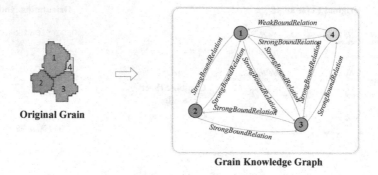

Grain Knowledge Graph

Fig. 5. Adjacent grains and the edges between them

Edge Between Grain Node and Size Attribute Node. In Sect. 3.1, we construct two types of attribute nodes. Here we associate the grain node with the attribute node to identify the property of the grain. First, we calculate the grain size category according to the Eq. 1, and then associate the corresponding grain node with the corresponding size attribute node to form the edge. As shown in the Fig. 6, we calculate the size categories {m, m, n, r} of the four grains {1, 2, 3, 4}, and then associate the corresponding grain node with the size attribute nodes to form the belonging relationship.

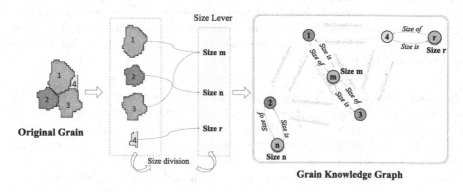

Grain Knowledge Graph

Fig. 6. Edge between grain node and size attribute node.

Edge Between Grain Node and Orientation Attribute Node. Similarly, we identify the orientation category for the grain node by associating it with the orientation attribute node. As shown in the Fig. 7, first we split the grains in the map, the bottom left of the picture shows the three Euler angles of the grains. Then we calculate the orientation category {i, j, k, l} of the grains {1, 2, 3, 4} according to Eq. 3. Finally, we associate the corresponding orientation attribute node with the corresponding grain node. Figure 7 shows the edges between the grain nodes and the orientation attribute nodes, reflecting the discrete orientation characteristics and orientation distribution of the grains.

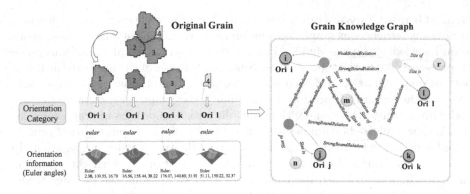

Fig. 7. Edge between grain node and orientation attribute node.

3.3   Grain Graph Convolutional Prediction Model

The structured grain knowledge graph can describe the microstructure of the material. Next, we build a graph feature convolution network (grain graph convolutional network) to embed the grain knowledge graph and realize graph feature extraction. Then, a feature mapping network based on a neural network is built to predict material properties with the graph feature. The complete model we built is shown in Fig. 8.

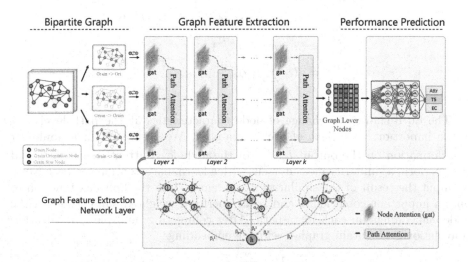

Fig. 8. Grain graph convolutional prediction model. The model is divided into graph feature extraction part and performance prediction part. The graph feature extraction network is composed of multiple node-level graph attention networks (gat) and a path-level attention aggregation network. The prediction network is a multilayer neural network. The graph feature network extracts graph-level features, and the prediction network maps graph-level features to material properties.

Grain Graph Convolutional Network. The graph features convolution network is a heterogeneous graph convolution network [15]. First, the heterogeneous grain knowledge graph is divided into multiple bipartite graphs and isomorphic graphs according to the type of edges. Next, the features of the nodes in the meta-path of subgraph are transferred and aggregated. Then the features of the same nodes of the subgraphs are fused, and finally, the graph-level characterization nodes are obtained through multiple convolutions. The process of graph convolution is shown at the bottom of Fig. 8. Specifically, the node aggregation process includes node-level feature aggregation and path-level feature aggregation. In node message transmission, we use node-level attention to learn the attention value of adjacent nodes on the meta-path. After completing the message transmission of all meta-paths, we use path-level attention to learn the attention value of the same nodes on different meta-paths. With the double-layer attention, the model can capture the influence factors of nodes and obtain the optimal combination of multiple meta-paths. Moreover, the nodes in the graph can better learn complex heterogeneous graphical and rich information. The Eq. 5 shows the feature aggregation transformation under the node-level attention. ι represents different paths/edges, and there are a total of p edges. α_{ij} represents the attention score between node i and node j. *LeakyReLU* and *Softmax* are activation functions, W is a learnable weight matrix, and $\vec{a}$ is a learnable weight vector. $\|$ represents concatenation, and $N(i)$ refers to all neighbor nodes of node i. h_i^{k+1} represents the $(k+1)$ layers embedding of node i.

$$\vec{z}_i^{k^{(\iota)}} = \boldsymbol{W}^{k^{(\iota)}} \cdot \vec{h}_i^{k^{(\iota)}}$$
$$e_{ij}^{k^{(\iota)}} = LeakyReLU(\vec{a}^{k^{(\iota)}} \cdot [\vec{z}_i^{k^{(\iota)}} \| \vec{z}_j^{k^{(\iota)}}])$$
$$\alpha_{ij}^{k^{(\iota)}} = Softmax_j(e_{ij}^{k^{(\iota)}}) \tag{5}$$
$$\vec{h}_i^{k+1^{(\iota)}} = \sigma(\sum_{j \in N(i)^{(\iota)}} \alpha_{ij}^{k^{(\iota)}} \cdot \vec{z}_i^{k^{(\iota)}})$$

The Eq. 6 shows the change of node characteristics at the path level, $\beta_{(\iota)}^k$ is the important coefficient of each meta-path. We first perform a nonlinear transformation on the output $\vec{h}_i^{k+1^{(\iota)}}$ of the node-level attention network, and then perform a similarity measurement with a learnable attention vector q. Next, we input the result of the similarity measurement into the Softmax function to obtain important coefficients, and finally perform weighted summation on the node embeddings on each meta-path. After completing multiple graph feature convolutions, we obtain graph-level node embeddings.

$$\beta_{(\iota)}^k = Softmax(\frac{1}{N(i)} \sum_{\iota \in N(i)} \vec{q} \cdot tanh(\boldsymbol{W}^k \cdot \vec{h}_i^{k+1^{(\iota)}} + \vec{b}))$$
$$\vec{h}_i^{k+1} = \sum_{\iota=1}^{p} \beta_{(\iota)}^k \cdot \vec{h}_i^{k+1^{(\iota)}} \tag{6}$$

Feature Mapping Network. The microstructure-performance relationship is usually qualitatively studied through statistical methods (e.g., statistics of grain size, orientation, and grain boundaries). The relationship between the microstructure and properties is difficult to obtain through comparative observation or direct calculation. However, Artificial neural networks can mine more essential characteristics of data and establish complex relationships between data [11]. Here, we have used the graph features convolution network to extract the features of the grain knowledge graph, so we use a feature mapping network based on a neural network to implement machine learning tasks. As shown in Eq. 7, $\vec{h}_i$ is the final graph-level node vector, fc is the mapping network. The network comprises a data normalization layer, a fully connected layer, an activation layer, and a random deactivation layer. Through this network, the feature of the grain knowledge graph can be mapped to the property of material.

$$prop = \frac{1}{n} \sum_{i=1}^{n} fc(\vec{h}_i) \tag{7}$$

4 Experiment

4.1 Dataset

The experimental data comes from the EBSD experimental data of 19 Mg metals and also includes the yield strength (ys), tensile strength (ts), and elongation (el) of the sample. The EBSD scan data contains a total of 4.46 million scan points. As a result, the number of nodes in all the constructed knowledge graphs is 40,265, and the number of edges reaches 389,210. We use EBSD knowledge graph representation as model input and mechanical properties as label.

4.2 Comparison Methods and Results

We design two different methods to compare with ours. They are traditional machine learning methods based on statistics, image feature extraction methods based on computer vision. We use traditional machine learning methods to directly calculate the attribute characteristics of all grains to obtain material properties. These methods include Ridge, SVR, KNN, ExtraTree. In addition, we use the pre-trained CNN model to directly learn visual features from the microstructure map and predict performance. The model is Resnet-50.

The model performance evaluation results are shown in Table 1. It can be seen that our method is superior to other methods. Our method obtained an R2 value of 0.74. This shows that our method can extract more effective features. Traditional machine learning methods and machine vision methods have obtained acceptable R2 values, which shows that both methods can obtain microstructure characteristics to a certain extent. However, compared with traditional machine learning, the method of machine vision does not show much superiority. It is because CNN training requires a larger amount of data, and our current data set is small.

Table 1. Results of model for ys prediction

Model	MSE	MAE	R2
Ridge	112.5	6.9	0.590
SVR	102.7	4.8	0.626
KNN	97.6	3.6	0.651
ExtraTree	105.9	5.6	0.610
Resnet50	94.8	**3.1**	0.667
Hetero_GAT (our)	**73.1**	5.9	**0.74**

5   Conclusion

This paper proposes a novel material organization representation and performance calculation method. First, we use the knowledge graph to construct the EBSD representation. Then, we designed a representation learning network to abstract the EBSD representation as graph-level features. Finally, we built a neural network prediction model to predict the corresponding attributes. The experimental results prove the effectiveness of our method. Compared with traditional machine learning methods and machine vision methods, our method is more reasonable and practical.

References

1. Carneiro, Í., Simões, S.: Recent advances in EBSD characterization of metals. Metals **10**(8), 1097 (2020)
2. Chen, C., Ye, W., Zuo, Y., Zheng, C., Ong, S.P.: Graph networks as a universal machine learning framework for molecules and crystals. Chem. Mater. **31**(9), 3564–3572 (2019)
3. Hamilton, W.L.: Graph representation learning. In: Synthesis Lectures on Artificial Intelligence and Machine Learning, vol. 14, no. 3, pp. 1–159 (2020)
4. Humphreys, F.: Characterisation of fine-scale microstructures by electron backscatter diffraction (EBSD). Scripta Mater. **51**(8), 771–776 (2004)
5. Jain, A., Ong, S.P., Hautier, G., Chen, W., Persson, K.A.: Commentary: the materials project: a materials genome approach to accelerating materials innovation. APL Mater. **1**(1), 011002 (2013)
6. Jang, J., Gu, G.H., Noh, J., Kim, J., Jung, Y.: Structure-based synthesizability prediction of crystals using partially supervised learning. J. Am. Chem. Soc. **142**(44), 18836–18843 (2020)
7. Kondo, R., Yamakawa, S., Masuoka, Y., Tajima, S., Asahi, R.: Microstructure recognition using convolutional neural networks for prediction of ionic conductivity in ceramics. Acta Mater. **141**, 29–38 (2017)
8. Lin, X., Quan, Z., Wang, Z.J., Ma, T., Zeng, X.: KGNN: knowledge graph neural network for drug-drug interaction prediction. In: Proceedings of the Twenty-Ninth International Joint Conference on Artificial Intelligence, IJCAI-2020, pp. 2739–2745. International Joint Conferences on Artificial Intelligence Organization (2020)

9. Pablo, J.D., et al.: New frontiers for the materials genome initiative. NPJ Comput. Mater. **5**(1) (2019)
10. Pokuri, B.S.S., Ghosal, S., Kokate, A., Sarkar, S., Ganapathysubramanian, B.: Interpretable deep learning for guided microstructure-property explorations in photovoltaics. NPJ Comput. Mater. **5**(1), 1–11 (2019)
11. Priddy, K.L., Keller, P.E.: Artificial Neural Networks: An Introduction, vol. 68. SPIE Press, Bellingham (2005)
12. Rekha, S., Raja, V.B.: Review on microstructure analysis of metals and alloys using image analysis techniques. In: IOP Conference Series: Materials Science and Engineering, vol. 197, p. 012010, May 2017
13. Sanyal, S., Anishchenko, I., Dagar, A., Baker, D., Talukdar, P.: ProteinGCN: protein model quality assessment using graph convolutional networks. BioRxiv (2020)
14. Wang, Q., Mao, Z., Wang, B., Guo, L.: Knowledge graph embedding: a survey of approaches and applications. IEEE Trans. Knowl. Data Eng. **29**(12), 2724–2743 (2017)
15. Wang, X., et al.: Heterogeneous graph attention network. In: The World Wide Web Conference, pp. 2022–2032 (2019)
16. Wang, Z.L., Adachi, Y.: Property prediction and properties-to-microstructure inverse analysis of steels by a machine-learning approach. Mater. Sci. Eng., A **744**, 661–670 (2019)
17. Wieder, O., et al.: A compact review of molecular property prediction with graph neural networks. Drug Discov. Today Technol. (2020)
18. Zhou, Q., Lu, S., Wu, Y., Wang, J.: Property-oriented material design based on a data-driven machine learning technique. J. Phys. Chem. Lett. **11**(10), 3920–3927 (2020)

Federated Knowledge Graph Embeddings with Heterogeneous Data

Weiqiao Meng[1], Shizhan Chen[1], and Zhiyong Feng[2]([✉])

[1] College of Intelligence and Computing, Tianjin University, Tianjin 300350, China
[2] College of Intelligence and Computing, Shenzhen Research Institute of Tianjin University, Tianjin University, Tianjin, China
zyfeng@tju.edu.cn

Abstract. Due to the problem of *privacy protection*, it is very limited to apply distributed representation learning to practical applications in the scenario of multi-party cooperation. Federated learning is an emerging feasible solution to solve the issue of data security. However, due to the heterogeneity of the data from multi-party platforms, it is not easy to employ federated learning directly to embed multi-party data. In this paper, we propose a new federated framework FKE for representation learning of knowledge graphs to deal with the problem of privacy protection and heterogeneous data. Experiments show that the FKE can perform well in typical link prediction, overcome the problem of heterogeneous data and have a significant effect.

Keywords: Federated learning · Knowledge graph embedding · Heterogeneous data

1 Introduction

With the rapid development of the Internet, there is a huge amount of information on the network in the era of the information explosion. The rich knowledge is hidden behind these massive amounts of information, and this knowledge will act in many related fields. In order to express faith in a way closer to human cognition, information is organized into knowledge graph. The knowledge graph is an essential foundation for the management, representation, and application of knowledge. In order to express the massive amounts of information in a way closer to human cognition, information is organized into knowledge graphs. Although the scale of the knowledge graph is constantly expanding, the knowledge graph is far from completion. Many methods [1] have been proposed to complete the knowledge graph. The existing methods of knowledge graph embedding show excellent performance on small-scale data. However, in the face of an oversize knowledge graph, it is difficult for the existing single-machine methods to construct a large-size representation learning model to process large-size data efficiently. Therefore, it is a huge challenge to improve the quality and efficiency of the representation learning model to complete a super large-scale knowledge graph.

© Springer Nature Singapore Pte Ltd. 2021
B. Qin et al. (Eds.): CCKS 2021, CCIS 1466, pp. 16–26, 2021.
https://doi.org/10.1007/978-981-16-6471-7_2

Distributed representation learning is faced with the privacy security problem which has been paid more and more attention. It is limited by the privacy protection policies of different countries and regions, and the data of each participant has obvious differences. Therefore, it is not trivial to explore richer semantic associations on a large-scale knowledge graph by distributed representation learning. To embed the large-scale knowledge graph, we have to address the following problems: (1) heterogeneous data; (2) privacy protection. Federated learning [12] needs further research in processing heterogeneous data, which mainly reflected in the bias between local and global models caused by the heterogeneity of data.

In this paper, we propose a new federated framework FKE for knowledge graphs representation learning to deal with the problem of privacy protection and heterogeneous data. Experiments show that Federated Knowledge Graph Embeddings can well avoid semantic loss and overcome the generalization problem of heterogeneous data.

In the remainder of the paper, we discuss related work in Sect. 2. In Sect. 3, we illustrate our proposed framework FKE. Experiments and analysis are shown in Sect. 4. Finally, we conclude our work and suggest some future works in Sect. 5.

2   Related Work

2.1   Knowledge Graph Embedding and Translation-Based Model

Framework design is closely related to representation learning model. There are great differences among various representation learning models, so it is not easy to support all the parts that need to be selected.

There are many types of representation learning models, and the following are some of them. (1) Translation-based models, which are distance models based on an additive formula, measure rationality through the distance between entities. Translation-based models started from transE [2] and formed a large transfamily [3,4]. (2) Semantic matching model, similarity model based on multiplication formula, measure rationality through the potential semantics of matching entities and the relationship contained in vector space representation, such as RESCAL model (bilinear model) [20], DistMult [6], HOLE [5], etc. (3) Neural network model, such as SME [7], NTN [8], and ProjE [9]. (4) Models with attention mechanism, such as KBAT [10]. The translation model is mature and classic. The graph model is new and has many varieties, so it is necessary to choose the appropriate representation model as the main body of the framework.

2.2   Federated Learning Research

Federated learning (FL) is a machine learning setting that enables multiple entities (or clients) to work together to solve a machine learning problem under the coordination of the central server or service provider. The original data of each client is stored locally, and there is no exchange or transmission among the original data. Instead, a specific update for model fusion is used to achieve the purpose of learning. As an encrypted distributed machine learning paradigm, federated learning can enable all parties to achieve the purpose of building a model without disclosing the original data. Federated learning connects data islands and establishes a common model with excellent performance without violating the data privacy protection regulations such as EU General Data Protection Regulation.

The basic concept of Federated learning was first proposed in a creative work about machine learning in mobile applications by Google in 2016 [13]. It started from machine learning applications in mobile application devices and edge computing scenarios. With personal privacy and data protection becoming the focus in recent years, federated learning has attracted more and more attention, both in research and application. Federal learning mainly includes cross device federal learning and cross silo federal learning. Currently, the existing federated application tools or frameworks include Tensorflow Federated [14] proposed by Google and FATE [15] by WeBank.

Federated learning involves a wide range of research directions, so the problems are not limited to machine learning, but more related to distributed systems, optimization algorithms, cryptography, information security, fairness, statistics and so on. On the whole, federal learning faces three main problems:

(a) Efficiency and effect: on the one hand, it is the core problem of optimization algorithm, from random gradient descent SGD to Federated average Fed-Avg (federalization of parallel SGD), considering how to design and implement a better optimization algorithm scheme in the federated scenario; on the other hand, it is communication and compression, in the federated scenario, the time cost of privacy preserving cluster communication is very high. It is necessary to be considered how to deal with data compression in communication. (b) Privacy protection: consider how to protect user privacy from four aspects: client, server, output model and deployment model. This is the aspect of secure multiparty computing (MPC) of Federated learning. Consider the application of security computing methods including homomorphic encryption [16], differential privacy [17], secure multiparty computing and trusted execution environment; (c) Robustness: it mainly considers how to deal with multi-level confrontation and how to deal with the fault tolerance and adaptability of non confrontation in application scenarios.

Federate average adopts different model aggregation methods to realize the federate design of distributed SGD and researches the federate training of IID data and non-IID data; FedMA [18] is re-designed by changing the average of parameters from index to layer; FedProx [19] uses the appropriate modification of the evaluation function to introduce the correction term to modify the parameters of the next iteration.

In this paper, the Federation scheme is used as privacy protection strategy to redesign the distributed representation learning with the parameter server [11], and optimize the performance of the model under the heterogeneous data of knowledge graph data, forming a federation knowledge graph representation learning framework.

3 Proposed Method

3.1 Overview

Figure 1 illustrates the overall framework of our proposed FKE, which consists of three modules: *Federated Configuration Module, Fedrated Trainer* and *Model Fusion Module.*

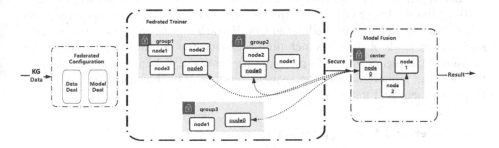

Fig. 1. Overall framework of FKE

- **Federated Configuration Module.** The federated configurator will process heterogeneous data building and configure federal training participants. Heterogeneous data processing is mainly used to construct the heterogeneity of RDF triple data. The federate training participant configuration module ensures the correct distribution of the constructed heterogeneous data, the construction of the local submodel, and the configuration of the federated cluster environment.
- **Federated Trainer.** In the federal training module, each cluster uses a decentralized way to train the local model. In this stage, each participant will not exchange or transmit any raw data. They only need to complete the training and update its sub-model. After communication with the federated model hybrid module in the encrypted environment, the necessary intermediate result is transmitted, and iterative training will finish.
- **Model Fusion Module.** The Federation model fusion module will complete the collection of submodels and update global models without original data. A central server independent of each participant will collect the necessary intermediate results to complete the model update. The intermediate results include sub-model parameter updating and constraint parameters based on data heterogeneity. In this module, the constraint parameters based on data heterogenoity are used to adjust the sub-model updating and model fusion to reduce the offset between the local and global models.

As a whole, the first part is the initialization of the federated environment, the second part is the common computing nodes as participants in the cluster, and the third part is the central server. Our federated knowledge graph embeddings with heterogeneous data focus on improving the updating process between the central server and computing nodes.

3.2   Federated Configuration Module

To complete the federated representation based on heterogeneous data, we will complete the necessary intermediate result preparation through the federation configuration module. On the one hand, the federated configurator will construct the heterogeneity of data. After checking the RDF triplet data and the working state of the device, we carry out the relation check, entity data mapping, and triple data mapping. Thus, Federated Configuration generates intermediate data such as two kinds of statistics including β-Entity, the distribution of entity data on relation category and β-Triplet, the distribution of triple data on relation, which is the prepare data of the sequent modules. On the other hand, we need to pre-segment and initialize the model.

3.3   Federated Local Updating Process

The main task of the federation training module is to update the local model (initialization and re-update), train the sub-model based on local data, and generate the necessary intermediate results for global updates. For the first iteration, the module will generate a sub-model suitable for its local data according to the preselected representation learning model, which is as small as possible as the global model covering all local data, to ensure that each participant consumes the appropriate computing load. Next, to reduce the communication pressure between local and global, the training of sub-model is carried out in the way of multiple iterations. In this way, the scale of sub-model, the size of related batch processing, the number of iterations e, the setting of the learning rate, the processing of intermediate results, and so on will affect the performance of Federated training. In the first mock exam, the intermediate results of Federated aggregation will be explained in conjunction with the next module, which includes data heterogeneity and the data needed to calculate the deviation.

3.4   Federated Fusion

The main work of the Federation aggregation module is to collect the sub-models generated by the Federation local training under the security environment of privacy protection, develop a new round of global models by using the appropriate aggregation method and complete the model distribution. The first is the setting of a privacy protection environment. To ensure the data security of each participant, the data related to model updating will be taken as the intermediate result, and feasible encryption strategies such as homomorphic encryption and differential privacy will be adopted for protection. Then, for the sub-models provided by each participant, considering the potential offset between the sub-model and the global model, we will use the regularization method based on data heterogeneity to constrain the sub-model to reduce the impact of knowledge graph embedding model offset as much as possible, which caused by the difference between local knowledge graph data and global data.

Devices usually generate and collect data on the network in different ways. The amount and characteristics of data across devices vary greatly. The data in the federal learning network is non-IID (independently identically distributed). The challenge of heterogeneous data features is the offset between the local and global models, leading to the degradation of accuracy and performance.

The beginning idea in the paper is to introduce data distribution characteristics in the whole design and adjust the aggregation stage of the model by measuring the distribution characteristics of non-IID data. Our research work is mainly to adjust the new parameters generated by each iteration in the Federation update phase according to the difference in data distribution characteristics between the local and global models. The specific implementation involves module 2 and module 3 of the federated framework.

In fact, in distributed representation learning, we usually calculate the data size of each node to weigh the sub-model based on frequency, which is a simple and intuitive way to deal with the heterogeneity of data size. However, data size is only the intuitive part of the difference between local knowledge graph data and global data. The goal of this part is to measure the difference between local knowledge graph data and global data more comprehensively. In statistics, variance measures the degree of dispersion of a random variable and it is consistent with our evaluation index obviously. And variance is used to measure the degree of deviation between a random variable and its mathematical mean in probability theory. So firstly, we consider a data distribution deviation degree based on variance to measure the data distribution difference of knowledge graph data between local and global parts. Different from the intuitive data of data scale, the calculation of distribution deviation degree of knowledge graph data is more complex.

First, we explain the judgment and setting of the heterogeneity of knowledge graph data. In the common task of image classification, the data heterogeneity is reflected in the difference in the number of images with a particular category in local data. Different from the conventional image classification task, the entity and relation of RDF data of knowledge data are different from the image data categories. One feasible scheme is to take relation as a class to consider data heterogeneity. For example, in data sets WN18, the number of relations is less. It can be divided based on relation number to allocate RDF data for all participants.

In terms of heterogeneous data construction, image classification can be constructed in 1-class and 2-class. Each client only contains 1 or 2 categories of image data, which is a typical non-IID data. In our research, we divide data based on relation and construct three groups of heterogeneous data, which are rand-relation, t-relation, and 2t-relation, among which t-relation only contains triplet data corresponding to t kinds of relation, and 2t-relation contains triplet data corresponding to 2t kinds of relation, The rand-relation is the random partition of all data the same as the distributed representation model.

Then about FedAvg, the FederatedAveraging algorithm has researched the federal training of IID data and non-IID data. Its main idea is shown as follows. However, it has poor performance on non-IID data at present. Fedavg has poor performance in non-IID data processing. To solve the problem of non-IID data, the specific methods include sharing data and regularization terms. Our research focuses on model regularization by considering data heterogeneity.

In the case of data partition based on relation, we need to generate a parameter to describe the different degrees of each local data distribution deviating from the global distribution to adjust the updating of the model. Based on the triple class distribution, the deviation parameter is set as β and based on taking relation as a category.

The calculation method of deviation degree based on category for knowledge graph data is:

$$\beta = \sum_{i \in T} \left| \frac{n_{li}}{n_l} - \frac{n_i}{n} \right|^2 \tag{1}$$

where T is the entity or relation set, for the items to be counted (the quantity distribution of entity data in relation category) , n_{li} is the local number, n_l is the local total number, n_i is the global number, and n is the global total number.

In the previous chapter, to solve the problem of data heterogeneity in federated knowledge representation, we proposed a deviation degree scheme based on the distribution characteristics of local data and global data. In this scheme, we measure the data distribution characteristics based on the variance and adjust the constraints in the aggregation phase of the Federation model based on the deviation degree to solve the deviation between the local model and the global model in the federated representation caused by data heterogeneity.

Therefore, we can consider a normalization scheme based on the measurement of data heterogeneity $\zeta = \text{F(D-local, D)}$. Starting from the heterogeneity of knowledge graph data, combined with the characteristics of data and related algorithms, the feasible normalization degree is generated from the characteristics of local data and global data through the evaluation function mapping. Then the influence of the local model on the global model is constrained based on the normalization method to reduce the deviation. Based on this idea, we can consider choosing different evaluation functions, localizing the evaluation function for federated knowledge representation, and evaluating the performance impact of different schemes.

(1) Scale scheme: Based on the conventional frequency based weighting method, combined with RDF triple data characteristics, the setting scheme is $f = f[y - e, y - tri]$ on the basis of the relationship as the category.

(2) Deviation scheme: Starting from the data distribution difference of knowledge graph between local and global parts, variance is used as the calculation prototype, and the scheme is set as $\text{F} = f\{\beta-\text{entity}, \beta-\text{tri}\}$ by combining representation model updating and triple data. For each participant's local data,

β-entity refers to the data distribution of entities on the relation categories, β-tri is the data distribution of triples in relation categories, and the final deviation degree is produced by entity distribution difference and triples distribution difference.

(3) Correlation scheme: Considering the correlation between the participant data and the overall data, the Pearson correlation coefficient is used as the calculation prototype for localization design. The local and global correlation degree is obtained as the final correlation degree.

In fact, based on the design, we can consider more evaluation functions that can be used here to measure the degree of data heterogeneity. After localization and appropriate adjustment of the evaluation function on the knowledge graph data, we can complete more new designs, which will be left to the future.

4   Experiments and Evaluations

4.1   Experimental Setting

In knowledge representation, data is organized in RDF triples, and the data scale of some common data sets is as follows:

The training tasks of knowledge graph representation learning include link prediction, relation extraction, and so on. We take a subset of data set fb15k as an example to illustrate the experimental setup. The triples include 14541 entities, 237 relationships, the size of training sets is 272115, and the size of the test set is 20466, nearly 50% parts. The basic processing of the data set is as follows. Heterogeneous data simulation is carried out based on the relation, that is, three kinds of basic data segmentation, including Rand relation data, t-relation data, and 2t-relation data. T-relation data means that the local data of each participant only contains the triplet data corresponding to the t kinds of relation, 2t-relation data means that the local data only contains the triplet data corresponding to the 2t kinds of relation. Rand relation data is the data generated by sequential segmentation after all triplet data are randomly scrambled. T-relation data and 2t-relation data are two kinds of heterogeneous knowledge graph data with obvious differences in heterogeneity degree. In contrast, Rand relation data is random data with uncertain heterogeneity.

In this experiment, the client scale n is set as 12, the single training client terminal set C (subset) scale K is set as 4, the statistical data heterogeneity is set as the above three levels of division, including heterogeneous data set 0, the training data sequence is divided into several clients, and the number of relationships is far more than 40; Heterogeneous data set 1, the relationship is divided into six parts, each with 40 kinds of relationship corresponding to the training data, each is evenly divided into two parts (each contains 40 types of relationship), and then allocated to n clients, that is, each device only contains 40 kinds of relationship; Heterogeneous data set 2: the relationship is divided into 12 pieces, each with 20 kinds of training data corresponding to the relationship, which are assigned to 12 clients, that is, each device only contains a specific number of relationships; The basic task is link prediction task. The main iteration number of the server is set to a maximum of 1000. The initial setting of the client learning rate, batch size, local training iteration number e, and other parameters adopts the default values on the dataset.

Experiments are implemented in a cluster running linux system by C++ and MPI. We evaluated the performance of the link prediction on WN18, and used Hit@10 as evaluation method under the settings of hyper-paparameters, with the dimension $d = 100$, the margin $m = 1$, and the learning rate $\eta = 0.001$.

4.2 Design and Implementation of Federated Framework

Table 1. Implementation of federated framework.

Method	Hit@10 (%)	Hit@10 (%)
Distributed	87.4	87.3
FKE-scheme0	87.2	86.7

In this table, we can see that the heterogeneity of knowledge graph data leads to a significant decline in the experimental results, which reflects the deviation of local model and global model caused by the difference of typical knowledge graph between local data and global data.

4.3 Federated Embeddings with Heterogeneous Data

Table 2. The comparison between KG FedAVG and 3 kinds of FKE schemes

Method	Hit@10 (%)	Hit@10 (%)
Base	87.2	86.7
FKE-scheme1	87.1	87.0
FKE-scheme2	87.9	87.4
FKE-scheme3	88.2	88.1

We performed some typical link prediction tasks on the knowledge graph data set WN18 to improve the performance Hit@10. As an evaluation index, it evaluates the differences between the basic realization of representation learning and the three patterns of federation representation based on data heterogeneity. The above experimental results show that the scale scheme starts from the smaller part of the data size, and produces a certain experimental effect, but it is not obvious; The deviation degree scheme starts from the difference of data distribution characteristics, and the similarity scheme starts from the similarity of local data and global data, which has produced obvious effect improvement.

5   Conclusion

In this paper, we propose a federated framework FKE for knowledge graph representation learning. Experiments show that our design can perform well in multiple tasks by dealing with the problem of privacy and heterogeneous data. Moreover, our framework achieves a significant acceleration effect. We believe that the framework FKE is also applicable to other representation learning models of knowledge base.

In fact, the current research of this paper is still limited to the preliminary exploration of the federated knowledge graph representation assumption and has been simplified to a certain extent in the aspects of data sets, evaluation indicators, comparative objectives and so on. In the follow-up work, we will conduct more experiments on other representation learning models to verify the generalization ability of the framework, more complete evaluation indexes, the natural heterogeneity of different knowledge graphs and more comparative experiments with existing Federation design.

Acknowledgments. The work described in this paper is supported by Shenzhen Science and Technology Foundation (JCYJ20170816093943197).

References

1. Gutierrez, C., Sequeda, J.: Knowledge graphs. Commun. ACM. **64**(3), 96–104 (2021)
2. Bordes, A., Usunier, N., Garcia-Duran, A., Weston, J., Yakhnenko, O.: Translating embeddings for modeling multi-relational data. In: 27th Annual Conference on Neural Information Processing Systems (NIPS), pp. 2787–2795 (2013)
3. Wang, Z., Zhang, J., Feng, J., Chen, Z.: Knowledge graph embedding by translating on hyperplanes. In: 28th AAAI Conference on Artificial Intelligence (AAAI), pp. 1112–1119 (2014)
4. Lin, Y., Liu, Z., Sun, M., Liu, Y., Zhu, X.: Learning entity and relation embeddings for knowledge graph completion. In: 29th AAAI Conference on Artificial Intelligence (AAAI), pp. 2181–2187 (2015)
5. Nickel, M., Rosasco, L., Poggio, T.: Holographic embeddings of knowledge graphs. In: 30th AAAI Conference on Artificial Intelligence (AAAI), pp. 1955–1961 (2016)
6. Yang, B., Yih, W., He, X., Gao, J., and Deng, L.: Embedding Entities and Relations for learning and inference in knowledge bases. In: 3rd International Conference on Learning Representations (ICLR), Poster (2015)
7. Glorot, X., Bordes, A., Weston, J., Bengio, Y.: A semantic matching energy function for learning with multi-relational data. Mach. Learn. **94**(2), 233–259 (2014)
8. Socher, R., Chen, D., Manning, C., Ng, A.: Reasoning with neural tensor networks for knowledge base completion. In: 27th Annual Conference on Neural Information Processing Systems (NIPS), pp. 926–934 (2013)
9. Shi, B., Weninger, T.: Embedding projection for knowledge graph completion. In: 31st AAAI Conference on Artificial Intelligence (AAAI), pp. 1236–1242 (2017)
10. Nathani, D., Chauhan, J., Sharma, C., Kaul, M.: Learning attention based embeddings for relation prediction in knowledge graphs. In: 57th Annual Meeting of the Association for Computational Linguistics on Proceedings (ACL), pp. 4710–4723 (2019)

11. Li, M., Andersen, D.G., Park, J.W., Smola, A.J.: Scaling distributed machine learning with the parameter server. In: 11th USENIX Symposium on Operating Systems Design and Implementation (OSDI), pp. 583–598 (2014)
12. Kairouz, P., et al.: Advances and open problems in federated learning. arXiv preprint arXiv:1912.04977 (2019)
13. Brendan, H., Moore, E., Ramage, D., Hampson, S., Agüera y Arcas, B.: Communication-efficient learning of deep networks from decentralized data. In: 20th International Conference on Artificial Intelligence and Statistics (AISTATS), pp. 54–55 (2017)
14. TensorFlow Federated. https://www.tensorflow.org/federated
15. Federated AI technology enabler. https://www.fedai.org/
16. Brakerski, Z., Gentry, C., Vaikuntanathan, V.: Leveled fully homomorphic encryption without bootstrapping. ACM Trans. Comput. Theory. $6(3)$, 1–36 (2014)
17. Cheu, A., Smith, A., Ullman, J., Zeber, D., Zhilyaev, M.: Distributed differential privacy via shuffling. In: 38th Annual International Conference on the Theory and Applications of Cryptographic Techniques, pp. 375–403 (2019)
18. Wang, H.: Federated learning with matched averaging. In: 8th International Conference on Learning Representations (ICLR), Poster (2020)
19. Li, T., Sahu, A. K., Zaheer, M., Sanjabi, M., Talwalkar, A., Smith, V.: Federated optimization in heterogeneous networks. In: Machine Learning and Systems, poster (2020)
20. Nickel, M., Tresp, V.: Tensor factorization for multi-relational learning. In: Machine Learning and Knowledge Discovery in Databases - European Conference, pp. 23–27 (2013)

Text-Guided Legal Knowledge Graph Reasoning

Luoqiu Li[1,2], Zhen Bi[1,2], Hongbin Ye[1,2], Shumin Deng[1,2(✉)], Hui Chen[3],
and Huaixiao Tou[3]

[1] Zhejiang University and AZFT Joint Lab for Knowledge Engine, Hangzhou, China
{luoqiu.li,bi_zhen,yehongbin,231sm}@zju.edu.cn
[2] Hangzhou Innovation Center, Zhejiang University, Hangzhou, China
[3] Alibaba Group, Hangzhou, China
{weidu.ch,huaixiao.thx}@alibaba-inc.com

Abstract. Recent years have witnessed the prosperity of legal artificial intelligence with the development of technologies. In this paper, we propose a novel legal application of legal provision prediction (LPP), which aims to predict the related legal provisions of affairs. We formulate this task as a challenging knowledge graph completion problem, which requires not only text understanding but also graph reasoning. To this end, we propose a novel text-guided graph reasoning approach. We collect amounts of real-world legal provision data from the Guangdong government service website and construct a legal dataset called LegalLPP. Extensive experimental results on the dataset show that our approach achieves better performance compared with baselines. The code and dataset are available in https://github.com/zjunlp/LegalPP for reproducibility.

1 Introduction

Legal Artificial Intelligence (LegalAI) mainly concentrates on applying artificial intelligence technologies to legal applications, which has become popular in recent years [16]. As most of the resources in this field are presented in text forms, such as legal provisions, judgment documents, and contracts, most LegalAI tasks are based on Natural Language Processing (NLP) technologies. In this paper, we introduce a novel application of **Legal Provision Prediction (LPP)** for LegalAI.

Legal Provision Prediction (LPP) aims to predict the related legal provisions of affairs. For example, given an affair "task_336: ......超出许可业务范围或无许可证的中介服务机构发布广告的处罚" (...Penalties for advertisements issued by intermediary service agencies that are beyond the scope of the licensed business or without a license), the task is to predict the most related legal provisions

---

L. Li, Z. Bi and H. Ye—Equal contribution and shared co-first authorship.

© Springer Nature Singapore Pte Ltd. 2021
B. Qin et al. (Eds.): CCKS 2021, CCIS 1466, pp. 27–39, 2021.
https://doi.org/10.1007/978-981-16-6471-7_3

such as "人才市场管理规定_004/026/001" (Talent Market Management Regulations_004/026/001) as the Table 1 shows. LPP is a real-world application that plays a significant role in the legal domain, as it can reduce heavy and redundant work for legal specialists or government employees.

Intuitively, there are many domain knowledge and concepts with well-defined rules in LegalAI, which cannot be ignored; we formulate the legal provision prediction task as a **knowledge graph completion** problem. We regard affairs and legal provisions as entities and utilize their well-defined schema structure as relations (e.g., base_entry_is, base_law_is, etc.). In such a way, the LPP problem becomes a link prediction task in the knowledge graph (e.g., whether there exists the base_entry_is relation between the affair entity and the legal provision entity). Numerous link prediction approaches [1,5,14] have been proposed for knowledge graph completion; however, there are still several non-trivial challenges for LPP:

Table 1. Legal provision prediction (LPP) task.

Type	Affair	Legal_Provision
Graph Vertex (Entity)	task_336	人才市场管理规定_004/026/001 (Talent Market Management Regulations_004/026/001)
Vertex Description	对人才中介服务机构超出许可业务范围发布广告、广告发布者为超出许可业务范围或无许可证的中介服务机构发布广告的处罚。(The punishment for talent intermediary service agencies to publish advertisements beyond the scope of the licensed business, and the advertisement publishers to publish advertisements for intermediary service agencies that are beyond the scope of the licensed business or without a license.)	人才中介服务机构通过各种形式、在各种媒体（含互联网）为用人单位发布人才招聘广告，不得超出许可业务范围... (Talent intermediary service agencies publish talent recruitment advertisements for employers in various forms and various media (including the Internet), and must not exceed the scope of the licensed business...

- *Text Understanding.* Many entities in the legal knowledge graph have well-formalized description information. For example, the legal provision "task_336" has the description "......为超出许可业务范围或无许可证的中介服务机构发布广告的处罚。" (...Penalties for advertisements issued by intermediary service agencies that are beyond the scope of the licensed business or without a license). Those texts provide enriched information for understanding the affairs and legal provisions, which is quite important, and utilizing that description is of great significance.
- *Legal Reasoning.* Some complex legal provisions may require sophisticated reasoning as legal data must strictly follow the rules well-defined in law. For example, given an affair "task_155: 市政府投资项目稽查" (Audit of

municipal government investment projects), human beings can quickly obtain the related legal provisions through *two-hop reasoning* as "task_155: 市政府投资项目稽查" (Audit of municipal government investment projects) is **following** "深圳经济特区政府投资项目管理条例" (Shenzhen Special Economic Zone Government Investment Project Management Article) and "深圳经济特区政府投资项目管理条例" (Shenzhen Special Economic Zone Government Investment Project Management Article) **has the provision of** "深圳经济特区政府投资项目管理条例第3节第1款" (Sect. 1, Paragraph 3 of the Shenzhen Special Economic Zone Government Investment Project Management Regulations).

The key to solving the issues mentioned above is combining text representation and structured knowledge with legal reasoning. To this end, we propose a **Text**-guided **G**raph **R**easoning (**T-GraphR**) approach for this task which bridges text representation with graph reasoning. Firstly, we utilize the pre-trained language model BERT [3] to represent entities with low dimension vectors. Then, we leverage graph neural networks (GNN) that assimilate generic message-passing inference algorithms to perform legal reasoning on the legal knowledge graph. We utilize two kinds of GNN, namely, R-GCN [11] and GAT [13]. Note that our approach is a model-agnostic method and is readily pluggable into other graph neural networks approaches. We collect legal provisions data from Guangdong government service website[1] and construct a dataset LegalLPP. Extensive experimental results show that our approach achieves significant improvements compared with baselines. We highlight our contributions as follows:

- We propose a new legal task, namely, legal provision prediction, which requires both text representation and knowledge reasoning.
- We formulate this task as a knowledge graph completion problem and introduce a novel text-guided graph reasoning approach that leveraging text and graph reasoning.
- Extensive experimental results demonstrate that our approach achieves better performance compared with baselines.
- We release the LegalLPP dataset, source code, and pre-trained models for future research purposes.

2 Data Collection

2.1 Data Acquisition

We collect all the data from the Guangdong government service website. We obtain about 140,482 raw affairs (including 1,552 unique affairs), and 4,042 laws with 269,053 legal provisions. We perform detailed analysis and conduct data preprocessing procedures to address those issues below:

---

[1] https://www.gdzwfw.gov.cn/.

Non-standard Text. There exist a huge discrepancy for the legal provisions and affairs, including: *Abbreviation*, such as "劳动法" (Labor Law) is the abbreviation of "中华人民共和国劳动法" (People's Republic of China Labor Law); *Missing*, such as missing of angle quotation mark ("《》") or the format of the version number (for example, the suffix of the "《广东省民用建筑节能条例》" ("Regulations on Energy Conservation of Civil Buildings in Guangdong Province" (Amended in 2014)). Those challenges make it difficult to establish the association between affairs and legal provisions. To handle those non-standard texts, we manually build a legal provision dictionary to normalize those non-standard texts.

Similar Affairs. From the raw data, we observe that a huge portion of affairs is very similar (with the same affair vertexes). Statistically, we find that the ratio of unique items to the total number of items is roughly 1:100. We analyze those similar affairs and find that most parts of them have the same content, while only the *time* in the affair is different. We merge those similar affairs in the prepossessing procedure.

No Legal Provisions. We observe that several affairs do not have any linking legal provisions, which implies no legal provisions. This problem is mainly due to outdated laws. As the legal provision will change over time, several old provisions may be deleted, making several affairs impossible to link. Also, there exists a little legal provision that does not have standard formats (in general, the standard format of the legal provision is XX law, chapter X, article X, paragraph X); thus, affairs cannot be linked to those legal provisions either. We filter out those no legal provision affairs in the prepossessing procedure.

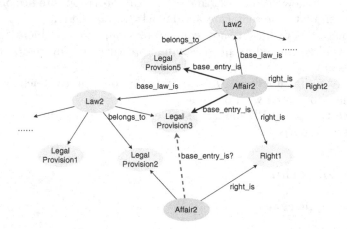

Fig. 1. Legal provision prediction as link prediction on legal knowledge graph. Best view in color. (Color figure online)

Table 2. Statistic of the legal knowledge graph. **base_entry_is** is the target relation.

Relation	Number	Description
base_entry_is	4,526	The legal provision is related to the affair
right_is	1,090	The affair has the right
base_law_is	2,152	The law is related to the affair
belongs_to	182,624	The legal provision belongs to the law

2.2 Legal Knowledge Graph

Our proposed legal task is a real-world application. As the fast updating of affairs and legal provisions, newly added affairs cannot be linked with existing legal provisions. We notice that there exist relations between affairs and laws following a well-defined schema. From two-hop reasoning on the legal knowledge graph, it is possible to judge whether there exists a relation between an affair and legal provisions. In this paper, we formulate the legal provision prediction task as the link prediction problem in the legal knowledge graph. We model the *legal provision, affair, law, right* as entities, as shown in Fig. 1. We detail the statistic of the legal knowledge graph in Table 2.

2.3 Dataset Construction

We randomly divide the triples with four type relations into the train, valid and test set with a ratio of 8:1:1, and we filter the triples with base_entry_is relation from the test set as the target test set. The detailed number of entities, relations, and triples of the LegalLPP dataset are shown in Table 3.

Table 3. Summary statistics of LegalLPP dataset.

Dataset	#Rel	#Ent	#Triple
Train (all)	4	151,746	152,307
Dev (all)	4	22,086	19,037
Test (all)	4	22,070	19,042
Test (target)	1	768	454

3 Methodology

3.1 Problem Definition

A knowledge graph G is a set of triplets in the form (h, r, t), $h, t \in \mathcal{E}$ and $r \in \mathcal{R}$ where $\mathcal{E}$ is the entity vocabulary and $\mathcal{R}$ is a collection of pre-defined relations as shown in Table 2. We are aimed at *predicting whether there exists the relation base_entry_is between affair entities and legal provision entities*. We construct positive triples with ground truth instances and negative triples with corrupted instances following [1].

3.2 Framework Overview

Our text-guided graph reasoning approach consists of two main components, as shown in Fig. 2. Our approach is not end-to-end as we firstly fine-tune the text representation and then leverage this feature to perform legal graph reasoning.

Text Representation Learning (Sect. 3.3). Given an affair and legal provision, we employ neural networks to encode the instance semantics into a vector. In this study, we implement the instance encoder with BERT [3]. We then apply an MLP layer to reduce the dimension of features (the dimension of BERT-base is 768, which is not convenient for training GNN) to obtain the text representations, which is more efficient for training and inference. We learn the text representation via fine-tuning with triple scores following TransE [1].

Legal Graph Reasoning (Sect. 3.4). After obtaining the learned text representations, we employ GNN to learn explicit relational knowledge. By assimilating generic message-passing inference algorithms with the neural-network counterpart, we can learn vertex embeddings with legal reasoning. Then we utilize a residual connection from the text representation to obtain the final representation. Finally, we utilize TransE [1], DistMult [14] and SimplE [5] as triple score functions.

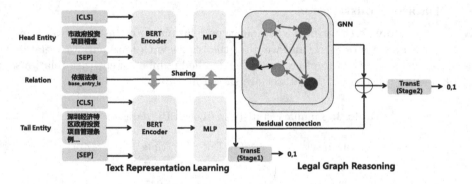

Fig. 2. Our approach of **T**ext-guided **G**raph **R**easoning (**T-GraphR**). TransE (Stage1) and TransE (Stage2) refers to the triple score function in text representation learning (Sect. 3.3) and legal graph reasoning (Sect. 3.4), respectively. Best view in color. (Color figure online)

3.3 Text Representation Learning

Given an input affair text h and legal provision text t, we utilize BERT [3] to obtain the text representations as follows:

$$m_h = \text{BERT}(h), m_t = \text{BERT}(t) \tag{1}$$

where h,t are raw input text and m_h, m_t are the output [CLS] embeddings of BERT. We then leverage an MLP layer to reduce dimension as follows:

$$v_h = ReLU(W^h * m_h + b^h), \qquad v_t = ReLU(W^t * m_t + b^t) \qquad (2)$$

where v_h, w_t are the final text representations which will then be fed into the GNN. To obtain more representative features, we finetune the text representation with TransE triple score function as:

$$\text{score}(h, r, t)_{transe} = \|v_h + v_r - v_t\|_p \qquad (3)$$

where $\text{score}(h, r, t)$ is the score of triple $< h, r, t >$, v_h, v_t, are entity representations from Eq. 2, v_r is **random initialized vectors**. We further analyze the empirical performance of other triple score function of DistMult and SimplE. The DistMult and SimplE score function are calculated as:

$$\text{score}(h, r, t)_{distmult} = \sum (v_h * v_r * v_t) \qquad (4)$$

$$\text{score}(h, r, t)_{simple} = \frac{\sum (v_h * v_r * v_t) + \sum (v_h * v_{r_{inv}} * v_t)}{2} \qquad (5)$$

DistMult is a simplified version of RESCAL [9] by using a diagonal matrix to encode relation. Different from DistMult, SimplE can handle asymmetric relations.

3.4 Legal Graph Reasoning

We feed the vertex representation v_i into a graph encoder to obtain the hidden vectors, which explicitly models the graph structure of the legal knowledge graph. We use an implementation of the GNN model following GAT [13] and R-GCN [11].

GAT. The GAT model uses multiple graph attention layer connections for encoding vertexs. Specifically, GAT calculates the attention weights of neighboring nodes for aggregation. The attention weights of node i and node j are calculated as follows:

$$\alpha_{ij} = \frac{\exp\left(\vec{\mathbf{a}}^T \left[\mathbf{W}v_i \| \mathbf{W}v_j\right]\right)}{\sum_{k \in \mathcal{N}_i} \exp\left(\vec{\mathbf{a}}^T \left[\mathbf{W}v_i \| \mathbf{W}v_k\right]\right)} \qquad (6)$$

where T represents transposition and $\|$ is the concatenation operation. Once obtained, the normalized attention coefficients are used to compute a linear combination of the features corresponding to them, to serve as the final output features for every node (after potentially applying a nonlinearity, σ):

$$v_i' = \sigma\left(\sum_{j \in \mathcal{N}_i} \alpha_{ij} \mathbf{W} v_j\right) \qquad (7)$$

where $\vec{v}'_i$ is the final graph node representation.

R-GCN. R-GCN utilizes multi-layer relational graph convolutional layer to represent node. The forward-pass update of an node denoted by v_i in a relational multi-graph is shown as follows:

$$v'_i = \sigma \left(\Sigma_{r \in R} \Sigma_{m \in N_i^r} \frac{1}{c_{i,r}} W_r v_i + W_0 v_i \right) \tag{8}$$

where $\vec{v}'_i$ is the final graph node representation, $\mathcal{N}_i^r$ denotes the set of neighbor indices of node i under relation $r \in \mathcal{R}$. $c_{i,r}$ is a problem-specific normalization constant that can either be learned or chosen in advance (such as $c_{i,r} = | \mathcal{N}_i^r|$). Note that, R-GCN considers the relation of triples in the convolution process and is able to learn different aggregation weights according to different relations.

Afterwards, we add a residual connection from the output of MLP layer to the graph node representation, denoted by:

$$v_i = v_i + GNN(v_i) \tag{9}$$

where v_i is the final entity representation of entities leveraging both text and graph reasoning.

Finally, we utilize the same score function of TransE, DistMult and SimpLE in the Sect. 3.3 to calculate triple scores. Note that, in the graph reasoning stage, v_h and v_t are combined with both text and graph features while v_r is initialized from **the tuned embedding in text representation learning** (Eq. 3 and Eq. 5). Though our approach is not end-to-end, the entity embeddings (e.g., legal provisions, affairs) can be pre-computed, which is quite efficient in for inference.

4 Experiments

4.1 Settings

We conduct experiments on the LegalLPP dataset. We use Pytorch [10] to implement baselines and our approach on single Nvidia 1080Ti GPU. We leverage Graph Deep Library[2] to implement all the GNN components. We utilize *bert-base-Chinese*[3] to represent text. We employ Adam [6] as the optimizer. In the text representation learning stage, the learning rate is 5e–5 with the warm-up proportion being 0.1; the batch size is 64, the maximum sequence length of each entity's text is 128. After 6 epochs of training, we generate 400-dimension text representations. In the legal graph reasoning stage, we set the learning rate of GNN to be 0.01. We train 4,000 epochs for GAT and R-GCN. We use TransE as the default triple score function. We evaluate the performance with Mean Rank (MR), Mean Reciprocal Rank (MRR), and HIT@N (N = 1,3,10).

---

[2] https://www.dgl.ai/.
[3] https://github.com/google-research/bert.

4.2   Baselines

We compare our approach with different kinds of baselines, as shown below:

No Reasoning. We conduct TransE [1] as an baseline. We also utilize two separate BERT encoders to represent the text with the TransE triple score function as a baseline.

Graph Only. We build the legal knowledge graph and leverage GNN approaches R-GCN and GAT without text features. The graph node representation is initialized randomly.

4.3   Evaluation Results

Table 4. Main results on LegalLPP dataset.

Model		MR	MRR	HIT@1	HIT@3	HIT@10
No reasoning	TransE	21615.832	0.179	0.121	0.196	0.258
	BERT	**404.308**	0.103	0.051	0.095	0.207
Graph only	GAT	14790.835	0.187	**0.137**	0.209	0.262
	R-GCN	35767.694	0.175	0.119	0.187	0.267
T-GraphR	GAT (TransE)	21339.555	**0.197**	0.133	**0.214**	**0.291**
	GAT (DistMult)	19546.152	0.047	0.011	0.041	0.119
	GAT (SimplE)	18164.057	0.094	0.062	0.099	0.145
	R-GCN (TransE)	1414.584	0.179	0.126	0.192	0.242

From Table 4, we observe:

1) Our approach T-GraphR with GAT achieves the best performance. We argue that our target task is to predict the base_entry_is relation between affairs and legal provisions, and there are only four relations in the graph; thus, GAT, which implicitly specifying different weights to different nodes in a neighborhood can obtain better performance.

2) Graph only approach achieves better performance than no reasoning methods BERT and TransE, which indicates that graph reasoning plays a vital role in legal provision prediction.

3) Our T-GraphR approach achieves the best performance and even obtain 12.8% hit@10 improvements compared with the text-only no reasoning model TransE.

4) The overall performance is still far from satisfactory (less than 0.3 with hit@10), and there is more room for future works.

We conduct experiments with different triple score function and report results in Table 4. We observe that TransE obtains better performance than DistMult

Table 5. Case studies.

Model	Affair	T-GraphR	BERT
Instance1	对占用道路、广场从事经营性车辆清洗活动的处罚 (Penalties for occupation of roads and plazas for cleaning vehicles)	肇庆市城区市容和环境卫生管理条例_005/053/001 (Regulations of Zhaoqing City on City Appearance and Environmental Sanitation_005/053/001)	中华人民共和国河道管理条例_003/035/001 (River Regulations of the People's Republic of China_003/035/001)
Instance2	对未按规定缴纳城市生活垃圾处理费的行政处罚 (Administrative penalties for failure to pay municipal solid waste disposal fees)	广东省城乡生活垃圾处理条例_004/037/001 (Guangdong Province Urban and Rural Domestic Waste Treatment Regulations_004/037/001)	广东省环境保护条例_004/056/001 (Guangdong Environmental Protection Regulations_004/056/001)

and SimplE. We argue that the TransE model represents relations as translations, which aims to model the **inversion** and **composition** patterns; the DistMult utilizes the three-way interactions between head entities, relations, and tail entities which aims to model the **symmetry** pattern, the SimplE model the **asymmetric** relations by considering two (head and tail) vectors only. In our LPP task, those inversion and composition patterns are common in the legal knowledge graph; thus, such translation assumption is advantageous.

4.4 Case Studies

We present some predicted instances obtained by our model to demonstrate the generalization ability in Table 5. Our method can predict correct legal provisions with complex surface contexts. Moreover, by reasoning on the legal knowledge graph, we can leverage the well-defined structure, which boosts performance. However, vanilla BERT only considers text, neglecting the structured knowledge in the legal knowledge graph, which results in unsatisfactory performance.

4.5 Entity Visualization

To further analyze the behavior of entity representations, we utilize T-SNE [8] to visualize five randomly selected entity embeddings. From Fig. 3, we find that entity embeddings of the graph only approaches have a compact data distribution, while with pre-trained LMs, entities of different types are **scattered**. To conclude, text features can enhance the vertex's discriminative ability to enhance node representations.

5 Related Work

Knowledge Graph Completion. In this paper, we formulate the LPP problem as a knowledge graph completion task by link prediction. A variety of approaches

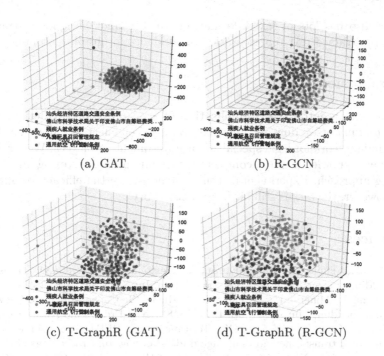

(a) GAT   (b) R-GCN

(c) T-GraphR (GAT)   (d) T-GraphR (R-GCN)

Fig. 3. T-SNE visualizations of entity (legal provision) embeddings. Figure (a) and (b) refer to the entity embeddings of **graph only** method, Figure (c) and (d) refer to our **T-GraphR** approach.

such as TransE [1], ConvE [2], Analogy [7], RotatE [12] have been proposed to encode entities and relations into a continuous low-dimensional space [15]. TransE [1] regards the relation r in the given fact (h, r, t) as a translation from h to t within the low-dimensional space. RESCAL [9] studies on matrix factorization based knowledge graph embedding models using a bilinear form as score function. DistMult [14] simplifies RESCAL by using a diagonal matrix to encode relation. [5] propose a simple tensor factorization model called SimplE through a slight modification of the Polyadic Decomposition model [4]. Since the relation of the legal knowledge graph is quite small, we utilize TransE [1], DistMult [14] and SimplE [5] as score functions of knowledge graph completion for computation efficiency.

Graph Neural Networks. Recently, graph neural network (GNN) models have increasingly attracted attention, which is beneficial for graph data modeling and reasoning. Some existing literature such as R-GCN [11], GAT [13] use GNN for structure learning. [11] introduces a relational graph convolutional networks (R-GCN) for knowledge base completion tasks that can deal with the highly multi-relational data. [13] propose a graph attention networks (GAT) that leveraging masked self-attentional layers based on neural graph networks. However, as legal provision also has lots of text information which cannot be ignored; thus, we leverage pre-trained text representation as guidance for graph reasoning.

6   Conclusion

In this paper, we introduce an application of legal provision prediction, which requires text understanding and knowledge reasoning. This task can reduce heavy and redundant work for legal specialists or government employees. We formulate this task as a knowledge graph completion task and propose a text-guided graph reasoning approach. Experimental results demonstrate the efficacy of our approach, however, the task is still far from satisfactory.

References

1. Bordes, A., Usunier, N., García-Durán, A., Weston, J., Yakhnenko, O.: Translating embeddings for modeling multi-relational data. In: NIPS, pp. 2787–2795 (2013)
2. Dettmers, T., Minervini, P., Stenetorp, P., Riedel, S.: Convolutional 2d knowledge graph embeddings. In: AAAI, pp. 1811–1818. AAAI Press, Palo Alto (2018)
3. Devlin, J., Chang, M.W., Lee, K., Toutanova, K.: BERT: pre-training of deep bidirectional transformers for language understanding. In: Proceedings of the 2019 Conference of the North American Chapter of the Association for Computational Linguistics: Human Language Technologies, vol. 1 (Long and Short Papers), pp. 4171–4186. Association for Computational Linguistics, Minneapolis, June 2019. https://doi.org/10.18653/v1/N19-1423
4. Hitchcock, F.L.: The expression of a tensor or a polyadic as a sum of products. J. Math. Phys. **6**(1–4), 164–189 (1927)
5. Kazemi, S.M., Poole, D.: Simple embedding for link prediction in knowledge graphs. In: Advances in Neural Information Processing Systems, pp. 4284–4295 (2018)
6. Kingma, D.P., Ba, J.: Adam: a method for stochastic optimization. CoRR abs/1412.6980 (2015)
7. Liu, H., Wu, Y., Yang, Y.: Analogical inference for multi-relational embeddings. In: Proceedings of the ICML, pp. 2168–2178 (2017)
8. Maaten, L.v.d., Hinton, G.: Visualizing data using t-SNE. J. Mach. Learn. Res. **9**, 2579–2605 (2008)
9. Nickel, M., Tresp, V., Kriegel, H.: A three-way model for collective learning on multi-relational data. In: Proceedings of ICML, pp. 809–816 (2011)
10. Paszke, A., et al.: Pytorch: an imperative style, high-performance deep learning library. In: Advances in Neural Information Processing Systems, pp. 8024–8035 (2019)
11. Schlichtkrull, M., Kipf, T.N., Bloem, P., van den Berg, R., Titov, I., Welling, M.: Modeling relational data with graph convolutional networks. In: Gangemi, A., et al. (eds.) ESWC 2018. LNCS, vol. 10843, pp. 593–607. Springer, Cham (2018). https://doi.org/10.1007/978-3-319-93417-4_38
12. Sun, Z., Deng, Z.H., Nie, J.Y., Tang, J.: Rotate: knowledge graph embedding by relational rotation in complex space. In: Proceedings of ICLR (2019)

13. Veličković, P., Cucurull, G., Casanova, A., Romero, A., Liò, P., Bengio, Y.: Graph attention networks. In: International Conference on Learning Representations (2018)
14. Yang, B., tau Yih, W., He, X., Gao, J., Deng, L.: Embedding entities and relations for learning and inference in knowledge bases. In: Proceedings of ICLR (2015)
15. Zhang, N., Deng, S., Sun, Z., Chen, J., Zhang, W., Chen, H.: Relation adversarial network for low resource knowledge graph completion. In: Proceedings of The Web Conference 2020, pp. 1–12 (2020)
16. Zhong, H., Xiao, C., Tu, C., Zhang, T., Liu, Z., Sun, M.: How does NLP benefit legal system: a summary of legal artificial intelligence. In: Jurafsky, D., Chai, J., Schluter, N., Tetreault, J.R. (eds.) Proceedings of ACL, pp. 5218–5230. Association for Computational Linguistics (2020)

Knowledge Acquisition and Knowledge Graph Construction

On Robustness and Bias Analysis of BERT-Based Relation Extraction

Luoqiu Li[1,2], Xiang Chen[1,2], Hongbin Ye[1,2], Zhen Bi[1,2], Shumin Deng[1,2], Ningyu Zhang[1,2(✉)], and Huajun Chen[1,2(✉)]

[1] AZFT Joint Lab for Knowledge Engine, Zhejiang University, Hangzhou, China
{luoqiu.li,xiang_chen,yehongbin,bizhen_zju,231sm,zhangningyu,
huajunsir}@zju.edu.cn
[2] Hangzhou Innovation Center, Zhejiang University, Hangzhou, China

Abstract. Fine-tuning pre-trained models have achieved impressive performance on standard natural language processing benchmarks. However, the resultant model generalizability remains poorly understood. We do not know, for example, how excellent performance can lead to the perfection of generalization models. In this study, we analyze a fine-tuned BERT model from different perspectives using relation extraction. We also characterize the differences in generalization techniques according to our proposed improvements. From empirical experimentation, we find that BERT suffers a bottleneck in terms of robustness by way of randomizations, adversarial and counterfactual tests, and biases (i.e., selection and semantic). These findings highlight opportunities for future improvements. Our open-sourced testbed **DiagnoseRE** is available in https:// github.com/zjunlp/DiagnoseRE.

1 Introduction

Self-supervised pre-trained language models (LM), such as the BERT [8] and RoBERTa [26], providing powerful contextualized representations, has achieved promising results on standard Natural Language Processing (NLP) benchmarks. However, the generalization behaviors of these types of models remain largely unexplained.

In NLP, there is a massive gap between task performance and the understanding of model generalizability. Previous approaches indicated that neural models suffered from poor **robustness** when encountering *randomly permuted contexts* [31], *adversarial examples* [20], and *contrastive sets* [13]. Moreover, neural models are susceptible to **bias** [32], such as *selection* and *semantic* bias. Concretely, models often capture superficial cues associated with dataset labels which are generally not useful. For example, the term , "airport," may indicate the output of the relation, "place_served_by_transport_hub," in the relation extraction (RE) task. However, this is clearly the result of a biased assumption.

Notably, there have been scant studies that analyzed the generalizability of NLP models [12,23,31]. This is surprising because this level of understanding

---

L. Li, X. Chen, H. Ye and Z. Bi—Equal contribution and shared co-first authorship.

© Springer Nature Singapore Pte Ltd. 2021
B. Qin et al. (Eds.): CCKS 2021, CCIS 1466, pp. 43–59, 2021.
https://doi.org/10.1007/978-981-16-6471-7_4

could not only be used to figure out missing connections in state-of-the-art models, but it could also be used to inspire important future studies while forging new ideas. In this study, we use RE as the study case and diagnose its generalizability in terms of robustness and bias. Specifically, we answer five crucial, yet rarely asked, questions about the pre-trained LM BERT [8].

Q1: Does BERT really have a generalization capability, or does it make shallow template matches? For this question, we leverage a randomization test for entity and context to analyze BERT's generalizability. Furthermore, we utilize data augmentation to determine whether this is beneficial to generalization. **Q2:** How well does BERT perform with adversarial samples in terms of RE? For this question, we introduce two types of adversarial methods to evaluate its performance. Then, we conduct experiments to understand how adversarial training influences BERT's generalizability. **Q3:** Can BERT generalize to contrast sets, and does counterfactual augmentation help? For this question, we evaluate whether the model can identify negative samples via contrastive sets (samples within a similar context but with different labels). We also propose a novel counterfactual data augmentation method that does not require human intervention to enhance generalization. **Q4:** Can BERT learn simple cues (e.g., lexical overlaps) that work well with most training examples but fail on more challenging ones? We conduct an in-depth analysis and estimate whether its tokens are prone to biased correlations. We also introduce a de-biased method to mitigate selection bias. **Q5:** Does semantic bias in the pre-trained LM hurt RE generalization? We attempt to identify whether these biases exist in BERT, and we introduce an entity-masking method to address this issue.

Main Contributions. This paper provides an understanding of BERT's generalization behavior from multiple novel perspectives, contributing to the field from the following perspectives. We first identify the shortcoming of previous RE models in terms of robustness and bias and suggest directions for improvement. Other tasks can benefit from the proposed counterfactual data augmentation method, which notably does not require human intervention. This research also enhances the generalization of two sampling approaches to bias mitigation. We also provide an open-source testbed, "DiagnoseRE," for future research purposes. Ours is the first approach that applies adversarial and counterfactual tests for RE. Our approach can be readily applied to other NLP tasks such as text classification and sentiment analysis.

Observations. We find that BERT is sensitive to random permutations (i.e., entities), indicating that fine-tuning pre-trained models still suffer from poor robustness. We also observe that data augmentation can benefit performance. BERT is found to be vulnerable to adversarial attacks that comprise legitimate inputs that are altered by small and often imperceptible perturbations. Adversarial training can help enhance robustness, but the results are still far from satisfactory. We find that model performance decays in the contrast setting, but counterfactual data augmentation does enhance robustness. BERT is susceptible to learning simple cues, but re-weighting helps to mitigate bias. There exists a semantic bias in the model that hurts generalization, but entity masking can slightly mitigate this.

2 Related Work

2.1 Relation Extraction

Neural models have been widely used for RE because they accurately capture textual relations without explicit linguistic analyses [24,42–44,48,51]. To further improve their performance, some studies have incorporated external information sources [16,19,46] and advanced training strategies [10,18,25,29,39,41,47,49, 50]. Leveraging the prosperity of pre-trained LMs, [36] utilized a pre-trained LM for RE: the OpenAI generative pre-trained transformer. [3] proposed a solution that could complete multiple entities RE tasks using a pre-trained transformer. Although they achieved promising results on benchmark datasets, the generalizability of RE was not well examined. To the best of our knowledge, we are the first to rigorously study the generalizability of RE.

2.2 Analyzing the Generalizability of Neural Networks

Most existing works [12] analyzed the generalizability of neural networks using parameters and labels and influencing the training process on a range of classification tasks. [4] examined the role of memorization in deep learning, drawing connections to capacity, generalization, and adversarial robustness. [11] developed a perspective on the generalizability of neural networks by proposing and investigating the concept of neural-network stiffness. [55] sought to understand how different dataset factors influenced the generalization behavior of neural extractive summarization models. For NLP, [1] introduced 14 probing tasks to understand how encoder architectures and their supporting linguistic knowledge bases affected the features learned by the encoder. [2] attempted to answer whether or not we have reached a performance ceiling or if there was still room for improvement for RE. The current study aims to better understand the generalizability of the fine-tuned pre-trained BERT models regarding robustness and bias.

3 Task, Methods, and Datasets

3.1 Task Description

Definition 1. Robustness is a measure that indicates whether the model is vulnerable to small and imperceptible permutations originating from legitimate inputs.

Definition 2. Bias is a measure that illustrates whether the model learns simple cues that work well for the majority of training examples but fail on more challenging ones (Table 1).

Table 1. Outline of our experiment designs.

Q.	Perspectives	Evaluation settings	Improved strategies
Q1	Randomization	Random permutation	Data augmentation (DA)
Q2	Adversarial	Adversarial attack	Adversarial training (Adv)
Q3	Counterfactual	Contrastive masking	Counterfactual data augmentation (CDA)
Q4	Selection	Frequent token replacement	De-biased training
Q5	Semantic	Entity-only	Selective entity masking

RE is usually formulated as a sequence classification problem. For example, given the sentence, "Obama was born in Honolulu," with the head entity, "Obama," and the tail entity, "Honolulu," RE assigns the relation label, "place_of_birth," to the instance. Formally, let $X = \{x_1, x_2, \ldots, x_L\}$ be an input sequence, $h, t \in X$ be two entities, and Y be the output relations. The goal of this task is to estimate the conditional probability, $P(Y|X) = P(y|X, h, t)$.

3.2 Fine-Tuning the Pre-trained Model for RE

To evaluate the generalization ability of RE, we leveraged the pre-trained BERT base model (uncased) [8]. Other strong models (e.g., RoBERTa [26] and XLNet [40]) could also be leveraged.

We first preprocessed the sentence, $\mathbf{x} = \{w_1, w_2, h, \ldots, t, \ldots, w_L\}$, for BERT's input form: $\mathbf{x} = \{[CLS], w_1, w_2, [E1], h, [/E1], \ldots, [E2], t, [/E2], \ldots, w_L, [SEP]\}$, where $w_i, i \in [1, n]$ refers to each word in a sentence and h and t are head and tail entities, respectively. [E1], [/E1], [E2], and [/E2] are four special tokens used to mark the positions of the entities. As we aimed to investigate BERT's generalization ability, we utilized the simplest method, i.e., [CLS] token, as the sentence-feature representation. We used a multilayer perceptron to obtain the relation logits, and we utilized cross-entropy loss for optimization.

3.3 RE Datasets for Evaluation

We conducted experiments on two benchmark datasets: Wiki80 and TACRED. The Wiki80 dataset[1] [17] was first generated using distant supervision. Then, it was filtered by crowdsourcing to remove noisy annotations. The final Wiki80 dataset consisted of 80 relations, each having 700 instances. TACRED[2] [53] is a large-scale RE dataset that covers 42 relation types and contains 106,264 sentences. Each sentence in two datasets has only one relation label. To analyze the RE model's generalization, we constructed our robust/de-biased test set based on Wiki80 and TACRED. We evaluate the generalization of BERT on those test sets. More details can be found in the following sections. We used the micro F1 score to evaluate performance.

---

[1] https://github.com/thunlp/OpenNRE.
[2] https://nlp.stanford.edu/projects/tacred/.

4 Diagnosing Generalization with *Robustness*

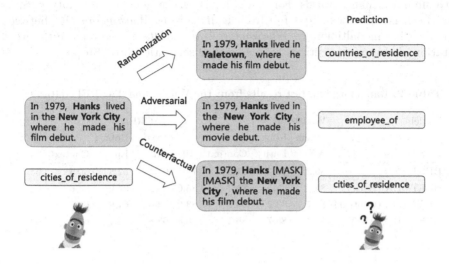

Fig. 1. Diagnosing generalization with robustness.

We studied the robustness of RE from three aspects: randomized [31], adversarial [20], and counterfactual [13]. For example, we would like to investigate the performance of fine-tuned models having diverse surface forms, adversarial permutations, and contrast settings, as shown in Fig. 1. The randomization test aims to probe the performance with random token permutations. Meanwhile, adversarial analysis is used to study its stability when encountering adversarial permutations. The contrast set is used to analyze whether the model has captured the relevant phenomena, compared with standard metrics from i.i.d. test data.

4.1 Randomization Test

To conduct the randomization test for RE, we utilized a random permutation for tokens in the test set to construct new robust sets. Note that the entity and context may provide different contributions to the performance of the RE task. Thus, we introduced two types of permutation strategies regarding entity and context as follows:

Entity Permutation is used to investigate the diversity of name entities for RE, which replaces the same entity mention with another entity having the same entity type. Thus, we can identify robust performance with different sentence entities. For example, in Fig. 1, given the sentence, "In 1979, *Hanks* lived in the *New York City*, where he made his film debut," we replace the entity "*New York City*" with the entity "Yaletown," having the same type, "LOCATION," to construct a new testing instance.

Context Permutation is used to investigate the impact of context. Unlike entity permutation, context permutation replaces each word between two entities with similar semantic words. For example, given the sentence, "Utility permits have been issued to extend a full from Baltimore to *Washington DC*, between Penn Station in Baltimore to *Washington Union Station*," we replace the word "Station" between two entities with a similar semantic word "Stop."

Table 2. Randomization test results from the Wiki80 and TACRED datasets.

Model	Wiki80				TACRED			
	Origin	Robust			Origin	Robust		
	All	Entity	Context	All	All	Entity	Context	All
BERT	86.2	78.4	81.6	79.1	67.5	57.8	60.4	58.3
BERT+DA (Entity)	85.9	85.5	85.7	85.6	64.3	64.6	64.6	64.6
BERT+DA (Context)	85.7	83.2	85.6	83.9	63.6	61.6	63.6	62.1
BERT+DA (All)	85.7	85.7	86.1	85.8	63.8	64.1	64.4	64.1

Specifically, we leveraged the CheckList behavioral testing tool[3] [31] to generate entity and context permutations, and we utilized an invariance test (INV) to apply label-preserving perturbations to inputs while expecting the model prediction to remain the same. We leveraged three methods to generate candidate token replacements. First, we used WordNet categories (e.g., synonyms and antonyms). We selected context-appropriate synonyms as permutation candidates. Furthermore, we used additional common fill-ins for general-purpose categories, such as named entities (e.g., common male and female first/last names, cities, and countries) and protected-group adjectives (e.g., nationality, religion, gender, and sexuality) for generating permutation candidates. These two methods can generate vast amounts of robust test instances efficiently. Additionally, we leveraged the pre-trained LM RoBERTa [26] to generate permutation candidates. We randomly masked tokens in the sentences and generated the mask token via a mask LM. For example, "New York is a [MASK] city in the United States" yields {"small", "major", "port", "large"}. We randomly selected the top-two tokens as permutation candidates and leveraged three strategies for robust set construction.

To ensure that the random token replacement was label-preserving, we manually evaluated the quality of the generated instances. We randomly selected 200 instances and found that only two permuted sentences had the wrong labels, indicating that our robust set was of high quality. In total, we generated 5,600/15,509 test instances of entity and context permutations on both datasets. We constructed a combined robust set (Table 2) with both entity and context permutations. We evaluated the performance of BERT with the original test set and the robust set. We also trained BERT with data augmentation regarding entity

---

[3] https://github.com/marcotcr/checklist.

(BERT+DA (Entity)) and context (BERT+DA (Context)) for evaluation. We employed Adam [21], and the initial learning rate was 2e−5. The batch size was 32, and the maximum epoch was 5. The hyperparameter was the **same** for different experiments.

Results and Analysis. From Table 2, we observe that the overall performance decayed severely in the robust set of entity and context permutations. BERT had a more remarkable performance decay with entity permutations, which indicates that the model was unstable with different head and tail entities. Furthermore, we found that BERT+DA achieved better performance in both the original and robust test sets. However, the robust test set's overall performance was still far from satisfactory. Thus, more robust algorithms are needed for future studies.

4.2   Adversarial Testing

In this section, we focus on the problem of generating valid adversarial examples for RE and defending from adversarial attacks with adversarial training. Given a set of N instances, $\mathcal{X} = \{X_1, X_2, \ldots, X_N\}$ with a corresponding set of labels, $\mathcal{Y} = \{Y_1, Y_2, \ldots, Y_N\}$, we have a RE model, $\mathcal{Y} = RE(\mathcal{X})$, which is trained via the input $\mathcal{X}$ and $\mathcal{Y}$.

The adversarial example X_{adv} for each sentence $X \in \mathcal{X}$ should conform to the requirements as follows:

$$RE\left(X_{\text{adv}}\right) \neq RE(X), \text{ and } \text{Sim}\left(X_{\text{adv}}, X\right) \geq \epsilon, \tag{1}$$

where Sim is a similarity function and ϵ is the minimum similarity between the original and adversarial examples. In this study, we leveraged two efficient adversarial attack approaches for RE: PWWS [30] and HotFlip [9].

PWWS, a.k.a., Probability Weighted Word Saliency, is a method based on synonym replacement. PWWS firstly find the corresponding substitute based on synonyms or entities and then decide the replacement order. Specifically, given a sentence of L words, $X = \{w_1, w_2, \ldots, w_L\}$, we first selected the important prediction tokens having a high score of I_{w_i}, which is calculated as the prediction change before and after deleting the word. Then, we gathered a candidate set with the WordNet synonyms and named entities. To determine the priority of words for replacement, we score each proposed substitute word w_i^* by evaluating the ith value of $\mathbf{S}(\mathbf{x})$. The score function $H\left(\mathbf{x}, \mathbf{x}_i^*, w_i\right)$ is defined as:

$$H\left(\mathbf{x}, \mathbf{x}_i^*, w_i\right) = \phi(\mathbf{S}(\mathbf{x}))_i \cdot P\left(y_{\text{true}} \mid \mathbf{x}\right) - P\left(y_{\text{true}} \mid \mathbf{x}_i^*\right) \tag{2}$$

where $\phi(\mathbf{z})_i$ is the softmax function, Eq. 2 determines the replacement order. Based on $H\left(\mathbf{x}, \mathbf{x}_i^*, w_i\right)$, all the words w_i in X are sorted in descending order. We then use each word w_i under this order. Specifically, we greedily select the substitute word w_i^* for w_i to be replaced which can make the final classification label change iteratively through the process until enough words have been replaced.

HotFlip is a gradient-based method that generates adversarial examples using character substitutions (i.e., "flips"). HotFlip also supports insertion and deletion operations by representing them as sequences of character substitutions. It uses the gradient from the one-hot input representation to estimate which individual change has the highest estimated loss efficiently. Further, HotFlip uses a beam search to find a set of manipulations that work together to confuse a classifier.

We conducted experiments based on OpenAttack[4] and generated adversarial samples of PWWS and HotFlip to construct a robust test set separately. We also conducted adversarial training experiments [35] to improve the robustness of machine-learning models by enriching the training data using generated adversarial examples. We evaluated the performance of the vanilla BERT and the version using adversarial training (BERT+Adv) in both the original and robust test sets.

Table 3. Adversarial test results from the Wiki80 and TACRED datasets. The former indicates the results of adversarial sets while the latter indicates the results of original sets.

Model	Wiki80	TACRED
BERT (Origin)	86.2	67.5
BERT (PWWS/Origin)	52.9/86.2	37.7/67.5
BERT (HotFlip/Origin)	56.3/86.2	49.2/67.5
BERT+Adv (PWWS/Origin)	86.4/86.4	72.1/65.9
BERT+Adv (HotFlip/Origin)	87.0/86.6	73.9/67.8

Results and Analysis. From Table 3, we observe that BERT achieved significant performance decay with PWWS and HotFip, revealing that fine-tuned RE models are vulnerable to adversarial attacks. We noticed that adversarial training helped achieve better performance. Conversely, the original set's evaluation results were slightly decayed, as was also found in [37]. We, therefore, argue that there is a balance between adversarial and original instances, and more reasonable approaches should be considered.

4.3   Counterfactual Test

Previous approaches [13,56] indicated that fine-tuned models, such as BERT, learn simple decision rules that perform well on the test set but do not capture a dataset's intended capabilities. For example, given the sentence "In 1979, Hanks lived in New York City, where he ...," we can classify the sentence into the label, "cities_of_residence," owing to the phrase "lived in." We seek to understand whether the prediction will change, given a sentence lacking such an indicating phrase. This sentence indicates the contrast set, as noted in [14].

---

[4] https://github.com/thunlp/OpenAttack.

Table 4. Important token generation with attention, integrated gradients and contrastive masking.

Method	Text	Label
Attention	[CLS] In 1979, Hanks lived in the New York City , ... [SEP]	cities_of_residence
Integrated Gradients	[CLS] In 1979, Hanks lived in the New York City , ... [SEP]	cities_of_residence
Constasive Masking	[CLS] In 1979, Hanks [MASK] [MASK] the New York City , ... [SEP]	**NOT**_cities_of_residence

We humans can easily identify that this instance does not contain the relation "cities_of_residence," via counterfactual reasoning. Motivated by this, we took our first step toward analyzing the generalization of RE in contrast sets. In this setting, we should generate examples of few permutations but with opposite labels. In contrast to the previous approach [14], which utilized crowdsourcing, we generated the contrast set automatically. Hence, we proposed a novel counterfactual data augmentation method lacking human intervention to generate contrast sets. We first generated the most informative tokens of the sentences regarding its relation labels, and we then introduced contrastive masking to obscure those tokens to generate the contrast set, as shown in Table 4.

Specifically, we leveraged *integrated gradients* [33] to generate informative tokens. We did not leverage attention scores [38] because [22] pointed out that analyzing only attention weights would be insufficient when investigating the behavior of the attention head. Furthermore, attention weights disregarded the hidden vector's values. Moreover, as shown in Table 4, we empirically observed that attention scores were not suitable for generating important tokens.

Intuitively, integrated gradients is a variation on computing the gradient of the prediction output w.r.t. features of the input, which simulate the process of pruning the specific attention head from the original attention weight, α, to a zero vector, α', via back-propagation [56]. Note that integrated gradients can generate attribution scores reflecting how much changing the attention weights will change the model's outputs. In other words, the higher of attribution score, the greater importance given to attention weights. Given an input, x, the attribution score of the attention head, t, can be computed using:

$$Atr(\alpha^t) = (\alpha^t - \alpha'^t) \otimes \int_{x=0}^{1} \frac{\partial F(\alpha' + x(\alpha - \alpha'))}{\partial \alpha^t} dx, \tag{3}$$

where $\alpha = [\alpha^1, \ldots, \alpha^T]$, and $\otimes$ is the element-wise multiplication. $Atr(\alpha^t) \in \mathbf{R}^{n \times n}$ denotes the attribution score, which corresponds to the attention weight α^t. Naturally, $F(\alpha' + x(\alpha - \alpha'))$ is closer to $F(\alpha')$ when x is closer to 0, and it is closer to α when x is closer to 1. We set the uninformative baseline α' as a zero vector and denote $Atr(\alpha_{i,j}^t)$ as the interaction from token $\mathbf{h}_i$ to $\mathbf{h}_j$. We approximate $Atr(\alpha^t)$ via a gradient summation function following [34,56]:

$$Atr(\alpha^t) ::= (\alpha^t - \alpha'^t) \odot \sum_{i=1}^{s} \frac{\partial F(\alpha' + i/s(\alpha - \alpha'))}{\partial \alpha'^t} \times \frac{1}{s}, \qquad (4)$$

where s is the number of approximation steps for computing the integrated gradients. We selected the top $k = 1,2$ informative tokens from instance and implemented contrastive masking by replacing informative tokens with unused tokens (e.g., [unused5]). We leveraged this procedure for both training and testing datasets, and the overall algorithm is:

Algorithm 1. Counterfactual Data Augmentation for RE

1: Train Relation Classifier RE with X,Y
2: **for** x in X **do**
3:     ig = IntegratedGradients(RE)
4:     attributes = ig.attribute(X)
5:     candidates = select_top_k(attributes)
6:     x^{mask} = mask(x,candidates)
7:     X ← X ∩ x^{mask} Y ← Y ∩ NA
8: Re-train RE with X,Y

We generated 15,509 samples to construct a robust contrast set. Since the Wiki80 dataset does not contain NA relation, we only evaluate results on the TACRED dataset with the performance of vanilla BERT and BERT with counterfactual data augmentation (BERT+CDA). Note that the contrast set of RE comprised instances with **NOT_such_relation (NA)** labels. Thus, we utilized the F1 score including NA.

Table 5. Counterfactual analysis results on the TACRED dataset (F1 score including **NA**). The former and the latter indicates the results of contrast and original sets, separately.

Model	TACRED
BERT (Origin)	87.7
BERT (Contrast Set, $k = 1$)	32.6
BERT (Contrast Set, $k = 2$)	45.1
BERT+CDA (Contrast Set/Origin, $k = 1$)	89.0/86.8
BERT+CDA (Contrast Set/Origin, $k = 2$)	95.0/87.2

Results and Analysis. From Table 5, we notice that BERT achieved poor performance on the robust set, which shows that fine-tuned models lack the ability of counterfactual reasoning. We also found that BERT+CDA achieved better results than BERT on a robust set, indicating that counterfactual data augmentation was beneficial. Note that, unlike previous counterfactual data augmentation approaches, such as [5,38], our method was a simple, yet effective, automatic algorithm that can be applied to other tasks (e.g., event extraction [6], text classification [7,52], sentiment analysis [28] and question answering [54]).

5   Diagnosing Generalization with *Bias*

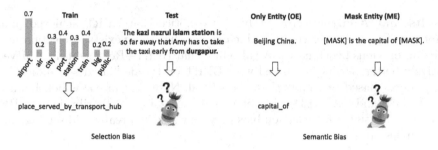

Fig. 2. Diagnosing generalization with bias.

5.1   Selection Bias

Selection bias emerges from the non-representative observations, such as when the users generating the training observations have different distributions than that in which the model is intended to be applied [32]. For a long time, selection bias (a.k.a. sample bias) has been a concern in social sciences, so much so that considerations of this bias are now considered primary considerations in research design. For the RE, we are the first to have studied selection bias. Given a running example, as shown in Fig. 2, owing to the high frequency of the token "airport," the fine-tuned model can memorize the correlation between the existence of the token and the relation "place_served_by_transport_hub" by neglecting the low-frequency words (e.g., "train station").

The origin of the selection bias is the non-representative data. The predicted output is different from the ideal distribution, for example, because the given demographics cannot reflect the ideal distribution, resulting in lower accuracy. To analyze the effect of selection bias for RE, we constructed a de-biased test set that replaced high-frequency tokens with low-frequency ones. We evaluated the performance of BERT on a de-biased set, and we introduced a simple method, BERT+De-biased, which masks the tokens based on the token frequency (neglecting the stop words and common words, such as "the," "when," and "none").

Table 6. Selection bias analysis results on the Wiki80 and TACRED datasets. The former indicates the results of de-biased sets while the latter indicates the results of orginal sets.

Model	Wiki80	TACRED
BERT (Origin)	86.2	67.5
BERT (De-biased)	80.2	64.7
BERT+De-biased (De-biased)	84.3/85.3	67.1/67.4

Results and Analysis. From Table 6, we notice that BERT achieved poor performance on the de-biased set, which shows that there exists a selection bias for RE in previous benchmarks. We also find that BERT+Re-weighting achieved relatively better results, compared with BERT on the de-biased set, indicating that frequent-based re-sampling was beneficial. Note that previous benchmarks (e.g., Wiki80 and TACRED) did not reveal the real data distribution for RE. Therefore, we argue that selection bias may be worse in a real-world setting, and more studies are required.

Table 7. Semantic bias analysis results on the Wiki80 and TACRED dataset.

Model	Wiki80				TACRED			
	Origin	OE	ME	De-biased	Origin	OE	ME	De-biased
BERT	86.2	66.5	52.4	67.0	67.5	42.9	36.6	57.4
BERT+ME (50%)	86.4	65.3	73.5	67.9	67.9	42.9	55.8	61.5
BERT+ME (100%)	86.0	62.9	74.7	68.1	67.4	43.6	55.9	60.5
BERT+ME (Frequency)	–	–	–	–	67.9	40.8	53.5	61.6

5.2 Semantic Bias

Embeddings (i.e., vectors representing the meanings of words or phrases) have become a mainstay of modern NLP, which provides flexible features that are easily applied to deep machine learning architectures. However, previous approaches [27] indicate that these embeddings may contain undesirable societal and unintended stereotypes (e.g., connecting medical nurses more frequently to female pronouns than male pronouns). This is an example of semantic bias.

The origin of the Semantic bias may be the parameters of the embedding model. Semantic bias will indirectly affect the outcomes and error disparities by causing other biases (e.g., diverging word associations within embeddings or LMs [32]). To analyze the effects of semantic bias, we conducted experiments with two settings, as inspired by [15]: a *masked-entity* (ME) setting, wherein entity names are replaced with a special token, and an *only-entity* (OE) setting, wherein only the names of the two entities are provided. We also constructed a de-biased test set in which instances were wrongly predicted in the OE setting.

We conducted experiments on these datasets and introduced a simple method of selective entity masking to mitigate semantic bias. We masked $K\%$ of the entities with unused tokens to guide the model to pay closer attention to the context (BERT+ME ($K\%$)). We intuitively selected K via entity-pair frequencies (BERT+ME (Frequency))[5].

Results and Analysis. From Table 7, we observe that the models suffered a significant performance drop with both the ME and OE settings. Moreover, it was surprising to notice that, in most cases, with only entity names can archive better performance than those of text only with entities masked. These empirical results illustrate that both entity names and text provided important information for RE, and entity names contributed even more, indicating the existence of semantic bias. It is contrary to human intuition since we identity relations mainly through the context between the given entities, whereas the models take more entity names into consideration. Furthermore, we noticed that BERT achieved poor performance on the de-biased set. We also found that BERT+ME (k) obtained better performance than BERT and BERT+ME (frequency) achieved the best results on the de-biased set, indicating that selective entity masking was beneficial.

6 Discussion and Limitation

Evaluation of NLP Models. Several models that leveraged pre-trained and fine-tuned regimes have achieved promising results with standard NLP benchmarks. However, the ultimate objective of NLP is generalization. Previous works [31] attempted to analyze this generalization capability using NLP models' comprehensive behavioral tests. Motivated by this, we took the CheckList paradigm a step further to investigate generalization via robustness and bias. We used RE as an example and conducted experiments. Empirically, the results showed that the BERT performed well on the original test set, but it exhibited poor performance on the robust and de-biased sets, as shown in Fig. 3. This indicates that generalization should be carefully considered in the future.

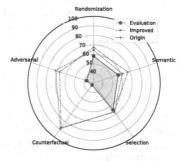

Fig. 3. Generalization analysis results of RE on TACRED. The *origin* and *evaluation* refer to the BERT performance on the original, robust/de-biased test set, respectively. The *improved* indicates the performance of our proposed methods.

Limitations. We only considered single-label classifications because there was only one relation for each instance. Arguably, there could exist multiple labels for each instance (e.g., multiple RE [45]). Moreover, apart from selection and semantic biases, label bias exists with the over-amplification of NLP [92]. Using the label bias as an example, the distribution of

---

the dependent variable in the train set may diverge substantially from the test, leading to the deterioration of performance. We leave this problem for future work.

7   Conclusion and Future Work

We investigated the generalizability of fine-tuned pre-trained models (i.e., BERT) for RE. Specifically, we diagnosed the bottleneck with regard to existing approaches in terms of robustness and bias, resulting in several directions for future improvement. We introduced several improvements, such as counterfactual data augmentation, sample re-weighting, which can be used to improve generalization. We regard this study as a step toward a unified understanding of generalization, and this offers hopes for further evaluations and improvements of generalization, including conceptual and mathematical definitions of NLP generalization.

References

1. Alt, C., Gabryszak, A., Hennig, L.: Probing linguistic features of sentence-level representations in neural relation extraction. arXiv preprint arXiv:2004.08134 (2020)
2. Alt, C., Gabryszak, A., Hennig, L.: Tacred revisited: a thorough evaluation of the tacred relation extraction task. arXiv preprint arXiv:2004.14855 (2020)
3. Alt, C., Hübner, M., Hennig, L.: Fine-tuning pre-trained transformer language models to distantly supervised relation extraction. arXiv preprint arXiv:1906.08646 (2019)
4. Arpit, D., et al.: A closer look at memorization in deep networks. arXiv preprint arXiv:1706.05394 (2017)
5. Chen, L., Yan, X., Xiao, J., Zhang, H., Pu, S., Zhuang, Y.: Counterfactual samples synthesizing for robust visual question answering. In: Proceedings of the IEEE/CVF Conference on Computer Vision and Pattern Recognition, pp. 10800–10809 (2020)
6. Deng, S., Zhang, N., Kang, J., Zhang, Y., Zhang, W., Chen, H.: Meta-learning with dynamic-memory-based prototypical network for few-shot event detection. In: Proceedings of the 13th International Conference on Web Search and Data Mining, pp. 151–159 (2020)
7. Deng, S., Zhang, N., Sun, Z., Chen, J., Chen, H.: When low resource NLP meets unsupervised language model: Meta-pretraining then meta-learning for few-shot text classification (student abstract). In: AAAI, pp. 13773–13774 (2020)
8. Devlin, J., Chang, M.W., Lee, K., Toutanova, K.: BERT: pre-training of deep bidirectional transformers for language understanding. In: Proceedings of NAACL, pp. 4171–4186. Association for Computational Linguistics, Minneapolis, June 2019
9. Ebrahimi, J., Rao, A., Lowd, D., Dou, D.: HotFlip: white-box adversarial examples for text classification. arXiv preprint arXiv:1712.06751 (2017)
10. Feng, J., Huang, M., Zhao, L., Yang, Y., Zhu, X.: Reinforcement learning for relation classification from noisy data. In: Proceedings of AAAI (2018)
11. Fort, S., Nowak, P.K., Jastrzebski, S., Narayanan, S.: Stiffness: a new perspective on generalization in neural networks. arXiv preprint arXiv:1901.09491 (2019)

12. Fu, J., Liu, P., Zhang, Q., Huang, X.: Rethinking generalization of neural models: a named entity recognition case study. In: AAAI, pp. 7732–7739 (2020)
13. Gardner, M., et al.: Evaluating NLP models via contrast sets. CoRR abs/2004.02709 (2020)
14. Gardner, M., et al.: Evaluating NLP models via contrast sets. arXiv preprint arXiv:2004.02709 (2020)
15. Han, X., et al.: More data, more relations, more context and more openness: a review and outlook for relation extraction. arXiv preprint arXiv:2004.03186 (2020)
16. Han, X., Liu, Z., Sun, M.: Neural knowledge acquisition via mutual attention between knowledge graph and text (2018)
17. Han, X., et al.: FewRel: a large-scale supervised few-shot relation classification dataset with state-of-the-art evaluation. In: Riloff, E., Chiang, D., Hockenmaier, J., Tsujii, J. (eds.) Proceedings of the 2018 Conference on Empirical Methods in Natural Language Processing, Brussels, Belgium, October 31 - November 4 2018, pp. 4803–4809. Association for Computational Linguistics (2018). https://doi.org/10.18653/v1/d18-1514
18. Huang, Y.Y., Wang, W.Y.: Deep residual learning for weakly-supervised relation extraction. arXiv preprint arXiv:1707.08866 (2017)
19. Ji, G., Liu, K., He, S., Zhao, J., et al.: Distant supervision for relation extraction with sentence-level attention and entity descriptions. In: Proceedings of AAAI, pp. 3060–3066 (2017)
20. Jin, D., Jin, Z., Tianyi Zhou, J., Szolovits, P.: Is BERT really robust? A strong baseline for natural language attack on text classification and entailment. arXiv pp. arXiv-1907 (2019)
21. Kingma, D.P., Ba, J.: Adam: a method for stochastic optimization. CoRR abs/1412.6980 (2015)
22. Kobayashi, G., Kuribayashi, T., Yokoi, S., Inui, K.: Attention module is not only a weight: analyzing transformers with vector norms. ArXiv abs/2004.10102 (2020)
23. Li, L., et al.: Normal vs. adversarial: salience-based analysis of adversarial samples for relation extraction. arXiv preprint arXiv:2104.00312 (2021)
24. Lin, Y., Shen, S., Liu, Z., Luan, H., Sun, M.: Neural relation extraction with selective attention over instances. In: Proceedings of ACL, vol. 1, pp. 2124–2133 (2016)
25. Liu, T., Wang, K., Chang, B., Sui, Z.: A soft-label method for noise-tolerant distantly supervised relation extraction. In: Proceedings of the 2017 Conference on Empirical Methods in Natural Language Processing, pp. 1790–1795 (2017)
26. Liu, Y., et al.: RoBERTa: a robustly optimized BERT pretraining approach. arXiv preprint arXiv:1907.11692 (2019)
27. Nadeem, M., Bethke, A., Reddy, S.: StereoSet: measuring stereotypical bias in pretrained language models. arXiv preprint arXiv:2004.09456 (2020)
28. Peng, H., Xu, L., Bing, L., Huang, F., Lu, W., Si, L.: Knowing what, how and why: a near complete solution for aspect-based sentiment analysis. In: AAAI, pp. 8600–8607 (2020)
29. Qin, P., Xu, W., Wang, W.Y.: DSGAN: generative adversarial training for distant supervision relation extraction. In: Proceedings of ACL (2018)
30. Ren, S., Deng, Y., He, K., Che, W.: Generating natural language adversarial examples through probability weighted word saliency. In: Proceedings of the 57th Annual Meeting of the Association for Computational Linguistics, pp. 1085–1097 (2019)

31. Ribeiro, M.T., Wu, T., Guestrin, C., Singh, S.: Beyond accuracy: behavioral testing of NLP models with checklist. In: Jurafsky, D., Chai, J., Schluter, N., Tetreault, J.R. (eds.) Proceedings of the 58th Annual Meeting of the Association for Computational Linguistics, ACL 2020, Online, 5–10 July 2020, pp. 4902–4912. Association for Computational Linguistics (2020). https://www.aclweb.org/anthology/2020.acl-main.442/

32. Shah, D., Schwartz, H.A., Hovy, D.: Predictive biases in natural language processing models: a conceptual framework and overview. arXiv preprint arXiv:1912.11078 (2019)

33. Sundararajan, M., Taly, A., Yan, Q.: Axiomatic attribution for deep networks. arXiv preprint arXiv:1703.01365 (2017)

34. Sundararajan, M., Taly, A., Yan, Q.: Axiomatic attribution for deep networks. In: Proceedings of the 34th International Conference on Machine Learning, ICML 2017, vol. 70. p. 3319–3328. JMLR.org (2017)

35. Tramèr, F., Kurakin, A., Papernot, N., Goodfellow, I., Boneh, D., McDaniel, P.: Ensemble adversarial training: attacks and defenses. arXiv preprint arXiv:1705.07204 (2017)

36. Wang, H., et al.: Extracting multiple-relations in one-pass with pre-trained transformers. arXiv preprint arXiv:1902.01030 (2019)

37. Wen, Y., Li, S., Jia, K.: Towards understanding the regularization of adversarial robustness on neural networks (2019)

38. Wiegreffe, S., Pinter, Y.: Attention is not not explanation. arXiv preprint arXiv:1908.04626 (2019)

39. Wu, Y., Bamman, D., Russell, S.: Adversarial training for relation extraction. In: Proceedings of EMNLP, pp. 1778–1783 (2017)

40. Yang, Z., Dai, Z., Yang, Y., Carbonell, J., Salakhutdinov, R.R., Le, Q.V.: XLNet: generalized autoregressive pretraining for language understanding. In: Advances in Neural Information Processing Systems, pp. 5753–5763 (2019)

41. Ye, H., Chao, W., Luo, Z., Li, Z.: Jointly extracting relations with class ties via effective deep ranking. In: Proceedings of ACL, vol. 1, pp. 1810–1820 (2017)

42. Ye, H., et al.: Contrastive triple extraction with generative transformer. arXiv preprint arXiv:2009.06207 (2020)

43. Yu, H., Zhang, N., Deng, S., Ye, H., Zhang, W., Chen, H.: Bridging text and knowledge with multi-prototype embedding for few-shot relational triple extraction. arXiv preprint arXiv:2010.16059 (2020)

44. Zeng, D., Liu, K., Chen, Y., Zhao, J.: Distant supervision for relation extraction via piecewise convolutional neural networks. In: Proceedings of EMNLP, pp. 1753–1762 (2015)

45. Zeng, D., Zhang, H., Liu, Q.: CopyMTL: copy mechanism for joint extraction of entities and relations with multi-task learning. In: AAAI, pp. 9507–9514 (2020)

46. Zeng, W., Lin, Y., Liu, Z., Sun, M.: Incorporating relation paths in neural relation extraction. In: Proceddings of EMNLP (2017)

47. Zeng, X., He, S., Liu, K., Zhao, J.: Large scaled relation extraction with reinforcement learning. In: Processings of AAAI, vol. 2, p. 3 (2018)

48. Zhang, N., et al.: OpenUE: an open toolkit of universal extraction from text. In: Proceedings of the 2020 Conference on Empirical Methods in Natural Language Processing: System Demonstrations, pp. 1–8 (2020)

49. Zhang, N., Deng, S., Sun, Z., Chen, J., Zhang, W., Chen, H.: Relation adversarial network for low resource knowledge graph completion. In: Proceedings of The Web Conference 2020, pp. 1–12 (2020)

50. Zhang, N., et al.: Long-tail relation extraction via knowledge graph embeddings and graph convolution networks. In: Proceedings of the NAACL, pp. 3016–3025 (2019)
51. Zhang, N., Deng, S., Sun, Z., Chen, X., Zhang, W., Chen, H.: Attention-based capsule networks with dynamic routing for relation extraction. In: Proceedings of EMNLP (2018)
52. Zhang, N., Jia, Q., Yin, K., Dong, L., Gao, F., Hua, N.: Conceptualized representation learning for Chinese biomedical text mining. arXiv preprint arXiv:2008.10813 (2020)
53. Zhang, Y., Zhong, V., Chen, D., Angeli, G., Manning, C.D.: Position-aware attention and supervised data improve slot filling. In: Proceedings of EMNLP, pp. 35–45 (2017)
54. Zhong, H., Wang, Y., Tu, C., Zhang, T., Liu, Z., Sun, M.: Iteratively questioning and answering for interpretable legal judgment prediction. In: Proceedings of the AAAI Conference on Artificial Intelligence, vol. 34, pp. 1250–1257 (2020)
55. Zhong, M., Wang, D., Liu, P., Qiu, X., Huang, X.: A closer look at data bias in neural extractive summarization models. arXiv preprint arXiv:1909.13705 (2019)
56. Zhou, P., Khanna, R., Lin, B.Y., Ho, D., Ren, X., Pujara, J.: Can BERT reason? Logically equivalent probes for evaluating the inference capabilities of language models. arXiv preprint arXiv:2005.00782 (2020)

KA-NER: Knowledge Augmented Named Entity Recognition

Binling Nie[1(✉)], Chenyang Li[2], and Honglie Wang[2]

[1] Digital Media Technology, Hangzhou Dianzi University, Hangzhou, Zhejiang, China
`binlingnie@hdu.edu.cn`
[2] Computer Science and Technology, Zhejiang University, Hangzhou, Zhejiang, China
`{licy_cs,wanghonglie}@zju.edu.cn`

Abstract. Most named entity recognition models comprehend words based solely on their contexts and man-made features, but neglect relevant knowledge. Incorporating prior knowledge from a knowledge base is straightforward but non-trivial due to two challenges: knowledge noise from unrelated span-entity mappings and knowledge gap between a text and a knowledge base. To tackle these challenges, we propose KA-NER, a novel knowledge-augmented named entity recognition model, in which sanitized entities from a knowledge base are injected to sentences as prior knowledge. Specifically, our model consists of two components: a knowledge filtering module to filter domain-relevant entities and a knowledge fushion module to bridge the knowledge gap when incorporating knowledge into a NER model. Experimental results show that our model achieves significant improvements against baseline models on different domain datasets.

Keywords: Named entity recognition · Knowledge filtering · Knowledge fushion

1 Introduction

Named entity recognition aims to detect terminology spans and classify them into pre-defined semantic categories such as person, location, organization, etc. It is a fundamental problem in natural language processing, particularly for knowledge graph construction and question answering. Existing named entity recognition methods adopt neural network architectures like LSTM-CRF with word-level and character-level representations [9,12,57]. When dealing with a text containing many specific domain terminologies, prevailing approaches solely capture word semantics based on plain texts and man-made features, which is far from sufficiency. Fortunately, high quality, human-curated knowledge contained in open knowledge bases (KB, e.g. Wikipedia [54]) is able to provide richer information for recognizing named entities. For example, given a text, we can utilize its prior knowledge to disambiguate boundaries and classify entity types. As shown in Fig. 1, the text span *diabetes insipidus* is associated with the KB

© Springer Nature Singapore Pte Ltd. 2021
B. Qin et al. (Eds.): CCKS 2021, CCIS 1466, pp. 60–75, 2021.
https://doi.org/10.1007/978-981-16-6471-7_5

entity *Diabetes_insipidus*. Therefore, the type of the text span is more likely to be predicted as the type of entity *Diabetes_insipidus*, i.e., *Disease*.

However, there are two main challenges when incorporating a knowledge base into named entity recognition models. (1) **Knowledge Noise:** taking all KB entities into consideration may cause noisy data problem. Given a text, we may find various span-entity mappings in KBs. Most of these entities are unrelated informative ones and simply leveraging all these entities may degrade the NER performance. Figure 2 illustrates an example from NCBI disease corpus [24], in which text spans *smooth muscles* and *neuromuscular diseases* are linked to KB entities *Smooth_musscle_tissue* and *Neuromuscular_disease* respectively. Since it is a disease name recognition task, the entity *smooth_muscle_tissue* with type */medicine/anatomical_structure* would mislead the model to disambiguate wrong boundary and classify incorrect entity type to the text span *smooth muscles*. (2) **Knowledge Gap:** the training procedure for language representation is quite different from the knowledge base representation procedure, which leads to two individual vector spaces. Given a text and its prior knowledge base, representations of a mention span in a text may have a low similarity score with the embeddings of its gold linked entity. These two type of feature are vector-space inconsistent, adding them directly may degrade the recognition performance of the model. Therefore, appropriately dealing with the gap of heterogeneous knowledge is crucial.

To address the above challenges, we propose a novel knowledge augmented named entity recognition model (KA-NER), which enhances named entity recognition with informative entities. To depress knowledge noise, we introduce a knowledge filter scheme based on KB entity types and word spans. For restraining domain-irrelevant knowledge noise, we design a rule-dependent entity selector. The intuition is that all entities which are not related to the target domain should be rejected even with high linked priors, while those relevant entities be retained. Afterwards, we apply a consistent knowledge fushion module to address

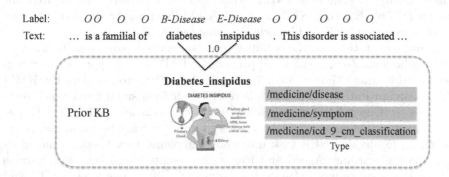

Fig. 1. An example of incorporating extra knowledge information for named entity recognition. The text span *diabetes insipidus* is linked to the KB entity *Diabetes_insipidus*. With knowing *Diabetes_insipidus* is */Medicine/Disease*, it is much easy to disambiguate the boundary and classify type *Disease* to *diabetes insipidus*.

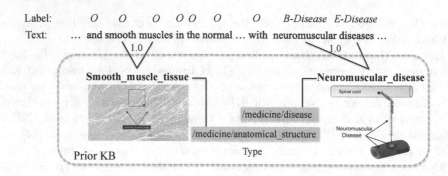

Fig. 2. An example of domain-irrelevant noise. Taking all linked entities into consideration may cause knowledge noise. Compared to the entity *Neuromuscular_disease*, the entity *Smooth_musscle_tissue* is out of the domain *Disease* and will mislead model to predict wrong labels.

the knowledge gap issue. Specifically, we construct a text-entity graph to bridge KB and plain texts, which models the connections between text spans and the matched entities. Based on the graph, a graph attention network embeds a word in a sentence by recursively aggregating node representations of its neighboring words and entities. Besides, we encode the graph structure of KBs with word embedding algorithms like GloVe [35] instead of utilizing triples facts in KBs directly and take the informative entity embeddings as the input of entity node in text-entity graph.

2   Related Work

2.1   Named Entity Recognition

Named Entity Recognition (NER) task has been widely investigated for decades [27,56]. Existing NER approaches can be roughly divided into two branches: common NER and knowledge enhanced NER.

Traditional statistic methods follow some supervised paradigms to learn to make predictions. Zhou and Su [58] proposed a HMM-based approach to recognize named entities. Malouf [33] added multiple features to compare the HMM with Maximum Entropy (ME). Schapire [46] combined small fixed-depth decision trees as binary AdaBoost classifiers for NER. Takeuchi and Collier [51] as well as Li et al. [28] adopted SVM models to improve NER performance. Ando and Zhang [5] divided NER task into many supplementary tasks by introducing structural learning. Agerri and Rigau [2] presented classifiers with various features in a semi-supervised way. For DrugNER [48], Shengyu et al. [49] and Rocktäschel [44] achieved promising results by adopting a CRF with different features. However, traditional statistic methods heavily rely on hand-crafted features or task-specific resources, which are cost-expensive and restricted by domains or languages.

Deep neural network architectures have also been vigorously studied. Researchers utilize different representations of characters, words, sub-word units or any combinations of these to assign proper categories to words in a sentence. The earliest research efforts on neural NER were based on feature vectors constructed from orthographic features, dictionaries and lexicons [11]. Later, researchers utilized word embeddings as input to Recurrent Neural Networks (RNN) instead of manually constructed features [8,42]. Meanwhile, character level architectures [23,41] advanced the study of deep neural networks on NER. Many researchers have proved that combining word contexts and the characters of a word could improve the overall performance with little domain specific knowledge or resources [9,12,29,30,34,38,57]. They studied different variants of this type model with a CNN/Bi-LSTM layer over characters of a word, a Bi-LSTM layer over the word representations of a sentence, and a softmax or CRF layer to generate labels. Furthermore, various studies focused on more elaborate deep neural network to improve the performance of NER. Xiao et al. [55] designed a similarity based auxiliary classifier to calculate the similarities between words and tags, and then compute a weighted sum of the tag vectors as a useful feature for NER tasks. Lu et al. Jiang et al. [20] studied differential neural architecture search method to recognize named entity for the first time. Arora et al. [6] presented a semi-markov structured support vector machine model for high-precision named entity recognition by assigning weights to different types of errors in the loss-augmented inference during training.

Recently, many efforts have been made to incorporate relevant knowledge to improve the performance of NER. Some researches pre-trained language models on biomedical literatures to improve the performance of biomedical named entity recognition [7,19,21,25]. Liu et al. [31] proposed to enrich pre-trained language models with relevant knowledge for downstream tasks, including NER. Liu et al. [32] extends a traditional, RNN-based neural language model, for unsupervised NER, which is the first unsupervised knowledge-augmented NER approach. Local Distance Neighbor (LDN) [3], substituted any external resource of knowledge such as the gazetteer to recognize emerging named entities in noisy user-generated text. There are also various methods incorporating gazetteer into a NER system [13]. [1] presented an incorporating token-level dictionary feature method which use a labeled dataset rather than an external dictionary and decoupled dictionary features from the external dictionary during the training stage. Meanwhile, growing experts started paying attention on utilizing knowledge bases to augment NER task. Rijhwani et al. [43] introduced soft gazetteer to extract available information from English knowledge bases for NER. A more recent work related to ours is KAWR [18], which designed a new recurrent unit (GREU) and a relation attention scheme to strengthen contextual embeddings from knowledge wise for downstream tasks, including NER. Nevertheless, KAWR only utilizes relation information to filter knowledge, which is far from sufficiency. By prior entity linking, each sentence obtains as many potential mentions as the linking model and each candidate mention would have various linked KB entities. The relations between potential mentions in a sentence can be simplified by taking entity types into consideration when filtering knowledge.

2.2   Other Related Work

There is a growing literature focusing on graph representation learning. These methods [15,17,22,39,52] aim to learn the low-dimensional representations of nodes in a graph. R-GCN [47] introduced convolutional neural networks that operate on graphs and applies spectral graph convolutions on multi-relational data to extract connectivity patterns in graph structure. Graph attention networks [53] performed node classification of graph-structured data by leveraging masked self-attention layers, which is adopted in our pipeline.

Entity linking maps an entity mention in texts to a KB entity. Most of existing methods compute the similarity of entity mentions and KB entities by modeling their local context [14,16,36,50]. Recently, KnowBERT-wiki [40] incorporated global information into entity linking by mention-span self-attention and achieved state-of-the-art performance. In this work, we adopt KnowBERT-wiki to generate candidate entities of sentences.

3   The Proposed Approach

In this section, we introduce a novel **K**nowledge **A**ugmented **N**amed **E**ntity **R**ecognition framework (KA-NER), which is able to effectively extract relevant informative entities from knowledge base for better named entity recognition.

3.1   Overview

The overall pipeline of the proposed framework is illustrated in Fig. 3. Our model KA-NER consists of two main components: the knowledge filtering module and the knowledge fushion module. The knowledge filtering problem is formulated as follows: given sentences and the related set of <KB entity, word span, KB entity type set> triples as $X = \{(e_1, s_1, T_1), (e_2, s_2, T_2), ..., (e_n, s_n, T_n)\}$, where e_i is the linked entity related to the word span ($s_i = (start_i, end_i)$) in sentences, and T_i (T_i can be empty) is the KB entity type set of the linked entity. The goal is to examine and determine which entity can provide strong signals for entity recognition and thus should be selected as a candidate one. Firstly, we filter those domain-irrelevant entities by an entity selector based on entity types. Then, the knowledge fushion problem is formulated as follows: given a sentence $<w_1, w_2, ..., w_n>$ and the selected candidate entities $<e_1, e_2, ..., e_m>$ with corresponding word spans $<s_1, s_2, ..., s_m>$, we construct a linked graph to integrate entities and the corresponding words into sentences, then use graph attention networks to model over the linked graph for knowledge fushion, and finally predict the word w_i's predefined semantic category y_i based on the linked graph.

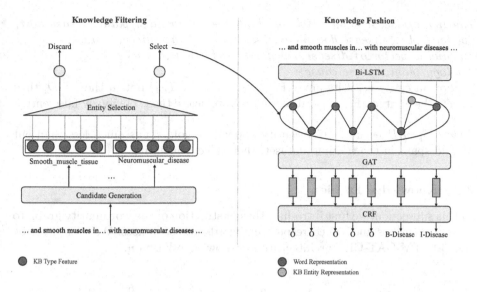

Fig. 3. The overall architecture of KA-NER. The entity selector chooses entities according to rules, and then the selected entities are used for knowledge fushion. For knowledge fushion, it adopts BiLSTM-GAT-CRF architecture to inject knowledge and assign predefined categories to words in a sentence.

3.2 Knowledge Filtering

Candidate Generation. For a mention $m \in \{x_i...x_j, 1 \leq i \leq j \leq n\}$ in X, our goal is to generate related KB entities $\mathcal{C}(m) = \{e_1, ..., e_k\}$. We first retrieve entities from KB by rules (like a WordNet lemmatizer) and heuristics (like string match). Then, we leverage KnowBERT [40], a powerful knowledge-aware language model to score every mention-entity pair with a prior and we only keep entities with priors above the threshold prior ρ.

Entity Selector. To encompass as wide knowledge as possible, we adopt an entity selector to select candidates from all possible entities. Specifically, given an input sentence and a KB, an entity linker takes the sentence as the input and returns a list of potential entities. Each entity is accompanied by a mention span and a KB entity type list. We introduce two rules to identify whether an entity is a candidate. The two rules can greatly depress the noises during inference.

Rule 1 Domain Dependent. Given a set of <KB entity, KB entity type set> triples as $X = \{(e_1, T_1), (e_2, T_2), ..., (e_n, T_n)\}$ and the predefined categories of the dataset C, find the corresponding KB entity types as a target type set $D = \{d_1, d_2, ..., d_l\}$. The dataset dependent rule is $\{\forall t_i \notin D, t_i \in T_i\} \Rightarrow \{reject se_i\}$. Take NCBI dataset [24] as an example, the predefined categories are $C = (B - Disease, I - Disease, E - Disease, S - Disease, O)$. Obviously, the target type list is $D = \{/medicine/disease, /medicine/disease\_cause,$

/*medicine/disease_stage*, /*biology/plant_disease*, /*biology/plant_disease_host*,
/*biology/plant_disease_documentation*, /*biology/plant_disease_cause*,
/*medicine/vector_of_disease*, /*medicine/infectious_disease*,
/*biology/plant_disease_cause*}.

For an entity e_i, if its any KB entity type $t_i \in T_i$ is not in the set D, then rejects the entitye_i. For type-absent entity, we delete those obviously false entity.

Rule 2 Span Dependent. If an entity appears in gold spans much less than out of gold spans in the training dataset, then reject the entity.

3.3   Knowledge Fushion

In this subsection, we first introduce the construction of the word-entity graph to integrate entities and the corresponding words into sentences. We then describe the BiLSTM-GAT-CRF architecture for knowledge Fushion.

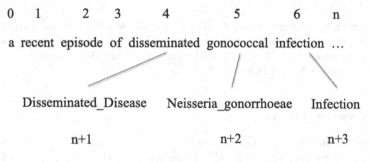

Fig. 4. Structure of a word-entity graph.

The Construction of Word-Entity Graph. To integrate entities and the corresponding words, we construct a word-entity interactive graph. The vertex set of this graph is made up of words in the sentence and linked prior entities. For example, as shown in Fig. 4, the vertex set is $V = \{a, recent, episode, of, disseminated, gonococcal, infection, ..., Disseminated\_Disease, Neisseria\_gonorrhoeae, Infection\}$. Adjacency matrix($\mathbb{A}$) is introduced to represent the edge set. As shown in Fig. 4, the word-entity graph can capture the boundaries and semantic information of entities. The word span $(4, 4)$ matches the entity $Disseminated\_Disease$, we will assign $\mathbb{A}_{4(n+1)} = 1$ and $\mathbb{A}_{(n+1)4} = 1$. Moreover, the word *gonococcal* is the nearest preceding or following word of a word *disseminated*, $\mathbb{A}_{45}$ and $\mathbb{A}_{54}$ will be assigned a value of 1.

BiLSTM. The input of the architecture is a sentence and all linked entities of the sentence. Given a sentence as $s = \{w_1, w_2, ..., w_n\}$ and its linked entities $e = \{e_1, e_2, ..., e_n\}$, a character BiLSTM is used to encode the character sequence within each word and automatically extract word level features w_i^c. Each word's representation w_i is the concatenation of a word embedding w_i^w by looking up from a pre-train word embedding matrix and the character sequence encoding hidden vector w_i^c.

To avoid changing the original sentence's meaning, a word BiLSTM is applied to $s = \{w_1, w_2, ..., w_n\}$. The words' contextual representation $\mathbb{H} = \{h_1, h_2, ..., h_n\}$ is the concatenation of the left-to-right and right-to-left LSTM hidden states.

$$\begin{aligned} \overrightarrow{h_i} &= \overrightarrow{LSTM}(w_i, \overrightarrow{h_{i-1}}) \\ \overleftarrow{h_i} &= \overleftarrow{LSTM}(w_i, \overleftarrow{h_{i-1}}) \end{aligned} \quad (1)$$

To represent the semantic information of entities, we look up entity embeddings e_i from a pre-trained entity GloVe embeddings. We directly concatenate word contextual representation and entity embeddings as the output of encoding layer, which is the initial node features of word-entity graph.

$$nf_0 = [h_1, h_2, ..., h_n, e_1, e_2, ..., e_m] \quad (2)$$

Graph Attention Network over Word-Entity Graph. We apply Graph Attention Network (GAT) to model over the word-entity graph. For a multi-layers GAT, the input to the i-th GAT layer is a set of node features, $nf_i = \{f_1, f_2, ..., f_{n+m}\}$, together with its adjacency matrix $\mathbb{A}$, $f_i \in \mathbb{R}^F$, $A \in \mathbb{R}^{(n+m) \times (n+m)}$, where $n+m$ is the number of nodes in word-entity graph, F is the dimension of node features. The i-th GAT layer outputs a new set of node features $nf_i\prime = \{f_1\prime, f_2\prime, ..., f_n\prime\}$. For K independent attention head, a GAT operation can be written as:

$$f_i\prime = \|_{k=1}^K \sigma(\sum_{j \in N_i} \alpha_{ij}^k \mathbb{W}^k f_j) \quad (3)$$

$$\alpha_{ij}^k = \frac{exp(LeakyReLU(a^T[\mathbb{W}^k f_i \| \mathbb{W}^K f_j]))}{\sum_{k \in N_i} exp(LeakyReLU(a^T[\mathbb{W}^k f_i \| \mathbb{W}^K f_k]))} \quad (4)$$

where $\|$ represents concatenation, α_{ij}^k are normalized attention coefficients by k-th attention head, and $\mathbb{W}^k$ is the corresponding input linear transformation's weight matrix. On the final layer of network, averaging is employed to get final output features.

$$nf_{final}' = \sigma(\frac{1}{K} \sum_{k=1}^K \sum_{j \in N_i} \alpha_{ij}^k \mathbb{W}^k h_j) \quad (5)$$

CRF. The inference layer takes a weighted sum of word sequence representations and final GAT output feature representations as input features (denoted as $R = \{r_1, r_2, ..., r_n\}$), which can avoid introducing noises to the original sentence, then assign labels to the word sequence. We adopt CRF to capture label dependencies by adding transition scores between neighboring labels. The probability of the ground-truth tag sequence $y = \{y_1, y_2, ..., y_n\}$ is

$$p(y|s) = \frac{exp(\sum_i (\mathbb{W}^{y_i} + \mathbb{T}_{(y_{i-1}, y_i)}))}{\sum_{y\prime} exp(\sum_i (\mathbb{W}^{y_i'} + \mathbb{T}_{(y_{i-1}', y_i')}))} \quad (6)$$

Therefore, the loss function of named entity recognizer is defined as:

$$\mathbb{L} = -\sum_{i=1}^{N} log(p(y_i|s_i))$$

(7)

4 Experimental Setup

In this section, we briefly describe the datasets across different domains and the baseline methods. Our experiments are conducted on four commonly-used datasets, namely NCBI [24], CoNLL2003 [45], Genia [37] and SEC [4]. We also give the hyper-parameter configuration used in our experiments.

For hyper-parameter settings, the hidden state size of LSTM is 200. The dropout rate is set as 0.2 and 0.5 for LSTM output and pretrained embedding respectively. We apply a two-layer GAT for knowledge fushion. The first layer consists of $K = 5$ attention heads computing $F = 30$ features each, followed by an exponential linear unit (ELU) [10] nonlinearity. The second layer is a single attention head that computes C features, where C is the number of classes in NER. The dropout rate is set as 0.1 and 0.4 for both GAT layers and pretrained entity embedding respectively. We use SGD optimizer in a mini-batch size of 16 with learning rate $\gamma = 1 \times 10^{-3}$ and L_2 regularization to $\lambda = 0.005$. In this paper, we adopt KnowBert-Wiki [40] to construct prior knowledge bases of three datasets from Wikidata. To evaluate the performance of knowledge fushion, we use entity GloVe embeddings obtained by a skip-gram like objective [35], pre-trained graph embeddings by using PBG [26], and one hot entity type embeddings. For entity types, we merge corresponding Freebase type system and Dbpedia type system to cover more entities.

5 Result and Analysis

5.1 Overall Performance

In this experiment, we compare our model with several state-of-the-art baselines, in terms of precision, recall and F1 score. We give the experimental results in Table 1.

- It can be observed that our method KA-NER significantly outperforms all baselines on all four datasets. Specifically, KA-NER achieves 2.79% on higher performance than KAWR+BiLSTM+FC (in terms of F1 score), which is the state-of-the-art model to leverage knowledge base into the NER task. Two advantages of KA-NER make its superiority of NER performance: (1) We introduce KB type schema to denoisy irrelevant knowledge for the first time. More accurate knowledge is the key point for NER. (2) We construct a word-entity graph to capture the boundary information of the mention spans. Besides, we introduce a GAT to learn the comprehensive node representations in the word-entity graph, which effectively handles the knowledge gap issue.

Table 1. Overall performance of KG4NER on NCBI (phrase-level F1), P: Precision, R: Recall, F: F1, AVG: Average value of four datasets.

Methods	Metrics	NCBI	CoNLL03	Genia	SEC	AVG
NCRF++	P	83.80	90.08	67.82	76.73	77.55
	R	83.89	91.71	77.22	81.47	82.97
	F	83.85	90.89	72.22	79.03	80.14
BERT+FC	P	75.56	90.08	67.82	76.73	77.55
	R	81.46	91.71	77.22	81.47	82.97
	F	78.40	90.89	72.22	79.03	80.14
BERT+BiLSTM+FC	P	77.21	90.99	68.13	76.95	78.32
	R	80.42	92.13	77.42	83.78	83.44
	F	78.78	91.56	72.48	80.22	80.76
KAWR+FC	P	77.13	91.10	68.40	77.27	78.45
	R	81.88	92.47	77.52	85.33	84.30
	F	79.43	91.78	72.68	81.1	81.25
KAWR+BiLSTM+FC	P	78.21	91.40	68.64	80.95	79.80
	R	83.02	92.20	77.25	85.33	84.45
	F	80.55	91.80	72.69	83.08	82.03
KA-NER	P	86.58	92.32	76.16	91.88	86.73
	R	85.67	92.47	75.03	79.28	83.11
	F	**86.12**	**92.40**	**75.59**	**85.16**	**84.82**

- KA-NER performs substantially better on biology domain datasets, NCBI and Genia, than universal domain datasets, CoNLL2003 and SEC. The reason is that domain-specific NER task are supposed to recognize more terminology. Terminology usually has been collected into knowledge base and belongs to specific types, which provide strong boundary and type signal for NER.
- Among all the baselines, the methods without BiLSTM module to represent word sequence has poorer performance on four datasets. It is mainly because that BiLSTM has the advantages to capture global features. Those global features contain more implicit contextual semantics to benefit the NER task.

5.2 Performance of Knowledge Filtering

We analyze performances of knowledge filtering on NCBI. As shown in Table 2, *prior* > 0 means all linked KB entities whose entity linking prior is greater than 0 are considered to construct the word-entity graph in KA-NER. As we analyzed before, many KB entities with low priors are domain-relevant. We can observe that the proportion of matched KB entities is advanced with the decrease of entity linking prior. At the same time, the number of noisy entities also increases with the increase of valid entities when in low priors. Therefore, knowledge filtering is needed. Table 2 illustrates that knowledge filtering improve performances

Table 2. The performance of knowledge filtering. Prior: the linked prior of candidate entities, Raw: considering all candidate entities, Filtered: only including valid candidate entities by knowledge filtering.

Datasets		P	R	F
NCBI ($prior > 0$)	Raw	0.8331	0.7887	0.8103
	Filtered	0.8585	0.8316	0.8448
NCBI ($prior > 25\%$)	Raw	0.8284	0.7981	0.8130
	Filtered	0.8658	0.8567	0.8612
NCBI ($prior > 50\%$)	Raw	0.8315	0.8107	0.8210
	Filtered	0.8652	0.8525	0.8588
NCBI ($prior > 75\%$)	Raw	0.8436	0.8399	0.8417
	Filtered	0.8628	0.8483	0.8555
NCBI ($prior = 100\%$)	Raw	0.8416	0.8536	0.8476
	Filtered	0.8552	0.8525	0.8539

Table 3. Comparison among four different fushion operators With/Without linked priors (F1). Weight = 1: Without linked priors, Weight = prior: with linked prior.

Method	None	Smul	Vmul	Concat
Weight = 1	84.52	**86.12**	83.43	85.88
Weight = priors	82.63	85.58	82.86	82.75

of KA-NER under all linking priors, showing the superiority of the filter based on KB entity type schema. $prior > 25\%$ gives the best F1 score of 86.12% on NCBI. The main reason is that our knowledge filtering module makes significant contributions on selecting domain-relevant entities only by semantic information and type features. Furthermore, entity linking prior also has the ability to filter noisy data (Table 3).

5.3 Performance of Knowledge Fushion

In this paragraph, we analyze the contributions and effects of knowledge fushion module. Firstly, we experiment with NCRF++ and KA-NER to measure the performance of encoding heterogeneous knowledge. Compared to KA-NER (BiLSTM-GAT-CRF), NCRF++ is a baseline without GATs to incorporate knowledge into NER, which only consists of BiLSTM layers and CRF layers for automatic feature extracting and inferencing respectively. From Table 1, we can observe that KA-NER consistently and effectively leverage knowledge into NER with a performance improvement. Besides, we evaluate KA-NER with different fusion operations on NCBI to evaluate whether the knowledge fushion module can avoid introducing noisy to the original sentence. In Fig. 3, *None*: directly take final GAT output as the input of CRF layer, *Smul*: take a weighted

Table 4. Comparison among three different entity embeddings With/Without linked priors. Weight = 1: Without linked priors (F1), Weight = prior: with linked prior.

Method	Glove	PBG	Type
Weight = 1	86.12	84.58	**83.34**
Weight = priors	86.06	85.32	82.68

sum of word sequence representations and final GAT output as the input of CRF layer, *Vmul*: a variant of smul, the shape of weight equals that of predefined categories, *Concat*: take the concatenation of word sequence representations and final GAT output as the input of CRF layer. Directly taking the output of GAT as the input of CRF layer gives lower F1-score than three other fusion operations, which proves that too much knowledge incorporation does hurt the meaning of original sentences (Table 4).

To further demonstrate the effects of linked prior, we conduct experiments with unweighted graph (weight = 1) and prior weighted graph (weight = prior). A GAT based on completely unweighted graph is equivalent to uniformly incorporating representations of linked KB entities for each word in sentence. Experiments in Fig. 3 show that prior weighted graph will hurt the performance of knowledge fushion module.

We further examine the performance of knowledge fushion module with different entity embeddings. Figure 4 shows the F1-scores of KA-NER on BC5CDR with three different entity embeddings. In Fig. 4, *GloVe*: entity GloVe embeddings, *PBG*: PBG embeddings, *Type*: entity type one hot embeddings. Compared with PBG embeddings and entity type one hot embeddings, KA-NER using entity GloVe embeddings give significant improvements. The GloVe 300-dimension embeddings performs better than PBG embeddings, which is consistent with the observation that entity GloVe embeddings fill the gap of structure knowledge in KB and semantics in plain texts.

6 Conclusions

In this paper, we propose KA-NER model for knowledge-enhanced NER task. Our method exploits an effective knowledge filter to directly capture the properties of KB types and word spans, leading to several advantages with respect to raw knowledge. We also empirically show that sentence representations that incorporate structural knowledge in knowledge base improves NER performance. Experiments on four challenging datasets demonstrate the effectiveness of our proposed model.

References

1. Mu, X., Wang, W., Xu, A.: Incorporating token-level dictionary feature into neural model for named entity recognition. Neurocomputing **375**, 43–50 (2020)
2. Agerri, R., Rigau, G.: Robust multilingual named entity recognition with shallow semi-supervised features. Artif. Intell. **238**(C), 63–82 (2016)
3. Al-Nabki, M.W., Fidalgo, E., Alegre, E., Fernández-Robles, L.: Improving named entity recognition in noisy user-generated text with local distance neighbor feature. Neurocomputing **382**, 1–11 (2019)
4. Alvarado, J.C.S., Verspoor, K., Baldwin, T.: Domain adaption of named entity recognition to support credit risk assessment. In: Proceedings of the Australasian Language Technology Association Workshop, pp. 84–90 (2015)
5. Ando, R.K., Zhang, T.: A framework for learning predictive structures from multiple tasks and unlabeled data. J. Mach. Learn. Res. **6**, 1817–1853 (2005)
6. Arora, R., Chen-Tse, T., Ketevan, T., Prabhanjan, K., Yang, Y.: A semi-Markov structured support vector machine model for high-precision named entity recognition. In: Proceedings of the Annual Meeting of the Association for Computational Linguistics, pp. 5962–5866 (2019)
7. Beltagy, I., Lo, K., Cohan, A.: SciBERT: a pretrained language model for scientific text. In: Proceedings of the Conference on Empirical Methods in Natural Language Processing and the International Joint Conference on Natural Language Processing, pp. 3606–3611 (2019)
8. Chalapathy, R., Borzeshi, E.Z., Piccardi, M.: An investigation of recurrent neural architectures for drug name recognition. In: Proceedings of the International Workshop on Health Text Mining and Information Analysis, pp. 1–5 (2016)
9. Chen, H., Lin, Z., Ding, G., Lou, J., Zhang, Y., Karlsson, B.: GRN: Gated relation network to enhance convolutional neural network for named entity recognition. In: Proceedings of the AAAI Conference on Artificial Intelligence, vol. 33, pp. 6236–6243 (2019)
10. Clevert, D.A., Unterthiner, T., Hochreiter, S.: Fast and accurate deep network learning by exponential linear units (ELUs). In: Proceedings of the International Conference on Learning Representations (2016)
11. Collobert, R., Weston, J.: A unified architecture for natural language processing: deep neural networks with multitask learning. In: Proceedings of the International Conference on Machine Learning, pp. 160–167 (2008)
12. Cui, L., Zhang, Y.: Hierarchically-refined label attention network for sequence labeling. In: Proceedings of the Conference on Empirical Methods in Natural Language Processing and the International Joint Conference on Natural Language Processing, pp. 4106–4119 (2019)
13. Ding, R., Xie, P., Zhang, X., Lu, W., Si, L.: A neural multi-digraph model for Chinese NER with gazetteers. In: Proceedings of the 57th Annual Meeting of the Association for Computational Linguistics (2019)
14. Ganea, O.E., Hofmann, T.: Deep joint entity disambiguation with local neural attention. In: Proceedings of the Conference on Empirical Methods in Natural Language Processing, pp. 2619–2629 (2017)
15. Grover, A., Leskovec, J.: node2vec: scalable feature learning for networks. In: Proceedings of the ACM International Conference on Knowledge Discovery and Data Mining, pp. 855–864 (2016)
16. Gupta, N., Singh, S., Roth, D.: Entity linking via joint encoding of types, descriptions, and context. In: Proceedings of the Conference on Empirical Methods in Natural Language Processing, pp. 2681–2690 (2017)

17. Hamilton, W., Ying, Z., Leskovec, J.: Inductive representation learning on large graphs. In: Proceedings of the Advances in Neural Information Processing Systems, pp. 1024–1034 (2017)
18. He, Q., Wu, L., Yin, Y., Cai, H.: Knowledge-graph augmented word representations for named entity recognition. In: Proceedings of the AAAI Conference on Artificial Intelligence, pp. 7919–7926 (2020)
19. Huang, K., Altosaar, J., Ranganath, R.: ClinicalBERT: modeling clinical notes and predicting hospital readmission (2019)
20. Jiang, Y., Chi, H., Tong, X., Chunliang, Z., Zhu, J.: Improved differentiable architecture search for language modeling and named entity recognition. In: Proceedings of the Conference on Empirical Methods in Natural Language Processing, pp. 3583–3588 (2019)
21. Johnson, A.E., et al.: MIMIC-III, a freely accessible critical care database. Sci. Data **3**(1), 1–9 (2016)
22. Kipf, T.N., Welling, M.: Semi-supervised classification with graph convolutional networks. In: Proceedings of the International Conference on Learning Representations (2017)
23. Kuru, O., Can, O.A., Yuret, D.: CharNER: character-level named entity recognition. In: Proceedings of the International Conference on Computational Linguistics, pp. 911–921 (2016)
24. Leaman, R., Lu, Z.: NCBI disease corpus: a resource for disease name recognition and concept normalization. J. Biomed. Inform. **47**, 1 (2014)
25. Lee, J., et al.: BioBERT: a pre-trained biomedical language representation model for biomedical text mining. Bioinformatics **36**(4), 1234–1240 (2020)
26. Lerer, A., et al.: PyTorch-BigGraph: a large-scale graph embedding system. In: Proceedings the Conference on Machine Learning and Systems (2019)
27. Li, J., Sun, A., Han, J., Li, C.: A survey on deep learning for named entity recognition. IEEE Trans. Knowl. Data Eng. (2020)
28. Li, Y., Bontcheva, K., Cunningham, H.: SVM based learning system for information extraction. In: Winkler, J., Niranjan, M., Lawrence, N. (eds.) DSMML 2004. LNCS (LNAI), vol. 3635, pp. 319–339. Springer, Heidelberg (2005). https://doi.org/10.1007/11559887_19
29. Lu, P., Bai, T., Langlais, P.: SC-LSTM: learning task-specific representations in multi-task learning for sequence labeling. In: Proceedings of the North American Chapter of the Association for Computational Linguistics, pp. 2396–2406 (2019)
30. Liu, K., Li, S., Zheng, D., Lu, Z., Li, S.: A prism module for semantic disentanglement in name entity recognition. In: Proceedings of the Annual Meeting of the Association for Computational Linguistics (2019)
31. Liu, W., Zhou, P., Zhao, Z., Wang, Z., Wang, P.: K-BERT: enabling language representation with knowledge graph. In: Proceedings of the AAAI Conference on Artificial Intelligence, pp. 2901–2908 (2020)
32. Liu, A., Du, J., Stoyanov, V.: Knowledge-augmented language model and its application to unsupervised named-entity recognition. In: Proceedings of the ACM International Conference on Knowledge Discovery and Data Mining, pp. 2901–2908 (2020)
33. Malouf, R.: Markov models for language-independent named entity recognition. In: Proceedings of the Conference on Natural Language Learning, vol. 20, pp. 1–4 (2002)
34. Mayhew, S., Gupta, N., Roth, D.: Robust named entity recognition with truecasing pretraining. In: Proceedings of the AAAI Conference on Artificial Intelligence (2020)

35. Mikolov, T., Sutskever, I., Chen, K., Corrado, G.S., Dean, J.: Distributed representations of words and phrases and their compositionality. In: Proceedings of the Advances in Neural Information Processing Systems, pp. 3111–3119 (2013)
36. Murty, S., Verga, P., Vilnis, L., Radovanovic, I., McCallum, A.: Hierarchical losses and new resources for fine-grained entity typing and linking. In: Proceedings of the Annual Meeting of the Association for Computational Linguistics, vol. 1, pp. 97–109 (2018)
37. Ohta, T., Tateisi, Y., Kim, J.D., Mima, H., Tsujii, J.: The GENIA corpus: an annotated research abstract corpus in molecular biology domain. In: Proceedings of the International Conference on Human Language Technology Research, pp. 82–86 (2002)
38. Peng, M., Zhang, Q., Xing, X., Gui, T., Fu, J., Huang, X.: Learning task-specific representation for novel words in sequence labeling. In: Twenty-Eighth International Joint Conference on Artificial Intelligence (2019)
39. Perozzi, B., Al-Rfou, R., Skiena, S.: DeepWalk: online learning of social representations. In: Proceedings of the ACM International Conference on Knowledge Discovery and Data Mining, pp. 701–710 (2014)
40. Peters, M.E., et al.: Knowledge enhanced contextual word representations. In: Proceedings of EMNLP-IJCNLP, pp. 43–54 (2019)
41. Pham, T.-H., Le-Hong, P.: End-to-end recurrent neural network models for vietnamese named entity recognition: word-level vs. character-level. In: Hasida, K., Pa, W.P. (eds.) PACLING 2017. CCIS, vol. 781, pp. 219–232. Springer, Singapore (2018). https://doi.org/10.1007/978-981-10-8438-6_18
42. Plank, B., Søgaard, A., Goldberg, Y.: Multilingual part-of-speech tagging with bidirectional long short-term memory models and auxiliary loss. In: Proceedings of the Annual Meeting of the Association for Computational Linguistics, vol. 2, pp. 412–418 (2016)
43. Rijhwani, S., Zhou, S., Neubig, G., Carbonell, J.: Soft gazetteers for low-resource named entity recognition. In: Proceedings of the Annual Meeting of the Association for Computational Linguistics, pp. 2901–2908 (2020)
44. Rocktäschel, T., Huber, T., Weidlich, M., Leser, U.: WBI-NER: the impact of domain-specific features on the performance of identifying and classifying mentions of drugs. In: Proceedings of the International Workshop on Semantic Evaluation, pp. 356–363 (2013)
45. Sang, E.T.K., Buchholz, S.: Introduction to the CoNLL-2000 shared task chunking. In: Proceedings of the Conference on Computational Natural Language Learning and the Second Learning Language in Logic Workshop, pp. 127–132 (2000)
46. Schapire, R.E.: Explaining AdaBoost. In: Schölkopf, B., Luo, Z., Vovk, V. (eds.) Empirical Inference, pp. 37–52. Springer, Heidelberg (2013). https://doi.org/10.1007/978-3-642-41136-6_5
47. Schlichtkrull, M., Kipf, T.N., Bloem, P., van den Berg, R., Titov, I., Welling, M.: Modeling relational data with graph convolutional networks. In: Gangemi, A., et al. (eds.) ESWC 2018. LNCS, vol. 10843, pp. 593–607. Springer, Cham (2018). https://doi.org/10.1007/978-3-319-93417-4_38
48. Segura-Bedmar, I., Martınez, P., Herrero-Zazo, M.: SemEval-2013 Task 9: extraction of drug-drug interactions from biomedical texts, Atlanta, Georgia, USA, vol. 3206, no. 65, p. 341 (2013)
49. Shengyu, L., Buzhou, T., Qingcai, C., Xiaolong, W.: Effects of semantic features on machine learning-based drug name recognition systems: word embeddings vs. manually constructed dictionaries. Information 6(4), 848–865 (2015)

50. Sil, A., Cronin, E., Nie, P., Yang, Y., Popescu, A.M., Yates, A.: Linking named entities to any database. In: Proceedings of the Joint Conference on Empirical Methods in Natural Language Processing and Computational Natural Language Learning, pp. 116–127 (2012)
51. Takeuchi, K., Collier, N.: Use of support vector machines in extended named entity recognition. In: Proceedings of the Conference on Natural language learning, vol. 20, pp. 1–7 (2002)
52. Tang, J., Qu, M., Wang, M., Zhang, M., Yan, J., Mei, Q.: LINE: large-scale information network embedding. In: Proceedings of the International Conference on World Wide Web, pp. 1067–1077 (2015)
53. Veličković, P., Cucurull, G., Casanova, A., Romero, A., Liò, P., Bengio, Y.: Graph attention networks. In: Proceedings of the International Conference on Learning Representations, pp. 4171–4186 (2018)
54. Vrandečić, D., Krötzsch, M.: Wikidata: a free collaborative knowledgebase. Commun. ACM **57**(10), 78–85 (2014)
55. Xiao, S., Yuanxin, O., Wenge, R., Jianxin, Y., Xiong, Z.: Similarity based auxiliary classifier for named entity recognition. In: Proceedings of the Conference on Empirical Methods in Natural Language Processing, pp. 1140–1149 (2019)
56. Yadav, V., Bethard, S.: A survey on recent advances in named entity recognition from deep learning models. In: Proceedings of the International Conference on Computational Linguistics, pp. 2145–2158 (2018)
57. Yang, J., Zhang, Y.: NCRF++: an open-source neural sequence labeling toolkit. In: Proceedings of the Demonstrations at the Annual Meeting of the Association for Computational Linguistics, pp. 74–79 (2018)
58. Zhou, G., Su, J.: Named entity recognition using an HMM-based chunk tagger. In: Proceedings of the Annual Meeting of the Association for Computational Linguistics, pp. 473–480 (2002)

Structural Dependency Self-attention Based Hierarchical Event Model for Chinese Financial Event Extraction

Zhi Liu[1,2,3,4], Hao Xu[1], Haitao Wang[1(✉)], Dan Zhou[1], Guilin Qi[5], Wanqi Sun[1], Shirong Shen[5], and Jiawei Zhao[1]

[1] Zhejiang Lab, Hangzhou, China
wanghaitao@zhejianglab.com
[2] Institutes for Robotics and Intelligent Manufacturing, Chinese Academy of Sciences, Shenyang, China
[3] University of Chinese Academy of Sciences, Beijing, China
[4] Shenyang Institute of Automation, Chinese Academy of Sciences, Shenyang, China
[5] School of Computer Science and Engineering, Southeast University, Nanjing, China

Abstract. Document-level event extraction (DEE) now draws a huge amount of researchers' attention. Not only the researches on sentence-level event extraction have obtained a great progress, but researchers realize that an event is usually described by multiple sentences in a document especially for fields such as finance, medicine, and judicature. Several document-level event extraction models are proposed to solve this task and obtain improvements on DEE task in recent years. However, we noticed that these models fail to exploit the entity dependency information of trigger and arguments, which ignore the dependency information between arguments, and between the trigger and arguments especially for financial domain. For DEE task, a model needs to extract the event-related entities, i.e., trigger and arguments, and predicts its corresponding roles. Thus, the entity dependency information between trigger and argument, and between arguments are essential. In this work, we define 8 types of structural dependencies and propose a document-level Chinese financial event extraction model called SSA-HEE, which explicitly explores the structure dependency information of candidate entities and improves the model's ability to identify the relevance of entities. The experimental results show the effectiveness of the proposed model.

Keywords: Structural dependency · Hierarchical event representation · Financial event extraction · Document event extraction

1 Introduction

Event Extraction (EE), a task of Nature Language Processing, purposes to extract a detailed event trigger and its arguments to form a structural event from unstructured sentences. In recent years, the utilization of EE assists practitioners to obtain valuable information from various industries, and to make appropriate decisions or avoid potential risks. Especially in the financial areas, a large number of financial announcements

B. Qin et al. (Eds.): CCKS 2021, CCIS 1466, pp. 76–88, 2021.
https://doi.org/10.1007/978-981-16-6471-7_6

with similar content, rigorous structure, and sufficiently comprehensive information are generated at every minute. In the case of high-performance digital calculation, valuable information can be extracted from thousands of financial announcements and news in a few seconds. Previous methods of EE mainly include pattern recognition [1, 2] and machine learning [3–5]. Compared with pattern recognition and machine learning, deep-learning-based event extraction requires a large amount of labeled data and computing resources. With the rapid development of big data and deep learning technology in the past decade, deep learning models have been widely used in event extraction.

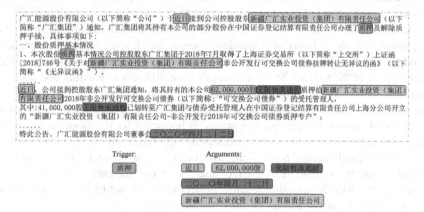

Fig. 1. An example of a Chinese financial announcement. Different entities are distinguished by colored rounded rectangles. The trigger entity and argument entity are listed below. We omit some content of the announcement for the clarity purpose. As we can see, the trigger and arguments scatter in the text. The argument "新疆广汇实业投资 (集团) 有限责任公司" appears at least 3 times in different parts of the announcement.

Event extraction mainly includes sentence-level EE (SEE) and document-level EE (DEE). Most of the current methods [6–8] focus on SEE. But in most cases, the event trigger and arguments are usually scattered across different sentences, and sometimes even more than one event can be found in one document. In the past few years, some encouraging achievements have been made in financial EE. Yang *et al.* propose a model called DCFEE [9] and explore DEE on ChFinAnn for the first time. This framework automatically generates labeled data based on distant supervision (DS) [10] and extracts events by using an LSTM-NER model and an arguments-completion strategy. Zheng *et al.* used a novel end-to-end model Doc2EDAG [11], which can generate an entity-based directed acyclic graph (EDAG) and simplify the hard table-filling task into sequential path-expanding sub-tasks. Doc2EDAG also considers the arguments-scattering and multi-event problems.

For the DEE task, however, a model needs to extract the event-related entities, i.e., trigger and arguments, and predict the corresponding roles of the extracted entities. Thus, the entity dependency information between trigger and argument, and between arguments are essential. For Chinese document-level financial event extraction, we notice the following facts: the event trigger may occur in multiple sentences throughout a document; one argument may occur in multiple sentences; different arguments may either occur in the same sentence simultaneously or occur in different sentences; just as the instance shown in Fig. 1. In the previous works, the proposed models capture the entity dependencies implicitly by the underlying language model (e.g., BERT, GPT-2, RoBERTa). These pretrained language models, however, are trained either to generate natural language or to obtain the bidirectional contextual representations. Thus, the previous works are failed to consider the entity dependency information properly and effectively in DEE task, which can aid the model to handle the issues of long-term dependency and coreference resolution properly.

As we can see from Fig. 1, the trigger entity "质押" appears in two sentences that are closer to each other. The argument entity "新疆广汇实业投资 (集团) 有限责任公司" scatters in multiple sentences in this document, so does the argument entity "近日". Intuitively, these kinds of dependencies indicate rich interactions among entities, and thereby provide informative priors for trigger and arguments extraction. Entities that appear in the same/different sentences have different contextual and semantic information. This different information of the same entity or different entity offers the dependency and semantic relationships between entities. This dependency information of the essential entities (trigger and arguments) is important for the model to extract the event. In this paper, we propose a structure dependency self-attention based hierarchical event extraction model, called SSA-HEE, to extract financial events from Chinese financial announcements and news. SSA-HEE explicitly explores the entity dependencies in a document to enrich the information that the model used to extract trigger and arguments. The contributions of this work can be summarized as follows:

- We summarized various types of entity dependencies that appeared in a document into a unified form. By incorporating these structural dependencies explicitly, the proposed model can be able to perform context and structure reasoning. In the end, the performance is improved.
- The proposed model achieves the state-of-art results on Chinese financial document-level event extraction.

2 Related Works

Challenges for sentence-level EE can be concluded as insufficient labeled data, multi-event extraction, and Multilingual event extraction. To address the problem of insufficient labeled data, Yang et al. separate the argument prediction in terms of roles to avoid the problem of overlapping roles and then edit prototypes to generate labeled data and select the samples according to the quality [12]. Sha et al. propose the DBRNN framework based on a recurrent neural network and enhanced by dependency bridges [13]. They also demonstrate the superior performance of simultaneously applying tree structure and sequence structure in RNN. Liu et al. pay attention to the situation of multiple events existing in the same sentence and exploit the JMEE framework to jointly extract multiple event triggers and arguments by implementing syntactic shortcut arcs in model graph information [14]. Subburathinam et al. focus on cross-lingual structure transfer techniques [15]. Relation- and event-relevant language-universal features are considered, and graph convolutional networks are utilized to train a relation or event extractor from the source language and apply it to the target language. Their experimental results show that the proposed model is comparable. In other studies, some significant improvements are obtained by the neural network structure improvements. LSTM and CNN are combined to capture sentence-level and lexical information in Chinese EE [16]. GAN is used for entity and event extraction with discriminators estimating proper rewards of the difference between the labels committed by the ground-truth (expert) and the extractor (agent) [17].

For DEE, Liao *et al.* [8] enforce the performance of ACE event extraction across document-level information. They consider information about different types of events rather than about the same type to predict if events and event arguments would occur in a text. Du *et al.* [18] propose a novel multi-granularity reader to aggregate the neural representations' information when learning at sentence-level or paragraph-level, which has been proved to be effective on the MUC-4 EE dataset. They are the first to investigate the effect of end-to-end neural sequence models when solving the document-level role filler extraction problem. Similarly, the end-to-end model has also exhibited improvements in capturing cross-event dependencies for DEE evaluated by [19]. Huang *et al.* probe a structured prediction algorithm named Deep Value Networks (DVN) and conducted experiments on ACE05, the results of which present that the proposed approach achieves higher computational efficiency and impressive performance to CRF-based models. For financial event extraction, DCFEE is the first probe to extract Chinese financial events proposed by Yang *et al.* [9]. This model can generate labeled data automatically and extract events from the whole documents in two stages: 1) a sequence tagging task for SEE, and 2) a key event detection and arguments completion for DEE. Doc2EDAG is another DEE model [11] designed for the Chinese financial EE task. By transforming tabular event data into entity-based directed acyclic graphs, this model has made further improvements in extracting multi-event and addressing event argument scattering in the context.

3   Methodology

3.1   Hierarchical Event Structure for Financial Event

The overall architecture of SSA-HEE is shown in Fig. 2. Different commonly event extraction datasets, such as ACE2005, KBP 2015, Chinese financial event extraction has the whole word tokenization and contains various subdivision event types, such as *Equity reduction* and *Equity increase* can be grouped into the category of *Trade*. The hierarchical event structure can express the relationships between similar event types and help the model to distinguish different event types. In [20], the proposed dynamic hierarchical event structure demonstrates its effectiveness. In this work, we define a hierarchical financial event (HFE). Part of the HFE is shown in Fig. 3. The details of the hierarchical events are shown in Appendix A.

Entity Dependency Structure based Joint Hierarchical Event Extraction Model. As shown in Fig. 2, the SSA-HEE model consists of a candidate argument extractor, a hierarchical event feature construction module, pedal attention mechanisms that obtain the semantic relationship and representation between candidate arguments and triggers,

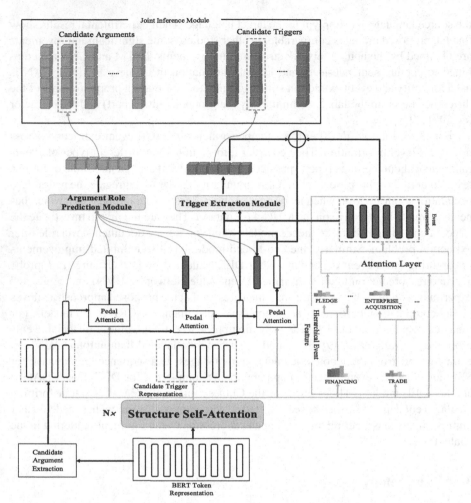

Fig. 2. The overall architecture of SSA-HEE model. The representation of the content of the document are obtained by the BERT. Then, the candidate arguments that contained in the document are extracted by the LTP tool. The contextual representation of the document is fed into the structure self-attention block to incorporate the dependency information. The candidate triggers then are obtained by the same way as candidate arguments. After that, the model can aggregate the candidate triggers and arguments through the pedal attention and the dependency-information-enhanced contextual representation and generate the argument-oriented and trigger-oriented representations. At the meantime, the model obtains the hierarchical financial event representation according to the pre-defined hierarchical financial event. The hierarchical financial event representation and the trigger-oriented are keep the detailed event information to aid the argument role prediction module when the argument role prediction module predicts the argument role for each candidate argument.

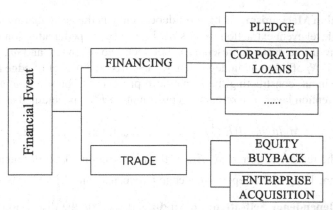

Fig. 3. A part of the hierarchical financial event type. For the samples in this work, we cooperate with financial experts to define a series of super class for the major events of public companies, such as FINANCING, TRADE and IPO RELATED. The event type of each sample is categorized into the pre-defined super classes.

a structural dependency self-attention module which is made up of a series of structural dependency self-attention layer, and a joint inference module which generates the joint probability for training and extraction.

Candidate Argument Extraction. Given the target financial document, there are no candidate arguments about the document. Thus, we need to extract the potential arguments at first. All target documents are annotated in the BIO schema. The candidate arguments are extracted from the target document by a BERT-based sequence annotation model. When this process is done, we obtain the candidate arguments $A = a_1, a_2, \ldots\ldots, a_K$.

Hierarchical Event Representation. The superordinate event type is important information in trigger classification. In this work, we make use of the hierarchical event attention proposed to construct each event type's feature. In this work, each event type T_i^e has a superordinate type T_i^s and a set of candidate argument A. We explore the scaled dot-product attention [21] to generate attention weights of T_i^e and T_i^s to each candidate argument. Then, each attention weight of T_i^e inherits the attention weight of its superordinate type (if it exists). The final attention weights $\widehat{W}_i^e$ of event type T_i^e for each candidate argument are recursively constructed in the following way:

$$\widehat{W}_i^e = \begin{cases} W_i^e, & T_i^e \text{ has no superordinate type} \\ (W_i^e + W_i^s)/2, & T_i^s \text{ is superordinate type of } T_i^e \end{cases}$$

After the attention weights are obtained, the hierarchical event representation F_i^e is computed as following,

$$F_i^e - \left[T_i^e, \sum_{j=1}^{K} \left(\widehat{W}_{i,j}^e (M_e E_{a_j} \mid b_e) \right) \right]$$

where E_{a_j} is the token representation of the argument a_j.

Pedal Attention Mechanism. Long-term dependency is the issue that must be solved in document-level event extraction. In this work, we use the pedal attention mechanism [20] to capture the semantic representation between two tokens ω_i and ω_j. It takes the adjacent words N_i of token ω_i in dependency parse tree and the set of edge D^i between ω_i and N_i as input, constructing the semantic representation between ω_i and ω_j by a multi-head attention layer. The semantic representation $F^p_{(i,j)}$ is obtained as following,

$$F^p_{(i,j)} = Multi\_head\,(E_j, \left[E_{N^i_1}, \ldots\ldots, E_{N^i_l}\right], \left[E_{D^i_1}, \ldots\ldots, E_{D^i_l}\right])$$

where E_j is the representation of ω_j, $E_{N^i_k}$ is the representation of k-th token in N_i, and $E_{D^i_k}$ is the embedding of the dependency edge between ω_j and N^i_k.

Structural Dependency Self-attention Module. Based on the discussion over entity dependency, how to explicitly structure the dependency information and combine it with the training samples is the key issue that to be solved. Inspired by [22], we define 8 types of entity-relationship according to the locations of the entities in a document. we use *intra/inter* to distinguish whether the entities appear in the same sentence and *coref/relate* to indicate whether the entities refer to the same entity. The 8 entity relationships are defined as follows:

- A trigger that appears in the same sentence multiple times is denoted as *T_intra + coref*.
- A trigger that appears in different sentences is denoted as *T_inter + coref*.
- An argument that appears in the same sentence multiple times is denoted as *A_intra + coref*.
- An argument that appears in different sentences is denoted as *A_inter + coref*.
- Trigger and arguments that occur in the same sentence are denoted as *intra + related*, which means this pair of distinctive entities are possibly related under certain predicates.
- Trigger and argument that occur in different sentence are denoted as *inter + related*.
- Relations between non-entity (NE) token and trigger/argument in the same sentence are denoted as *intraNE*.
- For other inter-sentences NE tokens and trigger/argument relations, we follow Xu's work which assumes there is no crucial dependency and denoted as *NA*.

Thus, the all 8 structure dependencies are formulated as 8 entity-centric adjacency matrices with all elements from a finite dependency set: {*T_intra + coref, T_inter + coref, A_intra + coref, A_inter + coref, intra + relate, inter + relate, intraNE, NA*} (see Fig. 4).

To combine the entity dependency into the end-to-end self-attention model, we instantiate each s_{ij} as neural layers with specific parameters. As a result, for each input structure S, we have a structured model composed of corresponding layer parameters. Then, we use biaffine transformation to incorporate query and key vector into a single-dimensional bias vector, and model the prior bias for each dependency independently. The biaffine transformation is computed as follows,

$$bias^l_{ij} = q^l_i A^l_{s_{ij}} {k^l_j}^T + b^l_{s_{ij}}$$

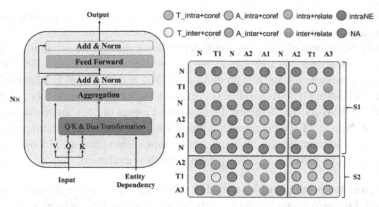

Fig. 4. Entity dependency structure and the structure self-attention (SSA). Each document has an entity dependency matrix of which each element indicates one type of entity dependency. In this case, a two sentences matrix is shown.

Where l means the l-th layer, $A_{s_{ij}}^l$ is the trainable layer of parameterized dependency s_{ij}.

Joint Inference Model. We add the global constraint to the whole model. We define association probability matrix W^{t2a} of which $W_{i,j}^{t2a}$ represents the probability that i-th type event contains a j-th type argument. The joint probability is defined as follows,

$$P(event|D) = P_t(k|\omega_i, D) \prod_{\omega_j \in A(\omega_i)} W_{i,j}^{t2a} P_{t,a}(r_j|\omega_i, \omega_j, D)$$

where $A(\omega)$ represents the argument set of ω, r_j represents the argument role of ω_j. We minimize the negative log-likelihood $-log(P(event|D))$ to optimize the model parameters during the training phase.

4 Experiments

4.1 Dataset and Comparison Models

We construct the Chinese financial event extraction dataset from two public datasets: DuEE-Fin[1] and DCFEE[2]. The dataset contains 8271 samples with 8 superordinate event types, 14 pre-defined subdivision event types, 62 pre-defined event-argument roles. For comparison models, we select the following state-of-the-art methods for comparison: (1) DMCNN [7] extracts event features by dynamic multi-pooling CNN; (2) DBRNN [13] extracts event triggers and arguments by dependency-bridge RNN; (3) DCFEE [9] explores sentence-level and document-level event features extraction, and arguments-completion strategy for event extraction. (4) Doc2EDAG [11] transforms the event into an entity-based directed acyclic graph (EDAG) and uses a named entity recognition (NER)

---

[1] https://aistudio.baidu.com/aistudio/competition/detail/65.
[2] https://github.com/yanghang111/DCFEE.

model to extract arguments at first. Then, the binary classifiers are used to determine the event type, and arguments are transformed into a directed acyclic graph.

To verify the effectiveness of structural dependency feature, we set up the following models for comparison: (1) BERT-base only uses the word representation output by BERT for trigger extraction and role prediction; (2) JHEE joins hierarchical event feature based on BERT-base; (3) SSA-N-JHEE is the model with both hierarchical event feature and various SSA layers. The N means the number of layers of SSA. In this work, we set N as 2 and 4. Noticed that all comparison models use the global constraint proposed in this work. For a fair comparison, all candidate arguments are generated by the candidate argument extraction module, and only the final result is evaluated.

Hyper-parameter Setting and Metric. For the input, we set the maximum document length 1024. During training, we set $\lambda_1 = 0.05$, $\lambda_2 = \lambda_3 = 0.95$ and$\gamma = 3$. We employ the Adam [23] optimizer with the learning rate$1e^{-5}$, train for at most 100 epochs and pick the best epoch by the validation score on the development set. We randomly split all samples into train, development, and test sets with the proportion of 6: 2: 2. All comparison models follow their optimal parameter settings as reported in its articles. The open-source dependency syntax analysis tool on Language Technology Platform[3] (LTP) is used to build the dependency syntax trees of all samples. The embedding dimension of the argument category and dependency syntax is set as 100. The hidden layer dimension of the multi-head attention is 256. We calculate the precision, recall, and F1 of the trigger and its arguments respectively as the criteria for judging the correctness of the predicted event.

Table 1. Overall performance on the test set.

	Trigger classification (%)			Argument classification (%)		
	P	R	F1	P	R	F1
DBRNN	87.5	79.3	83.2	76.7	77.7	75.6
DMCNN	80.8	75.2	77.9	73.7	75.7	74.7
DCFEE	49.5	38.8	43.5	–	–	–
Doc2EDAG	88.3	82.5	85.3	–	–	–
BERT-base	89.7	90.0	89.8	74.8	85.9	79.9
BERT-HEE	89.3	92.8	90.6	79.4	83.9	81.6
SSA-4-HEE	95	99.8	97.6	91.5	99.9	95.4
SSA-2-HEE	95.9	99.9	**97.8**	91.8	99.9	**95.6**

[3] https://github.com/HIT-SCIR/ltp.

4.2 Results

Overall Results. Table 1 shows the overall results of our financial event extraction dataset. As we can see, in both trigger and argument extraction tasks, SSA-HEE has achieved the best results among all the comparison models. BERT-base model achieves better results than the state-of-the-art models, which indicates the global constraints joint inference can enable the model to obtain a better performance in document-level financial domain event extraction. The Doc2EDAG obtains better performance than DBRNN, DMCNN, and DCFEE, which shows the effectiveness of pre-trained language models and the usefulness of the entity-based directed acyclic graph generation strategy. We will show the effectiveness of the hierarchical event representation and structural dependency self-attention. In addition, we notice that the performance of the DCFEE is extremely low compared with the other models. We make a 4-cross validation experiments to check DCFEE's performance. The results are the same. The reason that why the DCFEE obtained such performance may be that DFCEE assume one sentence contains the most event arguments and driven by a specific trigger is likely to be an event mention in an announcement. Thus, it performs the SEE first, then supplement the missing arguments. However, in our datasets, the triggers and arguments scatter throughout the document. The cases of one sentence contain most of the event elements (trigger and most arguments) are rare. In this scenario, the performance of SEE block is extremely limited, which lead to the poor performance of the entire model.

Without hierarchical event representation, the performance of the BERT-base is worse than BERT-HEE. By using the hierarchical event representation, the performance of BERT-HEE is improved both on trigger and argument classification. Compared with BERT-base, BERT-HEE achieves a 0.8% F1 increase on trigger classification and 1.7% F1 increase on argument classification. This shows that for trigger word and argument words, the hierarchical event representation leads to the correct classification and helps the classification of the roles of arguments correctly.

Effect of Structural Dependency Self-attention. As shown in Table 1, the models with SSA based on hierarchical event representation achieve at least F1 improvements of 7% and 13.8% on trigger and argument classification respectively. It proves that with the help of structural dependency information and structural self-attention, the model can extract the semantic relationship between trigger and arguments and keep the useful entities' information. Besides, we want to verify the influence of the number of SSA layers. We conduct the ablation experiments on SSA layers. We can notice that after we decrease the SSA layers, the model achieves a slightly F1 improvement of 0.2% both on trigger and arguments. It shows that the stability of SSA. This improvement of the performance may occur because of the insufficient training samples since 60% of our financial dataset only contains 4936 samples. The model with 4 layers of SSA may be too complex when trained on this training dataset. The model can slightly fix the overfitting problem on the training set when decreasing the complexity of the model.

5 Conclusion

In this work, we analyzed the Chinese financial documents and the models proposed in the previous works, noticing the entity dependencies that exist among the sentences in a document. Based on these facts of entity dependency, we defined 8 types of entity dependency among different types of essential entities and adopted the structural self-attention to incorporate the entity dependency into the token embeddings. We constructed a new document-level financial dataset that contains 8 superordinate event types: FINANCING, TRADE, EQUITY OVERWEIGHT/UNDERWEIGHT, FINANCIAL INDEX CHANGE, MULTI-PARTY COOPERATION, PERSONNEL CHANGE, IPO RELATED, and LAW ENFORCEMENT. The structural dependency self-attention based hierarchical event extraction model was proposed to capture the token-level semantic relation and the document-level entity dependencies. The experimental result proves the effectiveness of the proposed model.

In this work, only the entity dependency information (i.e., the entity dependency matrix that constructed according to the sentences and the its entities) are exploited for the document-level event extraction. The accuracy of the tokenized entity that the proposed model uses is the major pre-processing for the following operations. How to increase the accuracy of the tokenized entity is the key issue for SSA-HEE. As an alternative method, the combination of regular expression (RE) and neural networks had been studied and shown the promising performance in some NLP tasks. The regular expression is an effective tool for identify the specified unit in a sentence or a document. It could be a potential research direction that combing the RE with EE model. In our future work, we will study the possible method to combine the RE with the EE model.

Appendix A

The Definition of Hierarchical Event Structure

Table A shows the hierarchical relationship between financial events in the financial documents and announcement. The events in financial documents and announcement are divided into FINANCING, TRADE, EQUITY OVER/UNDER WEIGHT, FINANCIAL INDEX CHANGE, MULTI-PARTY COOPERATION, PSRSONNEL CHANGE, IPO RELATED, and LAW ENFORCEMENT.

Table A. The hierarchical Financial Event.

Category	Type
FINANCING	PLEDGE
	PLEDGE RELEASED
	CORPORATION LOANS
TRADE	EQUITY BUYBACK
	ENTERPRISE ACQUISITION
EQUITY OVER/UNDER WEIGHT	EQUITY UNDERWEIGHT
	EQUITY OVERWEIGHT
FINANCIAL INDEX CHANGE	DEFICIT
MULTI-PARTY COOPERATION	WIN BID

(continued)

Table A. (*continued*)

Category	Type
PSRSONNEL CHANGE	SENIOR MANAGER CHANGE
IPO RELATED	IPO
	BANKRUPTCY
LAW ENFORCEMENT	BE INTERVIEWED
	PUNISHMENT

References

1. Borsje, J., Hoge, F., Frasincar, F.: Semi-automatic financial events discovery based on lexico-semantic patterns. Int. J. Web Eng. Technol. **6**(2), 115–140 (2010)
2. Arendarenko, E., Kakkonen, T.: Ontology-based information and event extraction for business intelligence. In: Ramsay, A., Agre, G. (eds.) Artificial Intelligence: Methodology, Systems, and Applications, pp. 89–102. Springer Berlin Heidelberg, Berlin, Heidelberg (2012). https://doi.org/10.1007/978-3-642-33185-5_10
3. Ji, H., Grishman, R.: Refining event extraction through cross-document inference, pp. 254–262
4. Miwa, M., Sætre, R., Kim, J.-D., Tsujii, J. I.: Event extraction with complex event classification using rich features. J. Bioinform. Comput. Biol. **8**(01), 131–146 (2010)
5. Miwa, M., Thompson, P., Ananiadou, S.: Boosting automatic event extraction from the literature using domain adaptation and coreference resolution. Bioinformatics **28**(13), 1759–1765 (2012)
6. Nguyen, T.H., Cho, K., Grishman, R.: Joint event extraction via recurrent neural networks, pp. 300–309
7. Chen, Y., Xu, L., Liu, K., Zeng, D., Zhao, J.: Event extraction via dynamic multi-pooling convolutional neural networks, pp. 167–176
8. Liao, S., Grishman, R.: Using document level cross-event inference to improve event extraction, pp. 789–797
9. Yang, H., Chen, Y., Liu, K., Xiao, Y., Zhao, J.: Dcfee: A document-level Chinese financial event extraction system based on automatically labeled training data, pp. 50–55
10. Mintz, M., Bills, S., Snow, R., Jurafsky, D.: Distant supervision for relation extraction without labeled data, pp. 1003–1011
11. Zheng, S., Cao, W., Xu, W., Bian, J.: Doc2EDAG: An End-to-End Document-level Framework for Chinese Financial Event Extraction, pp. 337–346
12. Yang, S., Feng, D., Qiao, L., Kan, Z., Li, D.: Exploring pre-trained language models for event extraction and generation, pp. 5284–5294
13. Sha, L., Qian, F., Chang, B., Sui, Z.: Jointly extracting event triggers and arguments by dependency-bridge RNN and tensor-based argument interaction
14. Liu, X., Luo, Z., Huang, H.: Jointly multiple events extraction via attention-based graph information aggregation. arXiv:1809.09078 (2018)
15. Subburathinam, A., et al.: Cross-lingual structure transfer for relation and event extraction, pp. 313–325
16. Zeng, Y., Yang, H., Feng, Y., Wang, Z., Zhao, D.: A convolution BiLSTM neural network model for Chinese event extraction. In: Lin, C.-Y., Xue, N., Zhao, D., Huang, X., Feng, Y. (eds.) ICCPOL/NLPCC -2016. LNCS (LNAI), vol. 10102, pp. 275–287. Springer, Cham (2016). https://doi.org/10.1007/978-3-319-50496-4_23

17. Zhang, T., Ji, H., Sil, A.: Joint entity and event extraction with generative adversarial imitation learning. Data Intell. **1**(2), 99–120 (2019)
18. Du, X., Cardie, C.: Document-level event role filler extraction using multi-granularity contextualized encoding. arXiv:2005.06579 (2020)
19. Huang, K.-H., Peng, N.: Efficient End-to-end Learning of Cross-event Dependencies for Document-level Event Extraction. arXiv:2010.12787 (2020)
20. Shen, S., Qi, G., Li, Z., Bi, S., Wang, L.: Hierarchical Chinese Legal event extraction via Pedal Attention Mechanism, pp. 100–113
21. Vaswani, A., et al.: Attention is all you need. arXiv:1706.03762 (2017)
22. Xu, B., Wang, Q., Lyu, Y., Zhu, Y., Mao, Z.: Entity Structure Within and Throughout: Modeling Mention Dependencies for Document-Level Relation Extraction. arXiv:2102.10249 (2021)
23. Kingma, D.P., Ba, J.: Adam: A method for stochastic optimization. arXiv:1412.6980 (2014)

Linked Data, Knowledge Integration, and Knowledge Graph Storage Management

Integrating Manifold Knowledge
for Global Entity Linking
with Heterogeneous Graphs

Zhibin Chen[1] , Yuting Wu[1,2]([⊠]), Yansong Feng[1,2], and Dongyan Zhao[1,2]

[1] Wangxuan Institute of Computer Technology, Peking University, Beijing, China
{czb-peking,wyting,fengyansong,zhaodongyan}@pku.edu.cn
[2] The MOE Key Laboratory of Computational Linguistics, Peking University,
Beijing, China

Abstract. Entity Linking (EL) aims to automatically link the mentions in unstructured documents to corresponding entities in a knowledge base (KB), which has recently been dominated by global models. Although many global EL methods attempt to model the topical coherence among all linked entities, most of them failed in exploiting the correlations among various linking clues, such as the semantics of mentions and their candidates, the neighborhood information of candidate entities in KB and the fine-grained type information of entities. As we will show in the paper, interactions among these types of information are very useful for better characterizing the topic features of entities and more accurately estimating the topical coherence among all the referred entities within the same document. In this paper, we present a novel HEterogeneous Graph-based Entity Linker (HEGEL) for global entity linking, which builds an informative heterogeneous graph for every document to collect various linking clues. Then HEGEL utilizes a novel heterogeneous graph neural network (HGNN) to integrate the different types of information and model the interactions among them. Experiments on the standard benchmark datasets demonstrate that HEGEL can well capture the global coherence and outperforms the prior state-of-the-art EL methods.

Keywords: Entity linking · Heterogeneous graph · Graph neural network

1 Introduction

Entity Linking (EL) is the task of mapping entity mentions with specified context in an unstructured document to corresponding entities in a given Knowledge Base (KB), which bridges the gap between abundant unstructured text in large corpus and structured knowledge source, and therefore supports many knowledge-driven Natural Language Processing (NLP) tasks and their methods, such as question answering [29], text classification [26], information extraction [12] and knowledge graph construction [19].

© Springer Nature Singapore Pte Ltd. 2021
B. Qin et al. (Eds.): CCKS 2021, CCIS 1466, pp. 91–103, 2021.
https://doi.org/10.1007/978-981-16-6471-7_7

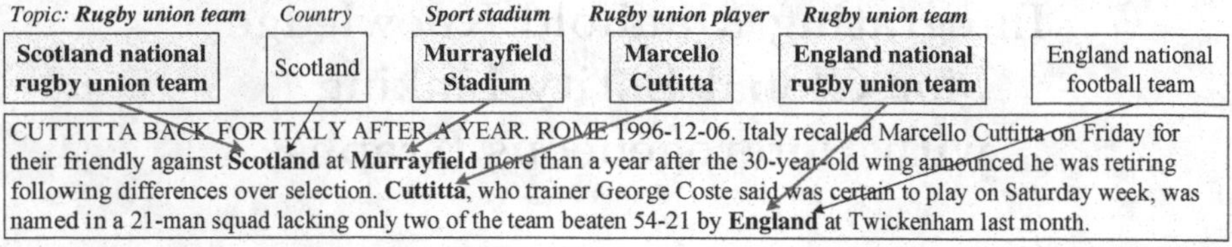

Topic: *Rugby union team*	*Country*	*Sport stadium*	*Rugby union player*	*Rugby union team*	
Scotland national rugby union team	Scotland	**Murrayfield Stadium**	**Marcello Cuttitta**	**England national rugby union team**	England national football team

CUTTITTA BACK FOR ITALY AFTER A YEAR. ROME 1996-12-06. Italy recalled Marcello Cuttitta on Friday for their friendly against **Scotland** at **Murrayfield** more than a year after the 30-year-old wing announced he was retiring following differences over selection. **Cuttitta**, who trainer George Coste said was certain to play on Saturday week, was named in a 21-man squad lacking only two of the team beaten 54-21 by **England** at Twickenham last month.

Fig. 1. The illustration example. By considering the topical coherence, an EL model can accurately link the mentions *"Scotland"*, *"Murrayfield"*, *"Cuttitta"* and *"England"* to their corresponding entities (in bold) that share the common topic "rugby".

Recently, EL task has been dominated by the global methods [1,5–8,13,16, 18,24,27,28], which model the topical coherence among the linked entities of mentions in the same document. Considering such global signal can help an EL model alleviate the biases from local contextual information. For instance, as shown in Fig. 1, for linking the mention *"England"*, it is difficult to decide between the candidate entities *England national football team* and *England national rugby union team* when only using the surrounding sports-related local context where there are the scores of matches or the name of stadium. However, if an EL model can capture the topical coherence of the common topic "rugby" among all the mentions *"Scotland"*, *"Murrayfield"*, *"Cuttitta"* and *"England"* in the current paragraph, the model can correctly link the mention *"England"* to the candidate *England national rugby union team*.

Although prior global EL approaches have greatly boosted the performance of local models, most of them do not simultaneously consider multiple types of useful information and the interactions among them, such as the semantics of mentions and their candidates, the neighborhood information of candidate entities in KB and the fine-grained type information of entities, when modeling the global coherence, and thus fail to precisely estimate the coherence among referred entities. As we will show in the paper, effectively modeling the interactions among above types of information can help to better model the topical coherence and achieving more accurate EL.

Most recently, some global methods [14,27] construct a document-level graph with candidate entities of the mentions as nodes and exploit Graph Convolutional Networks (GCN) [15] on the graph to integrate the global information, delivering promising results. Inspired by the effectiveness of using GCN to model the global signal,we present HEterogeneous Graph-based Entity Linker (**HEGEL**), a novel global EL framework designed to model the interactions among heterogeneous information from different sources by constructing a document-level informative heterogeneous graph and applying a heterogeneous architecture in GNN aggregation operation. We first construct a document-level informative heterogeneous graph with mentions, candidate entities, neighbors of entities and extracted keywords as nodes, and we create different types of edges to link these different types of nodes. Then we apply a designed heterogeneous graph neural network (HGNN) on the constructed heterogeneous graph to encode the global coherence, which allows information propagation along the informative graph structure and encourages sufficient interactions among different types of information.

Our contributions can be summarized as follows:

- We design a novel approach to construct a document-level informative heterogeneous graph to collect useful linking clues from different sources.
- We propose a meticulously designed heterogeneous graph neural network on the constructed graph, which integrates different sources of information and encourages sufficient interactions among them, more precisely characterizing the topic features of candidate entities and better capturing the topical coherence. To the best of our knowledge, this is the first work to employ a heterogeneous graph neural network in Entity Linking tasks.
- Extensive experiments and analysis on six standard EL datasets demonstrate that our HEGEL achieves state-of-the-art performance over mainstream EL methods.

2 Proposed Method: HEGEL

In addition to separately encoding the local features for every mention within a document as local models do, HEGEL constructs an informative heterogeneous graph for each document and then applies a heterogeneous GNN on it, which encodes the global coherence based on different types of information. Finally, HEGEL combines the local and global features and generates a final score for each mention-candidate pair.

2.1 Problem Formulation

Given a list of entity mentions $M = \{m_1, ... m_{|M|}\}$ in a document $\mathcal{D}$, the EL task can be formulated as linking each mention m_i to its corresponding entity $\tilde{e}_i$ from the entity collection $\mathcal{E}$ of KB or NIL (i.e. $\tilde{e}_i = NIL$, which means the mention m_i cannot be linked to any corresponding entity in $\mathcal{E}$ reasonably). Generally speaking, EL methods usually consist of two stages:

Candidate generation stage generates a small list of candidate entities $C_i = \{e_{i_1}, e_{i_2}, ..., e_{i_m}\} \subset \mathcal{E}$ for the mention m_i because of the unacceptable computation cost to traverse over the whole entity collection $\mathcal{E}$. For candidate generation, we use the method proposed in [7,16], which simply uses (1) computed mention-entity prior $\hat{p}(e|m)$ by averaging probabilities from mention entity hyperlink statistics of Wikipedia; and (2) the local context-entity similarity.

Candidate disambiguation stage assigns a score calculated in EL model to each candidate e_{i_k} and selects the top ranked candidate as the predicted answer, or predicts NIL under some specified situations. Most EL methods, including this work, focus on improving performance in this stage.

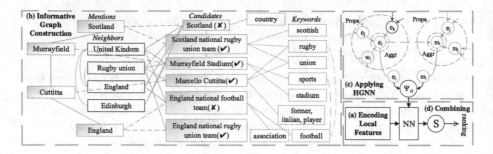

Fig. 2. The overall framework of our proposed model HEGEL

Figure 2 gives an overview of HEGEL that follows a four-stage processing pipeline: (a) encoding local features for each candidate independently, (b) informative graph construction for the document, (c) applying heterogeneous GNN on the graph, and (d) combining local and global features for scoring.

2.2 Encoding Local Features

Given a mention m_i in $\mathcal{D}$ and a candidate entity $e_{i_k} \in \mathcal{C}_i$, HEGEL computes three types of local features to encode the local mention-entity compatibility. These features consist of (a) the *Mention-Entity Prior* $P(e_{i_k}|m_i)$, which has been used in candidate generation stage, as referred in Sect. 2.1; (b) the *Context Similarity* $\Psi_C(e_{i_k}, c_{m_i})$, which utilizes an attention neural network to compute the similarity between candidate e_{i_k} and local context $c_{m_i} = \{w_1, ...w_{|c|}\}$ surrounding m_i by selecting K most relative words from c_{m_i}, eliminating noisy context words from computation; (c) the *Type Similarity* $\Psi_T(e_{i_k}, m_i)$, which estimates the similarity between the types (PER, GPE, ORG and UNK) of m_i and e_{i_k} by training a typing system proposed by [28].

2.3 Informative Heterogeneous Graph Construction

For the document $\mathcal{D}$, HEGEL builds an informative heterogeneous graph $\mathcal{G}_\mathcal{D}$ to collect different types of linking clues.

As shown in Fig. 2, $\mathcal{G}_\mathcal{D} =< V_\mathcal{D}, E_\mathcal{D} >$ contains three types of nodes: mention nodes V_{Ment}, entity nodes V_{Ent} and keyword nodes V_{Word}. Therefore, the node set $V_\mathcal{D} = V_{Ment} \cup V_{Ent} \cup V_{Word}$. V_{Ment} is naturally composed of all mentions m_i in $\mathcal{D}$. V_{Ent} contains two parts of entities: the mention candidates $V_{Ent,1} = \bigcup_{i=1}^{|M|} \mathcal{C}_i$, and the common neighbors in KB of at least two candidate entities in $V_{Ent,1}$, or formally $V_{Ent,2} = \{v|\exists v_1, v_2 \in V_{Ent,1}, v_1 \neq v_2, (v_1, r, v), (v_2, r, v) \in KB, v \notin V_{Ent,1}\}$. As reserving all neighbors in KB of $V_{Ent,1}$ is computationally unacceptable, we eliminate those nodes with only one neighbor because neighbors bridging two candidates are more informative for determining the relation between candidates, which is theoretically explained and experimentally proved in [18,21].

V_{Word} consists of the keywords extracted from the Wikipedia page of each candidate in $V_{Ent,1}$. We find that the first sentence in the Wikipedia page of an entity usually contains more fine-grained type information of the entity, which is a very useful linking clue. Therefore, for e in $V_{Ent,1}$, we extract the first sentence s from its Wikipedia page, find the first link verb in s, and pick the continuous phrase immediately after the link verb, which contains nouns, adjectives and conjunctions only. We regard the words in the picked phrase, except stopwords, as keywords characterizing the fine-grained type of e, and add them into V_{Word}.

After V_D generated, HEGEL creates heterogeneous edges between nodes of the same or different types by following rules: (a) the edges between two mention nodes $E_{MM} \subset V_{Ment} \times V_{Ment}$ are created between adjacent mentions (m_i, m_{i+1}) in D; (b) the edges between two entity nodes $E_{EE} \subset V_{Ent} \times V_{Ent}$ are created while there is a relation between them in KB; (c) the edges between two word nodes $E_{WW} \subset V_{Word} \times V_{Word}$ are created while the cosine similarity of two word embeddings is higher than a given threshold ϵ; (d) the edges from entities to mentions $E_{EM} \subset V_{Ent,1} \times V_{Ment}$ are consistent with the mention-candidate relation; (e) the edges from words to entities $E_{WE} \subset V_{Word} \times V_{Ent,1}$ are created while the word is one of the keywords for the entity. Note that (d) and (e) are unidirectional while (a)–(c) are bi-directional, and the performance of constructing bi-directional edges for (d) and (e) will be discussed later.

2.4   Heterogeneous Graph Neural Network

Given a constructed heterogeneous informative graph G_D, HEGEL applies a designed heterogeneous graph neural network (HGNN) on it to integrate different sources of information and encourage the interactions among them, generating information-augmented embeddings of V_{Ment} and $V_{Ent,1}$ for later scoring.

In order to avoid the requiring of expertise knowledge and information loss led by the former metapath-based HGNN methods, we design a novel metapath-free HGNN model. For the heterogeneous graph G_D, we represent an edge $e \in E_D$ from node $i \in V_D$ to node $j \in V_D$ with edge type r as (i, j, r). Note that in our informative graph, the node type (t_i, t_j) can exclusively determine the edge type r, and therefore we denote (t_i, t_j) as r in following explanation.

Node Embeddings. For a mention node $v_{m_i} \in V_{Ment}$, we use a text convolutional neural network (CNN) on the local context c_{m_i} surrounding m_i to compute the initial embeddings $h_{m_i}^0 \in R^{d_{cnn}+d_h}$:

$$c_{m_i} = \{w_1, ..., w_{|c|}\} \tag{1}$$

$$h_{m_i}^0 = [\frac{1}{len(m_i)} \sum_{w \in m_i} v_w; CNN(v_{w_1}, ..., v_{w_{|c|}})] \tag{2}$$

where $v_w, v_{w_i} \in R^{d_h}$ are corresponding word embeddings of mention surface words and its context c_{m_i} respectively, $[;]$ is concatenating operation. For the nodes in V_{Ent} and V_{Ment}, we naturally use the entity word embeddings $v_{e_i} \in R^{d_h}$ and $v_{w_i} \in R^{d_h}$ trained in [7] as initial embedding $h_{e_i}^0, h_{w_i}^0$.

Inter-Node Propagation. A node should receive different types information from its heterogeneous neighborhood in different ways. Motivated by previous work about metapath [4], HEGEL models the different information propagation with multiple feature transformations on different adjacent relations. Taking edge type $r = (t_i, t_j)$ into consideration, a node v_j with type t_j collects information from its neighborhood $N(v_j)$ with type t_i in l-th layer by a Graph Convolutional Network (GCN):

$$h_{v_j, t_i}^{l+1} = \frac{1}{Z} \sum_{v_i \in V_{t_i} \cap N(v_j)} W_{t_i, t_j}^l h_{v_i}^l \tag{3}$$

where $h_{v_i}^l \in R^{d_{l, t_i}}$ is v_i's embedding before l-th layer, $W_{t_i, t_j}^l \in R^{d_{l+1, t_j} \times d_{l, t_i}}$ is a trainable matrix in l-th layer, $h_{v_j, t_i}^{l+1} \in R^{d_{l+1, t_j}}$ is v_j's new embedding related to t_i, and Z is the normalization factor. Note that for edge types (t_i, t_i) connecting nodes with the same type, self-loop connections are added into its edge set.

Intra-Node Aggregation. In order to preserve the information from different types of relationship with neighborhoods, for the node v_j, HEGEL aggregates new embeddings to generate the input $h_{v_j}^{l+1}$ for next layer:

$$h_{v_j}^{l+1} = \sigma(f_{agg}(h_{v_j, t_i}^{l+1})) \tag{4}$$

where $f_{agg} : R^{d \times |\{t_i\}|} \rightarrow R^d$ is the aggregation function implemented as simple summation operation $f_{agg}(\{x_i\}) = \sum x_i$, σ is an activation function implemented as GELU($\cdot$) [10], $h_{v_j}^{l+1} \in R^{d_{l+1, t_j}}$ is the output embedding of l-th layer containing all types of one-hop neighborhood of v_j in heterogeneous graph structure. The L layers of inter-node propagation and intra-node aggregation encourage heterogeneous integrations and interactions among types of information, which are represented by the final output $h_{v_j}^L$.

Global Score Calculation. After getting the information-augmented embeddings $h_{m_i}^L$ for mention m_i and $h_{e_{i_k}}^L$ for corresponding candidate e_{i_k}, as we ensure that $d_{L, Ment} = d_{L, Ent}$, HEGEL applies a bi-linear similarity calculation to represent the global compatibility between the mention-candidate pair:

$$\Psi_G(e_{i_k}, m_i) = (h_{m_i}^L)^T \cdot D \cdot h_{e_{i_k}}^L \tag{5}$$

where $D \in R^{d_{L, Ment} \times d_{L, Ent}}$ is a trainable diagonal matrix.

2.5   Feature Combining and Model Training

HEGEL combines local features and the global compatibility score to compute the linking score for each candidate e_{i_k} of mention m_i:

$$S(m_i, e_{i_k}) = f([P(e_{i_k}|m_i); \Psi_C(e_{i_k}, c_{m_i}); \Psi_T(e_{i_k}, m_i); \Psi_G(e_{i_k}, m_i)]) \tag{6}$$

where f is a two-layered fully connect neural network. The candidate e_{i_k} with the highest final linking score $S(m_i, e_{i_k})$ is selected as the output linking result for m_i. HEGEL links m_i to NIL if and only if its candidate list $C_i = \emptyset$, or rather, there is no corresponding entity to m_i in KB entity set $\mathcal{E}$.

Following previous works, HEGEL attempts to make the ground truth entity $\tilde{e}_i$ ranking higher than other candidates, and therefore minimizes the following margin-based ranking loss:

$$L = \sum_{m_i \in \mathcal{D}} \sum_{e_{i_k} \in C_i} [\gamma - S(m_i, \tilde{e}_i) + S(m_i, e_{i_k})]_+ \tag{7}$$

where $\gamma > 0$ is the margin hyper-parameter, and $[x]_+$ is equal to x when $x > 0$, or equal to 0 otherwise.

3   Experiments and Analysis

3.1   Datasets

Following previous EL practice, we evaluate HEGEL on the benchmark dataset AIDA CoNLL-YAGO [11] for training, validation and the in-domain testing. To examine its cross-domain generalization ability, we use five popular datasets for cross-domain testing: MSNBC [3], AQUAINT [20], ACE2004 [24], CWEB [8] and WIKIPEDIA [8]. Table 1 shows the statistics and corresponding *recall* of candidate generation of all datasets used in our experiments.

Table 1. The statistics of used datasets. *Recall* represents the ratio of candidate lists containing corresponding ground truth entity.

Dataset	#Mentions	#Docs	#Ments / #Docs	Recall(%)
AIDA-train (train)	18448	946	19.50	100
AIDA-A (valid)	4791	216	22.18	97.72
AIDA-B (test)	4485	231	19.4	98.66
MSNBC	656	20	32.8	98.48
AQUAINT	727	50	14.54	94.09
ACE2004	257	36	7.14	91.44
CWEB	11154	320	34.86	91.90
WIKIPEDIA	6821	320	21.32	93.21

3.2   Experiment Settings

As we use the pre-trained Word2vec [17] word embeddings, and entity embeddings released by [7], the embedding dimension d_h is fixed to 300. The hyper-parameters are manually tuned based on the validation performance on AIDA-A.

CNN output dimension $d_{cnn} = 64$, all informative graph embedding dimensions $d_{l,t} = 32, l = 1, ..., L$, number of HGNN layers $L = 2$, margin $\gamma = 0.01$, $K = 40$, dropout rate is set to 0.5, E_{WW} threshold $\epsilon = 0.5$.

We use Adam optimizer to train HEGEL with a learning rate of $\alpha = 2e - 4$. The model is evaluated per 3 epochs, and the training process is terminated while highest validation performance is not exceeded in 10 evaluations.

To confine the graph size within a computable range, all documents with more than 80 mentions will be split into several documents as average as possible.

3.3 Compared Baselines

To illustrate the effect of modeling the interactions among different types of information, we evaluate and compare the performance of our HEGEL with 9 existing methods [1,2,6–8,11,14,16,24], which will be discussed in Sect. 4, on in- and cross-domain datasets.

It's worth noting that GNED claims they firstly construct a heterogeneous entity-word graph to model global information, but their nodes are not heterogeneous indeed as entity nodes share the same vector space with words. In addition, they do not apply any heterogeneous architecture in their GNN, as they regard all edges as the same type. Therefore, HEGEL is the first work to employ a heterogeneous GNN in EL tasks to our best knowledge.

3.4 Experiment Results

We report the performance of all the compared baselines and our HEGEL in Table 2. The top part shows the performance of non-GNN-based baselines, and other baselines are GNN-based.

The in-domain test dataset AIDA-B, which shares the similar data distribution with training dataset AIDA-train and validation dataset AIDA-A, is the most important benchmark. By modeling the latent relation between mentions and injecting entity coherence into it, which can be regarded as simply interaction between two types of information, Ment-Norm outperforms all baselines on AIDA-B. It shows that the interactions of heterogeneous information are beneficial for capturing global coherence. We observe that HEGEL, which integrates different types of information in a more interactive and effective way for capturing the global coherence, significantly outperforms the Ment-Norm method. The fact shows that our HEGEL can encourage more richer interactions among different types of information and greatly improve the performance.

It should be figured out that none of the models can consistently achieve the best F1-score on the all five cross-domain datasets. HEGEL outperforms the other two GNN-based method, NCEL and SGEL, on MSNBC and ACE2004. It shows that our HEGEL can handle cross-domain linking cases better than them in some extent.

Table 2. Performance on in-domain (AIDA-B) and cross-domain datasets.
We show the In-KB accuracy (%) for the in-domain and micro-F1 Score (%) for the cross-domain datasets respectively. For HEGEL we show std. deviation obtained over 3 runs.

Models	In-domain AIDA-B	Cross-domain MSNBC	AQUAINT	ACE2004	CWEB	WIKI
Prior $p(e\|m)$	71.51	89.3	83.2	84.4	69.8	64.2
AIDA [11]	-	79	56	80	58.6	63
GLOW [24]	-	75	83	82	56.2	67.2
RI [2]	-	90	90	86	67.5	73.4
WNED [8]	89	92	87	88	77	**84.5**
Deep-ED [7]	92.22	93.7	88.5	88.5	**77.9**	77.5
Ment-Norm [16]	93.07	93.9	88.3	89.9	77.5	78.0
GNED [14]	92.40	**95.5**	**91.6**	**90.14**	77.5	78.5
NCEL [1]	80	-	87	88	-	-
SGEL [6]	83	80	88	89	-	-
HEGEL	**93.65±0.1**	93.19±0.2	85.87±0.3	89.33±0.4	73.25±0.3	75.54±0.1
- w/o V_{Word}	91.94±0.2	93.18±0.2	85.35±0.4	88.40±0.5	71.95±0.5	74.70±0.2
- w/o $V_{Ent,2}$	92.22±0.1	92.93±0.4	85.07±0.7	88.93±0.4	72.57±0.5	75.45±0.2
HEGEL Local	91.03	91.97	84.06	86.92	71.45	74.79

Our HEGEL performs extremely well on in-domain cases by making full use of different types of linking clues for better capturing the global coherence, but it seems that there is no advantage on the cross-domain datasets. We find that the ground truth entities of cross-domain test sets are less popular, where the linking clues are sparse. To improve the generalization ability on such tough cases, the only effective way seems to be introducing large-scale corpus for training, aiming to more or less "see" the linking clues of cross-domain entities in the training stage. We will try to introduce large-scale pre-trained language models, such as BERT, to improve the generalization ability of our HEGEL in the future.

Comparing GNED with our simpler and effective way to extract keywords within the first sentence from the Wikipedia page of corresponding entity, they search on the whole Wikipedia KB to find the hyperlinks to corresponding entity and extract contexts in preprocessing stage, which have to iterate through all $|\mathcal{E}|$ entities and become very time-consuming. Even with less keyword evidence, our strategy still ourperforms GNED on in-domain dataset with lower time overhead. GNED accesses more additional linking clues and reach better performance on cross-domain datasets, and we suppose that these richer information can also improve the generalization ability of our HEGEL, and further boost our performance on cross-domain datasets.

3.5   Ablation Study

As shown in the bottom part of Table 2, HEGEL boosts the performance of local model with an average improvement of 1.77%, which shows that HEGEL is able to greatly enhance the local model.

To further examine the effect of our heterogeneous model, we remove the keyword nodes V_{Word} and neighbor nodes $V_{Ent,2}$ from $V_{\mathcal{D}}$ respectively, and therefore the related edges from $E_{\mathcal{D}}$ as well. After that, there is a significant drop in performance (0.89% and 0.61% on average respectively) across datasets, especially in-domain AIDA-B (1.71% and 1.43%). The results demonstrate the effectiveness of introducing the keyword (fine-grained type) information and neighborhood information of candidate entities and modeling the interactions among them, which can help to accurately capture the topical characteristics of candidates.

3.6   Analysis

The Impact of Edge Directions. As referred in Sect. 2.3, HEGEL only keeps one direction for E_{EM} and E_{WE}. We suppose that adding edges from V_{Ment} to $V_{Ent,1}$ and from $V_{Ent,1}$ to V_{Word} will lead to the over-smooth problem, as candidates to be disambiguated are related to the same mention and maybe the same keywords, where they might entangle with each other and make the disambiguation harder. As expected, the results shown in Table 3(a) prove that keeping these edges uni-directional can alleviate over-smooth and enhance the performance.

Table 3. Experiment results on (a) Changing the directionality of edges; (b) Scores in case study.

Models	AIDA-B	Cross-domain avg.
HEGEL	**93.65**	**83.44**
$+V_{Ent} \rightarrow V_{Word}$	93.15	82.93
$+V_{Ment} \rightarrow V_{Ent}$	92.64	83.03
$+$Both	91.21	81.81

Models	Scot.→ country	Scot.→ team	Eng.→ football	Eng.→ rugby
Gold	Low	**High**	Low	**High**
HEGEL	−0.162	**−0.144**	−0.147	**−0.145**
$-V_{Word}$	−0.336	**−0.309**	−0.312	−0.317
$-V_{Ent,2}$	**−0.176**	−0.187	−0.170	**−0.168**

Case Study. As shown in Fig. 2, HEGEL needs to map the mentions "Scotland", "Murrayfield", "Cuttitta" and "England" in the same document to corresponding entities. "Murrayfield" and "Cuttitta" are not ambiguous as they have only one candidate respectively. However, "Scotland" and "England" are linked to wrong candidates by local model, where our HEGEL outputs the right answers by correctly modeling the interactions among heterogeneous types of information, especially from the neighborhood around "Marcello Cuttitta" (a former rugby union player) and "Rugby Union", and from the respective keywords related to "rugby". Ablation score calculating results shown in Table 3(b) manifest that information from keyword nodes V_{Word} and neighbor nodes $V_{Ent,2}$ and correctly handling the information are both important for HEGEL to correctly capture the topical coherence and model the heterogeneous interactions.

4 Related Work

4.1 Entity Linking

Most existing models not only use local methods relying on local context of individual mentions independently [2,11,22,23], but also use global methods considering the coherence among the linked entities of all mentions by jointly linking on the whole document [8,24]. As the global coherence optimization problem is NP-hard, different approximation methods are often used. Apart from traditional methods like loopy belief propagation [7,16], several works approximate the problem into sequence decision problem [5] or graph learning [1,6,14,27]. Following the graph based neural network modeling methods, HEGEL expands the graph utilization in EL task to heterogeneous style, which not only enjoys the strong representation ability of heterogeneous graph structure, but also becomes effective enough because of avoiding other additional inference steps required in sequence-style models.

4.2 Graph Neural Networks

Graph Neural Network (GNN) is a strong and flexible framework to learn on data with graph structure. After the Graph Convolutional Network (GCN) [15] appeared, GNN is more and more widely used in many tasks, while several popular GNN architectures, such as GraphSAGE [9] and GAT [25], are proposed to learn the representation on graphs. The natural graph structure entailed in EL task becomes a favorable condition to apply GNN methods.

As the emergence of massive heterogeneous information, many works about Heterogeneous GNN have been proved to be effective. The mainstream of HGNN models is based on the construction of metapaths [4], but several HGNN architectures free of metapath are proposed recently [13]. Our HEGEL follows these works, and utilizes the heterogeneous structure to model the interactions among different types of linking information.

5 Conclusion

In this paper, we have presented HEGEL, a novel graph-based global entity linking method, which is designed to model and utilize the interactions among heterogeneous types of information from different sources. We achieve this aim by constructing a document-level informative heterogeneous graph and applying a heterogeneous GNN to propagate and aggregate information on the graph, which is hard to achieve by previous homogeneous architectures. Extensive experiments on standard benchmarks show that HEGEL achieves state-of-the-art performance in EL task.

Acknowledgments. This work is supported in part by the National Key R&D Program of China (2020AAA0106600) and the Key Laboratory of Science, Technology and Standard in Press Industry (Key Laboratory of Intelligent Press Media Technology).

References

1. Cao, Y., Hou, L., Li, J., Liu, Z.: Neural collective entity linking. arXiv preprint arXiv:1811.08603 (2018)
2. Cheng, X., Roth, D.: Relational inference for wikification. In: Proceedings of the 2013 Conference on Empirical Methods in Natural Language Processing (EMNLP 2013), pp. 1787–1796 (2013)
3. Cucerzan, S.: Large-scale named entity disambiguation based on Wikipedia data. In: Proceedings of the 2007 Joint Conference on Empirical Methods in Natural Language Processing and Computational Natural Language Learning (EMNLP-CoNLL), pp. 708–716. Association for Computational Linguistics (2007)
4. Dong, Y., Chawla, N.V., Swami, A.: metapath2vec: scalable representation learning for heterogeneous networks. In: Proceedings of the 23rd ACM SIGKDD International Conference on Knowledge Discovery and Data Mining, pp. 135–144 (2017)
5. Fang, Z., Cao, Y., Li, Q., Zhang, D., Zhang, Z., Liu, Y.: Joint entity linking with deep reinforcement learning. In: The World Wide Web Conference, pp. 438–447 (2019)
6. Fang, Z., Cao, Y., Li, R., Zhang, Z., Liu, Y., Wang, S.: High quality candidate generation and sequential graph attention network for entity linking. In: Proceedings of The Web Conference 2020, pp. 640–650 (2020)
7. Ganea, O.E., Hofmann, T.: Deep joint entity disambiguation with local neural attention. In: Proceedings of the 2017 Conference on Empirical Methods in Natural Language Processing, pp. 2619–2629 (2017)
8. Guo, Z., Barbosa, D.: Robust named entity disambiguation with random walks. Seman. Web **9**, 1–21 (2017)
9. Hamilton, W.L., Ying, R., Leskovec, J.: Inductive representation learning on large graphs. In: Proceedings of the 31st International Conference on Neural Information Processing Systems (NIPS 2017), pp. 1025–1035 (2017)
10. Hendrycks, D., Gimpel, K.: Gaussian error linear units (gelus). arXiv preprint arXiv:1606.08415 (2016)
11. Hoffart, J., et al.: Robust disambiguation of named entities in text. In: Proceedings of the 2011 Conference on Empirical Methods in Natural Language Processing, pp. 782–792 (2011)
12. Hoffmann, R., Zhang, C., Ling, X., Zettlemoyer, L., Weld, D.S.: Knowledge-based weak supervision for information extraction of overlapping relations. In: Proceedings of the 49th Annual Meeting of the Association for Computational Linguistics: Human Language Technologies, pp. 541–550 (2011)
13. Hong, H., Guo, H., Lin, Y., Yang, X., Li, Z., Ye, J.: An attention-based graph neural network for heterogeneous structural learning. In: Proceedings of the AAAI Conference on Artificial Intelligence, vol. 34, pp. 4132–4139 (2020)
14. Hu, L., Ding, J., Shi, C., Shao, C., Li, S.: Graph neural entity disambiguation. Knowl.-Based Syst. **195**, 105620 (2020)
15. Kipf, T.N., Welling, M.: Semi-supervised classification with graph convolutional networks. arXiv preprint arXiv:1609.02907 (2016)

16. Le, P., Titov, I.: Improving entity linking by modeling latent relations between mentions. In: Proceedings of the 56th Annual Meeting of the Association for Computational Linguistics (Volume 1: Long Papers), pp. 1595–1604, July 2018
17. Le, Q., Mikolov, T.: Distributed representations of sentences and documents. In: International Conference on Machine Learning, pp. 1188–1196. PMLR (2014)
18. Liu, M., Gong, G., Qin, B., Liu, T.: A multi-view-based collective entity linking method. ACM Trans. Inf. Syst. (TOIS) **37**(2), 1–29 (2019)
19. Luan, Y., He, L., Ostendorf, M., Hajishirzi, H.: Multi-task identification of entities, relations, and coreference for scientific knowledge graph construction. arXiv preprint arXiv:1808.09602 (2018)
20. Milne, D., Witten, I.H.: Learning to link with wikipedia. In: Proceedings of the 17th ACM Conference on Information and Knowledge Management (CIKM 2008,), pp. 509–518. Association for Computing Machinery, New York (2008)
21. Moreau, E., Yvon, F., Cappé, O.: Robust similarity measures for named entities matching. In: Proceedings of the 22nd International Conference on Computational Linguistics (Coling 2008). pp. 593–600. Coling 2008 Organizing Committee
22. Mulang', I.O., Singh, K., Prabhu, C., Nadgeri, A., Hoffart, J., Lehmann, J.: Evaluating the impact of knowledge graph context on entity disambiguation models. In: Proceedings of the 29th ACM International Conference on Information & Knowledge Management, pp. 2157–2160 (2020)
23. Raiman, J., Raiman, O.: Deeptype: multilingual entity linking by neural type system evolution. In: Proceedings of the AAAI Conference on Artificial Intelligence, vol. 32 (2018)
24. Ratinov, L., Roth, D., Downey, D., Anderson, M.: Local and global algorithms for disambiguation to Wikipedia. In: Proceedings of the 49th Annual Meeting of the Association for Computational Linguistics: Human Language Technologies, pp. 1375–1384. Association for Computational Linguistics, June 2011
25. Veličković, P., Cucurull, G., Casanova, A., Romero, A., Liò, P., Bengio, Y.: Graph attention networks. In: ICLR (2018)
26. Wang, J., Wang, Z., Zhang, D., Yan, J.: Combining knowledge with deep convolutional neural networks for short text classification. In: IJCAI, vol. 350 (2017)
27. Wu, J., Zhang, R., Mao, Y., Guo, H., Soflaei, M., Huai, J.: Dynamic graph convolutional networks for entity linking. In: Proceedings of The Web Conference 2020, pp. 1149–1159. WWW '20, Association for Computing Machinery (2020)
28. Xu, P., Barbosa, D.: Neural fine-grained entity type classification with hierarchy-aware loss. In: Proceedings of the 2018 Conference of the North American Chapter of the Association for Computational Linguistics: Human Language Technologies, pp. 16–25 (2018)
29. Yih, W.t., Chang, M.W., He, X., Gao, J.: Semantic parsing via staged query graph generation: question answering with knowledge base. In: Proceedings of the 53rd Annual Meeting of the Association for Computational Linguistics and the 7th International Joint Conference on Natural Language Processing (Volume 1: Long Papers), pp. 1321–1331 (2015)

Content-Based Open Knowledge Graph Search: A Preliminary Study with OpenKG.CN

Xiaxia Wang, Tengteng Lin, Weiqing Luo, Gong Cheng[(✉)], and Yuzhong Qu

State Key Laboratory for Novel Software Technology, Nanjing University, Nanjing, China
{xxwang,tengtenglin,wqluo}@smail.nju.edu.cn, {gcheng,yzqu}@nju.edu.cn

Abstract. Users rely on open data portals and search engines to find open knowledge graphs (KGs). However, existing systems only provide metadata-based KG search but ignore the contents of KGs, i.e., triples. In this paper, we present one of the first content-based search engines for open KGs. Our system CKGSE supports keyword-based KG search, KG snippet generation, KG profiling and browsing, all computed over KGs' (large) contents rather than their (small) metadata. We implement a prototype with Chinese KGs crawled from OpenKG.CN and we report some preliminary results about the practicability of such a system.

Keywords: Knowledge graph · Search engine · Snippet · Profiling

1 Introduction

Open data, especially knowledge graphs (KGs), plays an important role in scientific research and application development. These days, increasingly more research efforts for constructing reusable KGs bring about a large number of KG resources available on the Web, which motivates the development of data sharing platforms. Open data portals such as European Data Portal have made a promising start. Users are allowed to freely publicize their own KGs, or search and download published KGs from others. Recent research efforts have been paid to assist users to conveniently find the KG they want. Thus various systems have been developed, ranging from general search engines such as Google Dataset Search, to KG-centric systems like LODAtlas [14]. OpenKG.CN is a popular platform for Chinese open KGs, providing a keyword search service.

Limitations of Existing Work. The above systems provide search services based on *metadata*, i.e., meta-level data descriptions attached to a KG, such as authorship and provenance. This leads to two weaknesses. **(W1)** They cannot handle queries with keywords referring to the *content* (i.e., triples) of the target KG, such as a class, property, or entity. Queries of this kind are common in practice. Indeed, over 60% of KG queries contain keywords that refer to KG content [4]. **(W2)** Metadata cannot provide close-up views of the underlying KG content. Its utility for relevance judgment in search is limited [16].

© Springer Nature Singapore Pte Ltd. 2021
B. Qin et al. (Eds.): CCKS 2021, CCIS 1466, pp. 104–115, 2021.
https://doi.org/10.1007/978-981-16-6471-7_8

Our Work. To address the above two limitations, *content-based KG search* is needed but the *practicability* of such a new paradigm is unknown. In this paper, we present CKGSE, short for **C**hinese **KG S**earch **E**ngine, as our preliminary effort to build a content-based search system for open KGs. CKGSE has four components: *KG crawling and storage* for managing KGs and their metadata, *content-based keyword search* for parsing queries and retrieving relevant KGs based on their contents, *content-based snippet generation* for extracting a sub-KG from a KG to justify its query relevance, and *content profiling and browsing* for providing detailed information about a KG including abstractive and extractive summaries and a way to explore its content. To evaluate the practicability of CKGSE, we implement a prototype[1] based on Chinese KGs collected from OpenKG.CN. Our contributions are summarized as follows.

– We develop one of the first content-based search engines for open KGs.
– We report experimental results about the practicability of such a system.

Outline. Section 2 discusses related work. Section 3 and Sect. 4 describe an overview of CKGSE and its detailed implementation, respectively. Section 5 presents experimental results. Section 6 concludes the paper with future work.

2   Related Work

2.1   Open KG Portals and Search Engines

Today, hundreds of open data portals are available [13]. They manage collected resources with metadata using a vocabulary such as DCAT.[2] Google Dataset Search aims at navigating users to the original page of each data resource. It only indexes the metadata of each resource without considering its content. Current search engines for KGs are also based on metadata, e.g., LODAtlas [14].

To overcome the limitations of metadata-based systems, our CKGSE takes KG content into consideration. To support keyword search over KG content, complementary to metadata index, CKGSE also indexes the content of each KG. To facilitate relevance judgment of search results, CKGSE generates a snippet for each KG extracted from its content. To provide a close-up view for each KG, apart from its metadata, CKGSE offers content-based summaries and an exploration mechanism for easier comprehension.

2.2   KG Profiling

KG profiling is to interpret a KG according to various features [3,7,9]. In [7], by reviewing the literature over the past two decades, profiling features are divided into seven categories, most of which are metadata-oriented such as Provenance and Licensing. The Statistical category includes numbers and distributions of

---

[1] http://ws.nju.edu.cn/CKGSE.
[2] https://www.w3.org/TR/vocab-dcat-2/.

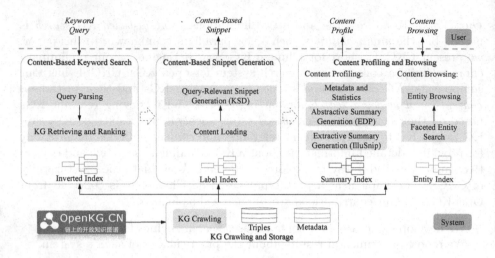

Fig. 1. Overview of CKGSE.

data elements such as the number of classes and properties instantiated in a KG [1]. The General category includes methods for choosing representative data elements, such as structural summaries [2,6,8,15] and pattern mining [20,21].

Benefited from these fruitful research efforts, our CKGSE incorporates several profiling techniques. In addition to metadata and statistics, CKGSE mines frequent entity description patterns (EDPs) in a KG as its abstractive summary [18,19], and illustrates its content with an extractive summary [5,10].

3   System Overview

An overview of CKGSE is presented in Fig. 1, including four main components. KG Crawling and Storage crawls, parses, and stores KGs and their metadata. Content-Based Keyword Search parses keyword queries and retrieves and ranks KGs. For each top-ranked KG, Content-Based Snippet Generation extracts a query-relevant sub-KG as a snippet presented in the search results page. For each KG selected by the user, Content Profiling and Browsing presents its profiled detailed information and supports exploring its content.

KG Crawling and Storage. As an offline process, KG Crawling retrieves all available metadata from OpenKG.CN through its CKAN API. Following the links in metadata, the dump files of all KGs are downloaded, parsed into triples, and stored in a local database. Four indexes are then built to support efficient online computation in downstream components, which will be detailed in Sect. 4.

Content-Based Keyword Search. Given an input (Chinese) keyword query, it is parsed by segmenting into words and removing stop words. The parsed query is executed over an inverted index to retrieve top-ranked KGs according to a relevance score function involving multiple fields of metadata and content.

Content-Based Snippet Generation. For each top-ranked KG, its content is loaded into memory from which a query-relevant snippet is online generated. The generation method, KSD [17], considers content representativeness and query relevance, and is supported by a label index. The output is an optimal snippet containing k top-ranked triples where k is a pre-defined size limit. It is presented as a node-link diagram in the search results page.

Content Profiling and Browsing. For each selected KG, its offline indexed profile is presented, including its metadata, statistics, and two content summaries. The abstractive summary contains the most frequent entity description patterns (EDPs) [18,19] in the KG, and the extractive summary generated by IlluSnip [5,10] contains an optimal sub-KG of k' triples in terms of content representativeness where k' is a pre-defined size limit. All entities in the KG can be explored in a faceted manner supported by an entity index, i.e., by filtering entities by their classes and properties. All triples describing a filtered entity can be browsed.

4   System Implementation

In this section we detail the implementation of each component of CKGSE.

4.1   KG Crawling and Storage

This component crawls and stores metadata and KGs from OpenKG.CN.

KG Crawling. Through the CKAN API, the metadata of all available resources are retrieved from OpneKG.CN, including KGs and non-KG resources which are not our focus. Among these resources, 40 are identified as KGs in formats such as RDF/XML, Turtle, and JSON-LD. Following the links in metadata, KG Crawling first downloads KG dump files, and then parses them into triples using Apache Jena 3.8.0. Metadata and triples are stored in a MySQL database.

Metadata. The metadata of all KGs is stored in a table, where each row represents a KG, and each column represents a field of metadata, such as `Title` and `Author`. Observe that incorrect/incomplete field values are common in practice since they are freely submitted by KG publishers.

Triples. The triples of each KG are stored in a table whose columns contain the subject, predicate and object of each triple. Moreover, each IRI in a KG is labeled by a human-readable textual form, including its local name, the value of its `rdfs:label` property, and the value of its associated literals. These labels will be used in downstream components.

4.2   Content-Based Keyword Search

Supported by an inverted index, this component parses the keyword query and retrieves a ranked list of relevant KGs.

Inverted Index. An inverted index is created for keyword-based KG search and has three fields: `Content`, `Title`, and `Other metadata`. For KG content which is not considered in existing systems, each triple is transformed into text by concatenating the labels of its subject, predicate, and object. The textual forms of all triples in a KG compose the field `Content`. Among metadata, the title of each KG is individually indexed in the field `Title` for its importance in KG search, while other metadata is combined into the field `Other metadata`.

Query Parsing. The keyword search component of CKGSE is developed over Apache Lucene 7.5.0, which provides only basic word segmentation modules incompetent for parsing Chinese queries. To address this problem, we incorporate the IK analyzer[3] which is an open-source tool for Chinese word segmentation.

KG Retrieving and Ranking. Receiving the parsed query from Query Parsing as a list of keywords, it is searched over the inverted index. Searches are performed using OR as the default boolean operator between keywords. A multi-field query parser provided by Apache Lucene is used to retrieve and rank the search results over the three fields. The relevance score function adopted by CKGSE, at the time of writing, is BM25 as default provided by Lucene, which could be replaced by more advanced functions for KGs in future work.

4.3   Content-Based Snippet Generation

To facilitate relevance judgement of KGs, this component outputs a snippet for each top-ranked KG to justify its query relevance and illustrate its main content. This unique feature distinguishes CKGSE from existing systems.

Label Index. For each KG, a label index maps each word to the IRIs whose textual forms contain the word. It is used to compute the query relevance of each triple, supporting query-relevant snippet generation. The query relevance of a triple is measured by the number of keywords contained by its textual form, i.e., contained by the textual form of its subject, predicate, or object.

Content Loading. Before snippet generation, for each KG in the search results, its triples are loaded into memory from the local triple store. During the loading process, the query relevance of each triple is computed with the support of the Label Index, as well as some scores measuring content representativeness such as the relative frequency of the class or property instantiated in the triple.

Query-Relevant Snippet Generation. KSD [17] formulates the computation of an optimal snippet for a KG as a weighted maximum coverage problem, where each triple is regarded as a set of elements including query keywords, classes, properties, and entities in the triple. Each element is assigned a weight of importance, e.g., relative frequency. KSD aims at selecting at most k triples maximizing the

---

[3] https://code.google.com/archive/p/ik-analyzer/.

total weight of covered elements. By applying a greedy strategy, it selects a triple containing the largest weight of uncovered elements in each iteration until reaching the size limit k or all elements are covered. We set $k = 10$. An example snippet is shown in Fig. 6. We refer the reader to [17] for details about KSD.

4.4   Content Profiling

This component presents an offline computed profile for each selected KG including metadata, statistics, an abstractive summary, and an extractive summary.

Metadata and Statistics. For each selected KG, its metadata is loaded from the database and presented to the user as a table. Each row contains a key-value pair as shown in Fig. 7. Statistics provide a high-level overview of the selected KG, including basic counts such as the number of triples, entities, the distributions of instantiated classes and properties presented in pie charts as shown in Fig. 8, and top-ranked entities by PageRank showing the central content of the KG.

Abstractive Summary Generation. Inspired by pattern mining techniques for KG profiling, CKGSE incorporates frequent entity description pattern (EDP) [18, 19] into the KG profile page. As each entity in the KG is described by a set of schema-level elements, i.e., classes and properties. EDPs retain the common data patterns shared among entities in the KG. Thus frequent EDPs can be regarded as an abstractive summary of the KG. Each of them consists of a set of classes, forward properties, and backward properties that describe an entity. In the profile page, each EDP is presented as a node-link diagram as in Fig. 9.

Extractive Summary Generation. Complementary to EDPs providing abstractive schema-level summaries, CKGSE also provides extractive summaries to exemplify the KG content. IlluSnip [5, 10] formulates the selection of triples as a combinatorial optimization problem, aiming at computing an optimal connected sub-KG containing the most frequently instantiated classes, properties, and the most important entities with the highest PageRank scores. Such a sub-KG can be viewed as an extractive summary of the KG. A greedy algorithm is applied to generate a sub-KG containing at most k' triples. We set $k' = 20$. The result is presented as a node-link diagram on the profile page as shown in Fig. 10.

Summary Index. Offline computed abstractive and extractive summaries are stored in a summary index.

4.5   Content Browsing

Beyond the profile page, CKGSE also enables the user to interactively explore the entities in the selected KG.

Entity Index. An entity index is created to support efficient filtering of entities. It consists of two parts, mapping classes and properties to their instance entities, respectively. The index is implemented with Lucene.

Table 1. Statistics of the KGs.

Triples		Classes		Properties		Entities		Dump files (MB)	
Median	Max	Median	Max	Median	Max	Median	Max	Median	Max
68,399	151,976,069	7	20,096	23	40,077	93,268	303,952,138	23	28,929

Table 2. Run-time of offline computation (hours).

Parsing	Inverted index	Label index	Statistics	Summary index	Entity index
20.27	5.10	2.52	2.97	31.38	33.21

Table 3. Disk use (MB).

Dump files	Triple store	Inverted index	Label index	Summary index	Entity index
54,133	35,756	1,816	9,708	165	1,129

Faceted Entity Search. As shown in Fig. 11, users are provided with a panel to choose classes and properties instantiated in the KG. The selected classes and properties are used as a filter for entities with AND as the boolean operator. Filtered entities are presented as a list. Each of them can be further browsed.

Entity Browsing. For each of the filtered entities, CKGSE retrieves all the triples describing it and presents them as a node-link diagram. Intuitively it shows the neighborhood of this entity in the original KG. By switching between entities, users are able to explore the KG according to her/his own interest.

5   Experiments

We implemented a prototype of CKGSE on an Intel Xeon E7-4820 (2GHz) with 100GB memory for JVM. Our experiments were focused on the practicability of CKGSE. We also presented a case study comparing CKGSE with the current version of OpenKG.CN to show the usefulness of the unique features of CKGSE.

5.1   Practicability Analysis

KG Crawling and Storage. Table 1 shows statistics about the 40 KGs crawled from OpenKG.CN. They vary greatly in size and schema. Some are very large.

As shown in Table 2, parsing all the 40 KGs affordably ended within 1 day, where 10% (4/40) were large and parsed in more than 1 h. Observe that the time for parsing the largest KG[4] reached 8 h.

Table 3 shows the disk use. The total size of the triple store and all the indexes is smaller than that of the original dump files, showing practicability. Figure 2 presents the disk use of each KG in ascending order. Compared with dump files, the indexes have relatively small sizes and are affordable.

---

[4] http://www.openkg.cn/dataset/zhishi-me-dump.

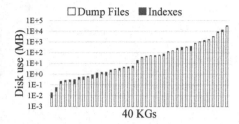

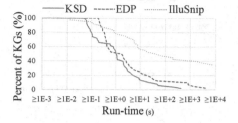

Fig. 2. Disk use of each KG.

Fig. 3. Cumulative distribution of run-time.

Content-Based Keyword Search. CKGSE spent 5 h to build an inverted index for all the 40 KGs to support keyword search over both metadata and KG contents (Table 2). Note that we did not build in parallel which would otherwise be much faster. The index only takes 1.8 GB, being small and affordable (Table 3).

Content-Based Snippet Generation. CKGSE spent 2.5 h to construct a label index to support query-relevant snippet generation (Table 2). About half of the time, i.e., 1.1 h, was spent on the largest KG. The index takes 9.7 GB (Table 3) where about half is for the largest KG.

To evaluate the performance of online snippet generation using KSD, we created 10 keyword queries containing 1–5 keywords and retrieved top-5 KGs with each query. We recorded the run-time of generating a snippet for each of these 50 KGs. Figure 3 shows the cumulative distribution of run-rime over all these KGs. The median run-time is only 1 s, while for 12% (6/50) the run-time exceeded 10 s. Further optimization of KSD is needed for large KGs.

Content Profiling. CKGSE spent 3 h to prepare statistics (Table 2), including about 2 h for computing PageRank to identify central entities.

The summary index only uses 165 MB (Table 3) but its computation spent 31 h (Table 2). Most time was for generating extractive summaries using IlluSnip. We used an anytime version of IlluSnip [10] and we allowed 2 h for generating a snippet. If needed, one could set a smaller time bound to trade snippet quality for generation time. In our experiments, the median run-time of IlluSnip was 31 s. By comparison, generating abstractive summaries (i.e., EDP) was much faster, being comparable with KSD as shown in Fig. 3.

Content Browsing. CKGSE spent 33 h to create an entity index to support faceted entity search (Table 2), although the index was as small as 1.1 GB upon completion (Table 3). Observe that there is much room for improving the performance of our trivial implementation of this index, e.g., using better index structures and/or faster algorithms.

(a) OpenKG.CN   (b) CKGSE

Fig. 4. Search results pages for the query "哈利波特 人物关系".

(a) OpenKG.CN   (b) CKGSE

Fig. 5. Search results pages for the query "格兰芬多 人物关系".

5.2 Case Study

We compare CKGSE with the current version of OpenKG.CN by a case study.

Keyword Search. Figure 4 shows the search results pages returned by OpenKG.CN and CKGSE for the query "哈利波特 人物关系". Both systems successfully find the target KG as both keywords in the query match the metadata of this KG.

However, for the query "格兰芬多 人物关系" where "格兰芬多" refers to an entity in the target KG, only CKGSE finds this KG as shown in Fig. 5, thanks to its content-based keyword search whose inverted index covers both metadata and KG content, distinguishing CKGSE from existing systems.

Snippet Generation. In the search results pages, while OpenKG.CN and other existing systems only show some metadata about each top-ranked KG, CKGSE further computes and presents an extracted sub-KG, as shown in Fig. 6. The

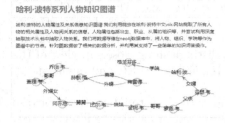

Fig. 6. Snippet (KSD).

Fig. 7. Metadata.

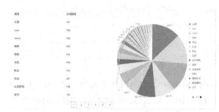

Fig. 8. Statistics.

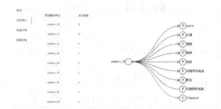

Fig. 9. Abstractive summary (EDP).

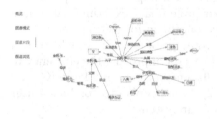

Fig. 10. Extractive summary (IlluSnip).

Fig. 11. Content browsing.

generation of this snippet is biased toward the keyword query, e.g., containing the "格兰芬多" entity mentioned in the query. Therefore, it can help the user quickly judge the relevance of the underlying KG to the query even before browsing its content which could be a time-consuming process.

Profiling and Browsing. When a KG is selected, both OpenKG.CN and CKGSE show its metadata in a profile page, as shown in Fig. 7. CKGSE further presents some statistics about the content of the KG, such as the distribution of all properties instantiated in the KG visualized as a pie chart shown in Fig. 8. Such statistics provide the user with a brief overview of the KG content.

Moreover, CKGSE provides two content summaries of the KG. The abstractive summary in Fig. 9 presents the most frequent EDPs in the KG to show how entities in the KG are described, i.e., by which combinations of classes and properties. The extractive summary in Fig. 10 presents an extracted sub-KG which is different from the snippet in the search results page. The sub-KG here is query-independent but illustrates the most frequent classes and properties in the KG with a few concrete entities and triples. Compared with metadata and statistics, our content summaries provide a distinguishing closer-up view of the KG content, thus assisting the user in comprehending the KG and further judging its relevance before downloading it.

Last but not least, in Fig. 11 CKGSE allows the user to interactively browse entities in the KG. The user can select classes and properties to filter entities. For each filtered entity, all its triples are visualized as a node-link diagram. With this simple yet effective browsing interface, for many users they do not need any other tools for KG browsing but can easily investigate the KG content.

6   Conclusion

We presented CKGSE, one of the first content-based search engines for open KGs. By incorporating triples into the inverted index, CKGSE can handle queries referring to the content of the target KG. Apart from metadata and statistics, CKGSE provides content snippets, summaries, and browsing capabilities to comprehensively assist users in relevance judgement. We implemented a prototype with KGs crawled from OpenKG.CN, and our preliminary experimental results demonstrated the practicability of such a new paradigm for KG search.

Our experiments also showed some shortcomings of CKGSE which we will address in future work. We will particularly focus on improving the efficiency of processing large KGs, and will improve the efficiency of browsing large KGs by using entity summarization techniques [11,12]. We also plan to conduct a user study to assess the usefulness and usability of our system.

Acknowledgement. This work was supported by the NSFC (62072224).

References

1. Auer, S., Demter, J., Martin, M., Lehmann, J.: LODStats - an extensible framework for high-performance dataset analytics. In: EKAW 2012, pp. 353–362 (2012). https://doi.org/10.1007/978-3-642-33876-2_31
2. Čebirić, Š, et al.: Summarizing semantic graphs: a survey. VLDB J. **28**(3), 295–327 (2018). https://doi.org/10.1007/s00778-018-0528-3
3. Chapman, A.: Dataset search: a survey. VLDB J. **29**(1), 251–272 (2019). https://doi.org/10.1007/s00778-019-00564-x
4. Chen, J., Wang, X., Cheng, G., Kharlamov, E., Qu, Y.: Towards more usable dataset search: from query characterization to snippet generation. In: CIKM 2019, pp. 2445–2448 (2019). https://doi.org/10.1145/3357384.3358096
5. Cheng, G., Jin, C., Ding, W., Xu, D., Qu, Y.: Generating illustrative snippets for open data on the web. In: WSDM 2017, pp. 151–159 (2017). https://doi.org/10.1145/3018661.3018670
6. Cheng, G., Jin, C., Qu, Y.: HIEDS: a generic and efficient approach to hierarchical dataset summarization. In: IJCAI 2016, pp. 3705–3711 (2016)

7. Ellefi, M.B., et al.: RDF dataset profiling - a survey of features, methods, vocabularies and applications. Semant. Web **9**(5), 677–705 (2018). https://doi.org/10. 3233/SW-180294
8. Khatchadourian, S., Consens, M.P.: ExpLOD: summary-based exploration of interlinking and RDF usage in the linked open data cloud. In: Aroyo, L., et al. (eds.) ESWC 2010, Part II. LNCS, vol. 6089, pp. 272–287. Springer, Heidelberg (2010). https://doi.org/10.1007/978-3-642-13489-0_19
9. Koesten, L., Simperl, E., Blount, T., Kacprzak, E., Tennison, J.: Everything you always wanted to know about a dataset: studies in data summarisation. Int. J. Hum. Comput. Stud. **135** (2020). https://doi.org/10.1016/j.ijhcs.2019.10.004
10. Liu, D., Cheng, G., Liu, Q., Qu, Y.: Fast and practical snippet generation for RDF datasets. ACM Trans. Web **13**(4), 19:1–19:38 (2019). https://doi.org/10. 1145/3365575
11. Liu, Q., Chen, Y., Cheng, G., Kharlamov, E., Li, J., Qu, Y.: Entity summarization with user feedback. In: Harth, A., et al. (eds.) ESWC 2020. LNCS, vol. 12123, pp. 376–392. Springer, Cham (2020). https://doi.org/10.1007/978-3-030-49461-2_22
12. Liu, Q., Cheng, G., Gunaratna, K., Qu, Y.: Entity summarization: state of the art and future challenges. J. Web Semant. **69**, 100647 (2021). https://doi.org/10. 1016/j.websem.2021.100647
13. Neumaier, S., Umbrich, J., Polleres, A.: Automated quality assessment of metadata across open data portals. ACM J. Data Inf. Qual. **8**(1), 2:1–2:29 (2016). https:// doi.org/10.1145/2964909
14. Pietriga, E., et al.: Browsing linked data catalogs with LODAtlas. In: Vrandečić, D., et al. (eds.) ISWC 2018, Part II. LNCS, vol. 11137, pp. 137–153. Springer, Cham (2018). https://doi.org/10.1007/978-3-030-00668-6_9
15. Song, Q., Wu, Y., Lin, P., Dong, X., Sun, H.: Mining summaries for knowledge graph search. IEEE Trans. Knowl. Data Eng. **30**(10), 1887–1900 (2018). https:// doi.org/10.1109/TKDE.2018.2807442
16. Wang, X., et al.: A framework for evaluating snippet generation for dataset search. In: Ghidini, C., et al. (eds.) ISWC 2019, Part I. LNCS, vol. 11778, pp. 680–697. Springer, Cham (2019). https://doi.org/10.1007/978-3-030-30793-6_39
17. Wang, X., Cheng, G., Kharlamov, E.: Towards multi-facet snippets for dataset search. In: PROFLILES & SemEx 2019, pp. 1–6 (2019)
18. Wang, X., Cheng, G., Lin, T., Xu, J., Pan, J.Z., Kharlamov, E., Qu, Y.: PCSG: pattern-coverage snippet generation for RDF datasets. In: ISWC 2021 (2021)
19. Wang, X., Cheng, G., Pan, J.Z., Kharlamov, E., Qu, Y.: BANDAR: benchmarking snippet generation algorithms for (RDF) dataset search. IEEE Trans. Knowl. Data Eng. (2021)
20. Zneika, M., Lucchese, C., Vodislav, D., Kotzinos, D.: RDF graph summarization based on approximate patterns. In: ISIP 2015. vol. 622, pp. 69–87 (2015). https:// doi.org/10.1007/978-3-319-43862-7_4
21. Zneika, M., Lucchese, C., Vodislav, D., Kotzinos, D.: Summarizing linked data RDF graphs using approximate graph pattern mining. In: EDBT 2016, pp. 684–685 (2016). https://doi.org/10.5441/002/edbt.2016.86

Natural Language Understanding
and Semantic Computing

Dependency to Semantics: Structure Transformation and Syntax-Guided Attention for Neural Semantic Parsing

Shan Wu[1,3(⊠)], Bo Chen[1], Xianpei Han[1,2], and Le Sun[1,2]

[1] Chinese Information Processing Laboratory, Beijing, China
{wushan2018,chenbo,xianpei,sunle}@iscas.ac.cn
[2] State Key Laboratory of Computer Science, Institute of Software,
Chinese Academy of Sciences, Beijing, China
[3] University of Chinese Academy of Sciences, Beijing, China

Abstract. It has long been known that the syntactic structure and the semantic representation of a sentence are closely associated [1,2]. However, it is still a hard problem to exploit the syntactic-semantic correspondence in end-to-end neural semantic parsing, mainly due to the partial consistency between their structures. In this paper, we propose a neural dependency to semantics transformation model – Dep2Sem, which can effectively learn the structure correspondence between dependency trees and formal meaning representations. Based on Dep2Sem, a dependency-informed attention mechanism is proposed to exploit syntactic structure for neural semantic parsing. Experiments on GEO, JOBS, and ATIS benchmarks show that our approach can significantly enhance the performance of neural semantic parsers.

1 Introduction

Semantic parsing is the task of transforming natural language utterances to their formal meaning representations [3–7], such as lambda calculus, FunQL [8], and structured SQL queries. For example in Fig. 1, a semantic parsing system transforms the sentence *"what river traverses the most states?"* to its FunQL representation $most(river(traverse\_2(state(all))))$.

In recent years, due to the strong representation learning ability and simple end-to-end architecture, neural semantic parsers have achieved significant progress [9–12]. Most neural semantic parsers follow the classical attention-based encoder-decoder architecture. Dong and Lapata [14] first propose a Seq2Seq model, where structural logical forms are linearized and generated as sequences during decoding. Noticing the hierarchical structure of logical forms, SEQ2TREE model [14] is proposed to exploit the structure of logical forms. In Xiao et al. [15], Sequence-based Structured Prediction model is used to predict derivation sequence, which consists of derivation steps relative to its underlying grammar. In Chen et al. [16,18], Seq2Action model is proposed to predict action sequence for semantic graph construction, so that the structure of semantic graph can be exploited. We can see that, although these approaches can exploit the

© Springer Nature Singapore Pte Ltd. 2021
B. Qin et al. (Eds.): CCKS 2021, CCIS 1466, pp. 119–133, 2021.
https://doi.org/10.1007/978-981-16-6471-7_9

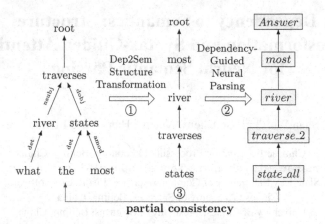

Fig. 1. To exploit the syntactic-semantic correspondence: 1) A Dep2Sem transformation model is proposed, so dependency trees are probabilistically transformed to structures which are consistent with meaning representations; and 2) A dependency-informed attention is designed to guide neural semantic parsing. In this way 3) the partial consistency problem can be effectively resolved.

structure of meaning representations, all of them simply model input sentences as token sequences and ignore their syntactic structures.

It has long been known that the syntactic structure and the semantic representation of a sentence are closely associated. For example, in Fig. 1 the dependency tree and the logical form share many substructures, e.g., the dependency relation "*river* $\xrightarrow{nsubj}$ *traverses*" can reveal the semantic relation between *river* and *traverse_2*. Such a syntactic-semantic structure correspondence has been heavily exploited in traditional grammar-based semantic parsers. Reddy et al. [19] directly transform dependency parse tree to lambda calculus representation using a manually designed grammar in DEPLAMBDA; Liang et al. [20] propose a dependency-based semantic representation – DCS, which uses trees to represent formal semantics. By explicitly exploiting syntactic structures, the complex syntactic phenomenons such as nested subordinations and quantifier scopes can be better addressed, and many long-range dependency problems can be avoided. We believe that neural semantic parsers can also be enhanced by exploiting the syntactic-semantic structure correspondence effectively.

However, it is not trivial to exploit the syntactic-semantic correspondence in end-to-end neural semantic parsing. The main challenge is that the syntactic structure and the meaning representation of a sentence are mostly only partially consistent, rather than isomorphic. For example, in Fig. 1, the word "*most*" has a long dependency path to "*river*", but the operation *most* is directly linked to the type predicate *river* in logical form. To resolve this partial consistency problem, traditional grammar-based methods usually rely on specially designed grammar rules, such as the mark-execute construction rule in Liang et al. [20], and the designed transformation rules in DEPLAMBDA [19]. Unfortunately, since neural semantic parsers are trained end-to-end and logical forms are predicted in sequence to sequence, it is hard to apply traditional solutions in neural semantic parsers.

To exploit syntactic-semantic correspondence for neural semantic parsing, this paper proposes a neural dependency to semantics transformation model – Dep2Sem, which can effectively model the structure correspondence between dependency trees and logical forms. Based on Dep2Sem, we enhance current neural semantic parsers by designing a dependency-informed attention mechanism, which can guide the decoding of logical forms to be more accurate. Figure 1 shows the framework of our method. Specifically, to resolve the partial structure consistency problem, our Dep2Sem model predicts whether the semantic derivations of two words form a parent-child relation in logical form, based on the dependency path between them. For example in Fig. 1, our Dep2Sem model will predict the type predicate $river$ as the child of the operation $most$ in logical form based on the dependency path "$most \xrightarrow{amod} \xrightarrow{dobj} \xleftarrow{nsubj} river$". Based on the Dep2Sem model, our dependency-informed attention guides the decoding of neural semantic parsers using syntactic-semantic correspondence. Using the above parent-child structure prediction between the derivations of words "$river$" and "$most$", our dependency-informed attention ensures that the word "$river$" is attended when decoding the children of the operation $most$.

We conduct experiments on three standard datasets GEO, JOBS, and ATIS. Experimental results show that, by modeling and exploiting syntactic-semantic correspondence, our method can significantly and robustly enhance current neural semantic parsers on different datasets with different meaning representations.

2 Dependency to Semantic Structure Transformation Model

As described above, the main challenge to exploit syntactic-semantic correspondence lies in the partial consistency between their structures. To resolve this problem, we propose a Dep2Sem structure transformation model (Sect. 2), which can transform a dependency tree into a new form which is consistent with its logical form. Then, we introduce how to incorporate such structure correspondence into neural semantic parsing via a dependency-informed attention mechanism (Sect. 3).

To transform dependency trees into logical form-consistent structures, two main problems need to be addressed:

1. How to represent the transformation rules from dependency to semantics. That is, we need an effective and compact way to model the structural correspondence between them, and the rules need to be easily represented, learned and exploited.
2. Because different semantic representation formalisms (such as lambda calculus, FunQL, Prolog, etc.) have different structures, the transformation rules need to be flexible and learnable from data. In previous studies, Reddy et al. [19] manually write transformation rules. However, manual rules need expert knowledge and thus are difficult to be generalized to different meaning representations/domains/datasets.

In this paper, given a sentence $\mathbf{x} = w_1, w_2, ..., w_n$, we model the Dep2Sem transformation rule as the probability of whether the logical forms derived by words w_i and w_k will form a parent-to-child relation in the final logical form of sentence $\mathbf{x}$, based on the dependency path between w_i and w_k: $D_{ik} = d_1, d_2, ..., d_{|D|}$. For example in Fig. 1, our Dep2Sem model predicts the parent-to-child relation between ($river, traverse\_2$)

using the dependency path pattern "$river \xrightarrow{nsubj} traverses$", and the parent-to-child relation between $(most, river)$ based on the dependency path between words "$most$" and "$river$":

$$most \xrightarrow{amod} \xrightarrow{dobj} \xleftarrow{nsubj} river$$

In this way, our Dep2Sem model can effectively model the dependency-semantic structure correspondence, and is flexible enough for different semantic representation formalisms. That is, our model can learn different transformation rules for different meaning representations.

By predicting the parent-to-child probabilities between any two words in a sentence, our Dep2Sem model can reconstruct a transformed dependency tree, which reflects the structure of its meaning representation. In this way, the partial structure consistency problem can be effectively resolved. For example, in Fig. 1 the transformed tree has the same structure with its FunQL representation.

To this end, this paper formally models the above transformation rules with a neural network. Given the dependency parse of a sentence, we first find the dependency paths between any two words. Then an RNN is learned to score the parent-to-child relation between two words using the dependency path between them. Finally, our Dep2Sem model predicts the child of each word based on the above scores. Given the dependency path from word w_i to word w_k as $D_{ik} = \mathbf{d}_1, \mathbf{d}_2, ..., \mathbf{d}_{|D|}$, we embed each item $\mathbf{d}_t = [\mathbf{l}_t; \mathbf{c}_t]$, where $\mathbf{l}_t$ is the dependency label embedding and $\mathbf{c}_t$ is the dependency direction embedding. The dependency path embeddings $\mathbf{d}_1, \mathbf{d}_2, ..., \mathbf{d}_{|D|}$ are then fed into a RNN which returns the hidden state sequence $\mathbf{e}_1, \mathbf{e}_2, ..., \mathbf{e}_{|D|}$. The final hidden state is used to represent D_{ik}, i.e. $\mathbf{E}_{ik} = \mathbf{e}_{|D|}$. For our RNN encoder, the GRU transition is used:

$$\mathbf{r}_t = \sigma(\mathbf{W}_r \mathbf{d}_t + \mathbf{U}_r \mathbf{e}_{t-1} + \mathbf{b}_r) \tag{1}$$
$$\mathbf{u}_t = \sigma(\mathbf{W}_u \mathbf{d}_t + \mathbf{U}_u \mathbf{e}_{t-1} + \mathbf{b}_u) \tag{2}$$
$$\tilde{\mathbf{e}}_t = \tanh(\mathbf{W}\mathbf{d}_t + \mathbf{U}[\mathbf{r}_t \odot \mathbf{e}_{t-1}] + \mathbf{b}) \tag{3}$$
$$\mathbf{e}_t = (1 - \mathbf{u}_t) \odot \mathbf{e}_{t-1} + \mathbf{u}_t \odot \tilde{\mathbf{e}}_t, \tag{4}$$

where $\mathbf{W}, \mathbf{W}_u, \mathbf{W}_r, \mathbf{U}, \mathbf{U}_u, \mathbf{U}_r, \mathbf{b}, \mathbf{b}_u, \mathbf{b}_r$ are GRU parameters.

The above neural Dep2Sem model can be easily extended with other useful features, the full Dep2Sem model also uses the hidden state of the head and tail words of the dependency path. Let $\mathbf{O}_{ik}$ be additional features, we calculate the score between word w_i and w_k as:

$$score_{ik} = \phi([\mathbf{E}_{ik}; \mathbf{O}_{ik}]) \tag{5}$$

where ϕ is a linear transformation outputting a scalar value. We finally normalize all scores of a word w_i to get its children probabilities as:

$$\mathbf{M}_{ik}^R = \frac{\exp(score_{ik})}{\sum_k \exp(score_{ik})} \tag{6}$$

here $\mathbf{M}_{ik}^R$ denotes the probability of whether w_k is the child of w_i in logical form. As Fig. 2(b) shows, the sentence structure organized by the parent-to-child probability

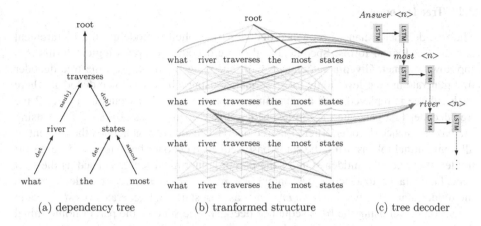

| (a) dependency tree | (b) tranformed structure | (c) tree decoder |

Fig. 2. Using Dep2Sem model, the transformed structure (b) from the dependency tree (a) is consistent with the meaning representation. The dependency-informed attention will guide the tree decoder (c) with more accurate attention using the transformed structure.

matrix $\mathbf{M}^R$ is expected to be fully consistent with the structure of its meaning representation.

The above neural Dep2Sem transformation model leverages the strong representation learning and prediction ability of neural networks, and is more tolerant to syntactic parsing errors and has better generalization ability than hard dependency path rules. Furthermore, the neural Dep2Sem model can be jointly learned with neural parsing models in an end-to-end manner, without the need for any manual rules or grammars.

3   Dependency-Informed Attention for Neural Semantic Parsing

In this section, we enhance neural semantic parsing by designing a dependency-informed attention mechanism. To better exploit syntactic-semantic structure correspondence, this paper uses SEQ2TREE [14] as our base model, which can explicitly model the structures of meaning representations using a hierarchical tree decoder. By exploiting syntactic information, our dependency-informed attention can guide the tree decoder with more accurate attention. In following we first briefly describe the sequence encoder and the tree decoder of the SEQ2TREE model, then we introduce our dependency-informed attention in detail.

3.1   Sequence Encoder

The SEQ2TREE model regards input sentence as a sequence, $\mathbf{x} = w_1, w_2, ..., w_n$. Follow previous approaches, we encode $\mathbf{x}$ using a bidirectional RNN, which learns hidden representations of $\mathbf{x}$ in both forward and backward directions. The hidden states in two directions are concatenated to form source word representation $\mathbf{h}_t = [\overrightarrow{\mathbf{h}}_t; \overleftarrow{\mathbf{h}}_t]$, and the encoding of the whole sentence is: $\mathbf{h}_1, \mathbf{h}_2, ..., \mathbf{h}_n$.

3.2   Tree Decoder

The tree decoder enhances sequence decoder by explicitly modeling the hierarchical structure of logical forms. Specifically, the tree decoder generates logical forms in a top-down manner. Given a sentence $\mathbf{x}$ with encoding $\mathbf{h}_1, \mathbf{h}_2, ..., \mathbf{h}_n$, the tree decoder first generate the top-level logic token sequence using a base sequence decoder, where each subtree is replaced with a nonterminal token $<n>$. For example, in Fig. 2 the tree decoder first generate the top-level logic token sequence as $Answer <n>$ using the base sequence decoder. Then, the tree decoder recursively generates the content of all nonterminal tokens using the same base sequence decoder but with a different start hidden state, i.e., the hidden state of the parent nonterminal $<n>$ is used as the start state. For instance, to generate the content of $<n>$ in $Answer <n>$, tree decoder uses the hidden state of $<n>$ in $Answer <n>$ as start state, and generates $most <n>$ in the second level using the base sequence decoder. The state of the parent nonterminal is also concatenated with the input during the expansion of nonterminal $<n>$.

Specifically, let $\mathbf{s}_{<n>}$ be the hidden state of nonterminal $<n>$ to be expanded, the hidden state $\mathbf{s}_j$ is computed as:

$$\mathbf{s}_j = \text{LSTM}(\mathbf{s}_{j-1}, [\mathbf{s}_{<n>}; \mathbf{y}_{j-1}]) \tag{7}$$

During decoding, attention mechanism plays a critical role, which is used to find which words in encoder-side will be used to generate the next logical token. The attention score between the current hidden state $\mathbf{s}_j$ and each source word encoding $\mathbf{h}_t$ is:

$$\alpha_{jt} = \frac{\exp(\mathbf{s}_j \cdot \mathbf{h}_t)}{\sum_{t=1}^{n} \exp(\mathbf{s}_j \cdot \mathbf{h}_t)} \tag{8}$$

Then we summarize the context vector $\mathbf{c}_j$ as the weighted sum of token encoding:

$$\mathbf{c}_j = \sum_{t=0}^{n} \alpha_{jt} \mathbf{h}_t \tag{9}$$

Finally, $\mathbf{c}_j$ is used to generate the next logic token $\mathbf{y}_j$ using the conditional probability:

$$p(y_j|y_{<j}, \mathbf{x}) \propto \exp(\mathbf{W}_o \tanh(\mathbf{W}_1 \mathbf{s}_j + \mathbf{W}_2 \mathbf{c}_j)) \tag{10}$$

We can see that, the above original attention only relies on training data to capture such correspondence, ignores the syntactic-semantic correspondence which would provide helpful information, too. Therefore if we can design a more accurate attention by further exploiting syntactic-semantic correspondence, we can improve the tree-decoder by providing more accurate context.

3.3   Dependency-Informed Attention

In the tree decoder, the logical tokens expanded from a nonterminal $<n>$ are all its direct children in logical form structure. For instance, when expanding $<n>$ in $Answer <n>$, all logical tokens such as $most <n>$ are children of $Answer$. Meanwhile, our Dep2Sem transformation model can predict whether the logical form derived by word w_i is the child of the logical form derived by word w_k using the dependency path between them. Therefore, our Dep2Sem model can help to provide more accurate attention based on the syntactic-semantic structure correspondence. For example, for expanding $<n>$ in *Answer* $<n>$, our Dep2Sem will predict "*most*" should be attended because the word "*most*" is the child of word "root" in the transformed dependency structure.

Based on the above observations, we design a dependency-informed attention. Specifically, given a sentence $\mathbf{x}$ and its dependency parse, we first compute its parent-to-child probability matrix using our Dep2Sem model, where $\mathbf{M}_{ik}^R$ is the probability that the logical token derived by w_i will be the parent of the logical token derived by of w_k. Based on $\mathbf{M}^R$, if we know the attended words in the current state, their children words in $\mathbf{M}^R$ will be more likely to be attended when decoding its child logical tokens. For example, in Fig. 2 if word "*most*" is currently attended, our dependency-informed attention will predict the next attended word will be "river" using the confident parent-to-child transformation rule "*most* $\xrightarrow{amod}$ $\xrightarrow{dobj}$ $\xleftarrow{nsubj}$ *river*".

To exploit transformation matrix $\mathbf{M}^R$ in SEQ2TREE model, we observed that when generating the next logic token y_j, its parent logic token y_{j_p} is always given. For example, when generating the content of the nonterminal node in *most* $<n>$, all logic tokens have the same parent logic token $most$. Then, let α_{j_p} be the attention vector of its parent logic token y_{j_p} on source words, we can compute the dependency-informed attention vector of current state by multiplicating the parent-to-child probability matrix $\mathbf{M}^R$ with α_{j_p}:

$$\alpha_j^{dep} = \mathbf{M}^R \alpha_{j_p} \tag{11}$$

here α^{dep} can be seen as a prior attention score guided by the syntactic-semantic structure correspondence. And we get the final attention score by combining the dependency-guided attention score α_j^{dep} and the original attention score α_j^{orig} computed in Sect. 3.2 using a probability p_m:

$$\alpha_j = p_m \alpha_j^{dep} + (1 - p_m) \alpha_j^{orig} \tag{12}$$

here p_m can be a fixed probability, or estimated using the parent logic token information. This paper estimates p_m using a multi-layer perceptron, its input is the concatenation of parent token's hidden state and the attentive context vector $[\mathbf{s}_{j_p}; \mathbf{c}_{j_p}]$. The multi-layer perceptron will be jointly learned during training.

3.4 Model Learning

Our approach contains two components: one is the Dep2Sem transformation model which models syntactic-semantic structure correspondence; the other is the enhanced SEQ2TREE semantic parsing model with dependency-informed attention. This section describes how to jointly learn the Dep2Sem model, dependency-informed attention and neural semantic parsing model in an end-to-end manner.

In semantic parsing, the training corpus is a set of training instances. Each instance $(\mathbf{x}, \mathbf{y})$ contains a sentence $\mathbf{x}$ and its labeled logical form $\mathbf{y} = [y_1, y_2, ..., y_m]$ where y_j

is logical token. For example, (*"what river traverses the most states"*, $Answer(most$ $(river(traverse\_2(state(all))))))$ is a training instance. Given training corpus $\mathbf{D}$, the objective function can be written as:

$$J = - \sum_{(\mathbf{x},\mathbf{y})\in\mathbf{D}} \sum_{j=1}^{m} \log p(y_j|\mathbf{x}, y_{<j}, y_{j_p}) \tag{13}$$

The two components interact with each other. To avoid the local optimization problem, this paper employs the widely used incremental training strategy. Firstly, we train the original SEQ2TREE model without using dependency-informed attention, i.e., we set the mix probability $p_m = 0$. Then we train the full model by initializing it using the trained original SEQ2TREE model. Furthermore, in order to ensure that the second training stage only fine-tune parameters, we add an L_1-norm regularization term in the loss function to impose restricts to dependency-informed attention:

$$\lambda_{att}(\mathbf{x}, y_j) = ||\boldsymbol{\alpha}_j^{dep} - \boldsymbol{\alpha}_j^{orig}||_1 \tag{14}$$

Then we can train the full model with the loss function: $J + \sum_{(\mathbf{x},\mathbf{y})} \sum_{j=1}^{m} \lambda_{att}(\mathbf{x}, y_j)$.

4   Experiments.

This section assesses the performance of our method and compares it with previous systems.

4.1   Experimental Settings

Datasets. We conduct experiments on three standard datasets: GEO, JOBS, and ATIS, which use different meaning representations and different domains.

ATIS. This is a large dataset, which contains 5,410 queries to a flight booking system. Each question is annotated with a lambda calculus query. Following Zettlemoyer and Collins [37], we use the standard 4,473/491/448 train/validation/test instance splits in our experiments.

GEO. This is a semantic parsing benchmark about U.S. geography [3]. The variable-free semantic representation FunQL [8] is used in this dataset. We follow the standard 600/280 train/test instance splits in our experiments.

JOBS. This dataset contains 640 queries to a database of job listings. Natural language questions are labeled with Prolog-style queries. We follow the standard 500/140 train/test splits [4] in our experiments.

In all our experiments, the standard accuracy is used to evaluate systems. The accuracies on all datasets are obtained as the same as Jia and Liang [7].

Table 1. Performances on GEO dataset. † represents the system uses lambda calculus expressions as meaning representations. For fair comparison, we also present the performance of the FunQL version of SEQ2TREE

Method	Accuracy
Non-Neural Models	
Kate et al., 2005 [8]	71.1
Lu et al., 2008 [6]	76.8
Kwiatkowski et al., 2010 [33]	82.1
Jones et al., 2012 [35]	79.3
Lu, 2015 [34]	86.8
Neural Models	
Jia and Liang, 2016 [7]†	85.0
Jia and Liang, 2016 [7]† (+data)	89.3
Dong and Lapata, 2016 [14]†: Seq2Seq	84.6
Dong and Lapata, 2016 [14]†: Seq2Tree	87.1
Susanto and Lu, 2017 [11]†	**90.0**
Xu et al., 2018 [21]† (+syntactic encoder)	88.1
Chen et al., 2018b [16]†	88.9
Jie and Lu, 2018 [22]	89.3
Our Models	
Seq2Tree(Base)	88.2
Dep2Sem	**90.0**
- Dep2Sem(WithoutDep)	88.6
- Dep2Sem(Fixed)	88.2

Data Preprocessing. Following Dong and Lapata [14], we handle entities with *Replacing* mechanism, which replaces identified entities with their types and IDs. Because tree decoder works better on variable-free meaning representations and different meaning representations can be transformed between each other, we preprocess different datasets as follows. For GEO, we directly adopt the functional query language FunQL [42] as its semantic formalism, which is variable-free itself. For JOBS, we transform Prolog-style queries into their variable-free forms using heuristic rules. For ATIS, we transform the lambda calculus to λ-DCS [41], which is more compact and with fewer variables.

System Settings. Our model uses 200 hidden units and 100-dimensional word vectors for sentence encoding. We initialize all parameters by uniformly sampling within [−0.1, 0.1]. The dimensions of dependency embeddings are 30, and the dimensions of dependency label embeddings and dependency direction embeddings are 15. We use Adam algorithm to update parameters, with the batch size set to 20. We train our model with 20 epochs and an initial learning rate of 0.001. The first 5 epochs train the original model, following 5 epochs learn the dependency-informed attention, final 10 epochs learn the full model. The beam size of the decoder is set to 10. We use the Stanford CoreNLP tool[1] [27] for dependency parsing. Our model is implemented in PyTorch, which will be released together with its configurations on github.com/lingowu/Dep2Sem.

4.2 Overall Results

We compare our method with previous methods on all three datasets. We train our model with three settings: the full model; **WithoutDep** – the transformation model trained using only head/tail words, without dependency path; **Fixed** – the dependency-informed attention is combined with original attention using a fixed probability p_m, which is tuned on the training corpus. For fair comparison with base models, we also report the performance of Seq2Tree(Base) with the same settings and preprocessing. The results are shown in Tables 1 and 2. we can see that:

1. Exploiting syntactic structure can effectively enhance neural semantic parsing. On all three datasets, our Dep2Sem achieves competitive performance. This verifies the effectiveness of combining neural models and syntactic structures. Furthermore, because all previous methods don't exploit syntactic-semantic structure correspondence, our method can be further enhanced by combing with other improvements. For instance, Xu et al. [21] enhance sentence encoding using both constituent parse and dependency parse, which can be incorporated into our model for further performance improvement.
2. By exploiting syntactic-semantic structure correspondence, our method can effectively enhance previous neural parsers. Combined with the base SEQ2TREE model, our method achieves performance improvements on all three datasets.
3. Dependency-informed attention is an effective way to exploit syntactic-semantic correspondence. Compared with the three base attention settings – original attention of SEQ2TREE, DepPath, and Fixed, the full dependency-based attention can improve our method on all three datasets.

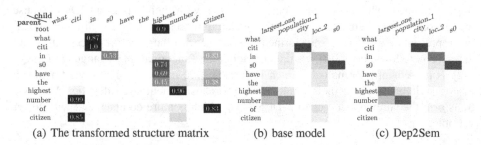

(a) The transformed structure matrix   (b) base model   (c) Dep2Sem

Fig. 3. (Left) The transformed structure matrix $\mathbf{M}^R$ of the sentence *"what citi in s0 have the highest number of citizen"*, whose logical form is $largest\_one(population\_1(city(loc\_2(s0))))$. (Middle&Right) The example demonstrating the difference between the attention alignments of base model and our Dep2Sem.

---

[1] https://stanfordnlp.github.io/CoreNLP/.

Table 2. Evaluation Accuracies on JOBS and ATIS.

Method	JOBS	ATIS
Non-Neural Models		
Zettlemoyer and Collins, 2007 [37]	79.3	84.6
Kwiatkowski et al., 2011 [39]	-	82.8
Liang et al., 2013 [38]	90.7	-
Wang et al., 2014 [40]	-	**91.3**
Zhao and Huang, 2015 [36]	85.0	84.2
Neural Models		
Jia and Liang, 2016 [7]	-	76.3
Jia and Liang, 2016 [7] (+data)	-	83.3
Dong and Lapata, 2016 [14]: Seq2Seq	87.1	84.2
Dong and Lapata, 2016 [14]: Seq2Tree	90.0	84.6
Xu et al., 2018 [21] (+syntactic encoder)	**91.2**	85.9
Chen et al., 2018 [16]	-	85.5
Our Models		
Seq2Tree(Base)	90.0	84.4
Dep2Sem	90.7	86.2
- Dep2Sem(WithoutDep)	90.0	84.8
- Dep2Sem(Fixed)	88.6	84.2

Table 3. The examples of the learned tranformation rules on GEO in FunQL, JOBS in Prolog, ATIS in λ-DCS. Our Dep2Sem model can learn different transformation rules for different meaning representations.

	Rule	Example	Logical form
GEO	$\overleftarrow{ROOT} \xrightarrow{nsubj} \overleftarrow{amod}$	ROOT $\overleftarrow{ROOT}$ is $\xrightarrow{nsubj}$ river $\overleftarrow{amod}$ largest	$Answer(largest(river(...)))$
	$\xrightarrow{amod}$	many $\xrightarrow{amod}$ cities	$count(city(...))$
	$\xrightarrow{nsubj} \xrightarrow{nmod} \xrightarrow{case}$	cities $\xrightarrow{nsubj}$ are $\xrightarrow{nmod}$ state $\overleftarrow{case}$ in	$city(loc\_2(state(...)))$
JOBS	$\overleftarrow{ROOT} \xrightarrow{nsubj}$	ROOT $\overleftarrow{ROOT}$ require $\xrightarrow{nsubj}$ jobs	$Answer(job(req\_deg(...)))$
	$\overleftarrow{nmod} \xrightarrow{case}$	jobs $\xrightarrow{nmod}$ areaid0 $\overleftarrow{case}$ in	$job(area(areaid0),...)$
	$\xrightarrow{dobj}$	use $\xrightarrow{dobj}$ languageid0	$language(languageid0)$
ATIS	$\xrightarrow{case}$	from $\xrightarrow{case}$ ci0	$From.ci0$
	$\overleftarrow{nmod} \xrightarrow{case}$	fare $\xrightarrow{nmod}$ al0 $\overleftarrow{case}$ on	$Fare.Airline.al0$
	$\overleftarrow{ROOT} \xrightarrow{nsubj}$	ROOT $\overleftarrow{ROOT}$ fly $\xrightarrow{nsubj}$ airline	$Answer.Airline$

4.3   Effects of Dep2Sem Transformation Model

To analyze the effect of our Dep2Sem transformation model, the transformation rules which are automatically learned by Dep2Sem in different datasets are shown in Table 3. And an example of the transformed structure matrix $\mathbf{M}^R$ and attention alignments is shown in Fig. 3.

In Table 3, the average score of Eq. 5 for each dependency path in training dataset are calculated, the top 3 transformation rules in different datasets are shown.

Figure 3 (a) shows the probability of each word as a child node of the other words. According to the transformed structure matrix $\mathbf{M}^R$, a sentence can be organized into a tree structure. Figure 3 (b&c) shows the attention alignments of base model and our Dep2Sem. Benefit from the parent-child relation "$citi \leftarrow in$" captured in $\mathbf{M}^R$, the attention of meaning representation token loc_2 can be improved by its parent node $city$.

We can see that:

1. Our Dep2Sem model can effectively learn the syntactic-semantic structure correspondence. For example, in Table 3 our model can capture that the parent-child relation in dependency "$subject \xleftarrow{amod} adjective$" is reversed in FunQL using the second rule learned from GEO dataset. And Fig. 3 shows that Dep2Sem transforms the dependency tree to the structure which is consistent with its target meaning representation.
2. By learning transformation rules from datasets, rather than manually designing them, our Dep2Sem model can easily adapt to different meaning representations. From Table 3, we can see that our method learns different rules for different meaning representations. For example, our model learns quite different transformation rules for GEO in FunQL and ATIS in λ-DCS datasets.

5 Related Work

In recent years, neural semantic parsers have achieved significant progress. Neural parsers model semantic parsing as a sentence to logical form translation task [24,26,28,31,32]. Xiao et al. [15] and Jia and Liang [7] translate word sequences to linearized logical forms. Chen et al. [13] proposed two sentence rewriting methods to resolve the structure mismatch problem in semantic parsing. Bian et al. [17] transferred structural knowledge to text to enhance commonsense question answering. To better exploit the structure of meaning representations, constrained decoding is often used, such as SEQ2TREE [14] which exploits the hierarchical structure of logical forms; Krishnamurthy et al. [26] used type constraints to filter illegal tokens; Liang et al. [43] used a Lisp interpreter to produce valid tokens; Iyyer et al. [31] adopted type constraints to generate valid actions.

Knowing that the syntactic information can help semantic parsing, many grammar-based techniques have been investigated to exploit syntactic information. Debusmann et al. [30] manually designed a syntax-semantics interface. B'edaride and Gardent [29] associated dependency with semantic representations using a semantic calculus. Reddy et al. [19] proposed to use lambda calculus for deriving logical forms from dependency trees. Reddy et al. [25] mapped natural language to logical forms using the grammar in DEPLAMBDA. Liang et al. [38] proposed a Dependency-based Compositional Semantics. Because neural parsers don't use grammars and its training and decoding are in an end-to-end manner, these traditional techniques cannot be used in neural parsers.

To exploit syntactic information in neural semantic parsers, the syntactic information have been exploited in neural methods [44,45]. In neural semantic parsing, Jie and

Lu [22] designed a specialized dependency-based hybrid tree representation to jointly capture both words and semantics. Xu et al. [21] improved sentence encoding with syntactic information. All these methods don't exploit the syntactic-semantic structure correspondence. Chen et al. [23] proposed a tree-to-tree neural network for program translation, but which can only exploit the regular structure correspondence between formal program languages. Compared to program language, in natural language the syntactic structure and the semantic structure are only partial consistent, this is why we design the neural Dep2Sem transformation model.

6   Conclusions and Future Work

This paper proposes Dep2Sem – an approach which can effectively exploit syntactic-semantic structure correspondence for neural semantic parsing. Specifically, a neural Dep2Sem model is proposed to capture the structure correspondence between dependency parses and meaning representations, and a dependency-informed attention is designed to guide the decoder with more accurate attention. Experiments verify the effectiveness of our method. We believe our approach provides a good start for exploiting linguistic structure in neural models, which can potentially benefit many other NLP tasks. For future work, we want to develop new techniques which can exploit syntactic-semantic structure correspondence in other manners, e.g., structure rewriting.

Acknowledgment. This work is supported by the National Natural Science Foundation of China under Grants no. 61906182 and 62076233. Moreover, we sincerely thank the reviewers for their valuable comments.

References

1. Montague, R.: Universal Grammar. In: ACL 2016, Berlin, Germany (1970)
2. Mark, S.: The Syntactic Process. Berlin, Germany (2000)
3. Zelle, J.M., Mooney, R.J.: Learning to parse database queries using inductive logic programming. In: AAAI, IAAI, Portland, Oregon, USA (1996)
4. Zettlemoyer, L.S., Collins, M.: Learning to map sentences to logical form: structured classification with probabilistic categorial grammars. In: UAI, Edinburgh, Scotland (2005)
5. Wong, Y.W., Mooney, R.: Learning synchronous grammars for semantic parsing with lambda calculus. In: ACL, Prague, Czech Republic (2007)
6. Lu, W., Ng, H.T., Lee, W.S., Zettlemoyer, L.: A generative model for parsing natural language to meaning representations. In: EMNLP, Honolulu, Hawaii, USA (2008)
7. Robin, J., Percy, L.: Data recombination for neural semantic parsing. In: ACL, Berlin, Germany (2016)
8. Kate, R.J., Wong, Y.W., Mooney, R.J.: Learning to transform natural to formal languages. In: AAAI and IAAI, Pittsburgh, Pennsylvania, USA (2005)
9. Li, D., Mirella, L.: Coarse-to-fine decoding for neural semantic parsing. In: ACL, Melbourne, Australia, (2018)
10. Jianpeng, C., Siva, R., Vijay, S., Mirella, L.: Learning structured natural language representations for semantic parsing. In: ACL, Vancouver, Canada (2017)
11. Susanto, R.H., Lu, W.: Semantic parsing with neural hybrid trees. In: AAAI, San Francisco, California, USA (2017)

12. Wu, S., et al.: From paraphrasing to semantic parsing: unsupervised semantic parsing via synchronous semantic decoding. arXiv preprint arXiv:2106.06228 (2021)
13. Han, B.C.L.S.X., An, B.: Sentence rewriting for semantic parsing. In: ACL (1) (2016)
14. Li, D., Mirella, L.: Language to logical form with neural attention. In: ACL, Berlin, Germany (2016)
15. Xiao, C., Dymetman, M., Gardent, C.: Sequence-based structured prediction for semantic parsing. In: ACL, Berlin, Germany (2016)
16. Chen, B., Sun, L., Han, X.: Sequence-to-action: end-to-end semantic graph generation for semantic parsing. In: ACL, Melbourne, Australia (2018)
17. Bian, N., Han, X., Chen, B., Sun, L.: Benchmarking knowledge-enhanced commonsense question answering via knowledge-to-text transformation. arXiv preprint arXiv:2101.00760 (2021)
18. Chen, B., Han, X., He, B., Sun, L.: Learning to map frequent phrases to sub-structures of meaning representation for neural semantic parsing. In: AAAI, pp. 7546–7553 (2020)
19. Reddy, S., et al.: Transforming dependency structures to logical forms for semantic parsing. In: ACL, Berlin, Germany (2016)
20. Liang, P., Jordan, M.I., Klein, D.: Learning dependency-based compositional semantics. In: ACL, Portland, Oregon, USA (2011)
21. Xu, K., Wu, L., Wang, Z., Yu, M., Chen, L., Sheinin, V.: Exploiting rich syntactic information for semantic parsing with graph-to-sequence model. In: EMNLP, (2018)
22. Zhanming, J., Wei, L.: Dependency-based hybrid trees for semantic parsing. In: EMNLP, (2018)
23. Chen, X., Liu, C., Song, D.: Tree-to-tree neural networks for program translation. In: NeurIPS 2018, Montréal, Canada (2018)
24. Rabinovich, M., Stern, M., Klein, D.: Abstract syntax networks for code generation and semantic parsing. In: ACL, Vancouver, Canada (2017)
25. Reddy, S., Täckström, O., Petrov, S., Steedman, M., Lapata, M.: Universal semantic parsing. In: EMNLP, Copenhagen, Denmark, (2017)
26. Krishnamurthy, J., Dasigi, P., Gardner, M.: Neural semantic parsing with type constraints for semi-structured tables. In: EMNLP, Copenhagen, Denmark (2017)
27. Manning, C.D., Surdeanu, M., Bauer, J., Finkel, J.R., Bethard, S., McClosky, D.: The Stanford CoreNLP natural language processing toolkit. In: ACL, Baltimore, MD, USA, System Demonstrations (2014)
28. Yih, S.W.T., Chang, M.W., He, X., Gao, J.: Semantic parsing via staged query graph generation: question answering with knowledge base. In: ACL, Beijing, China (2015)
29. Bédaride, P., Gardent, C.: Deep semantics for dependency structures. In: CICLing, Tokyo, Japan, (2011)
30. Debusmann, R., Duchier, D., Koller, A., Kuhlmann, M., Smolka, G., Thater, S.: A relational syntax-semantics interface based on dependency grammar. In: COLING, Geneva, Switzerland (2004)
31. Iyyer, M., Yih, W.T., Chang, M.W.: Search-based neural structured learning for sequential question answering. In: ACL, Vancouver, Canada (2017)
32. Chen, B., An, B., Sun, L., Han, X.: Semi-supervised lexicon learning for wide-coverage semantic parsing. In: COLING, pp. 892–904 (2018)
33. Kwiatkowksi, T., Zettlemoyer, L., Goldwater, S., Steedman, M.: Inducing probabilistic CCG grammars from logical form with higher-order unification. In: EMNLP, Massachusetts, USA (2010)
34. Lu, W.: Constrained semantic forests for improved discriminative semantic parsing. In: ACL, Beijing, China (2015)
35. Jones, B., Johnson, M., Goldwater, S.: Semantic parsing with bayesian tree transducers. In: ACL, Jeju Island, Korea (2012)

36. Zhao, K., Huang, L.: Type-driven incremental semantic parsing with polymorphism. In: NAACL HLT 2015, Denver, Colorado, USA, (2015)
37. Zettlemoyer, L., Collins, M.: Online learning of relaxed CCG grammars for parsing to logical form. In: EMNLP-CoNLL, Prague, Czech Republic (2007)
38. Liang, P., Jordan, M.I.: Klein, D.: Learning dependency-based compositional semantics. In: EMNLP-CoNLL 2007, Prague, Czech Republic (2013)
39. Kwiatkowski, T., Zettlemoyer, L., Goldwater, S., Steedman, M.: Lexical generalization in CCG grammar induction for semantic parsing. In: EMNLP, Edinburgh, UK (2011)
40. Wang, A., Kwiatkowski, T., Zettlemoyer, L.: Morpho-syntactic lexical generalization for CCG semantic parsing. In: EMNLP, Doha, Qatar (2014)
41. Liang, P.: Lambda dependency-based compositional semantics. In: EMNLP, 2014, Doha, Qatar (2013)
42. Zelle, J.M., Moriarty, D., Thompson, C., Hermjakob, U., Konvisser, J.: Using inductive logic programming to automate the construction of natural language parsers. In: EMNLP, 2014, Doha, Qatar (1995)
43. Liang, C., Berant, J., Le, Q., Forbus, K.D., Lao, N.: Neural symbolic machines: learning semantic parsers on freebase with weak supervision. In: ACL, Vancouver, Canada (2017)
44. Roth, M., Lapata, M.: Neural semantic role labeling with dependency path embeddings. In: ACL, Berlin, Germany, Vol. 1: Long Papers (2016)
45. Strubell, E., Verga, P., Andor, D., Weiss, D., McCallum, A.: Linguistically-informed self-attention for semantic role labeling. In: EMNLP, Brussels, Belgium (2018)

Research on Chinese-Korean Entity Alignment Method Combining TransH and GAT

Cheng Jin, Rongyi Cui[✉], and Yahui Zhao

Intelligent Information Processing Laboratory, Yanbian University, Yanji, China
{cuirongyi,yhzhao}@ybu.edu.cn

Abstract. Cross-language entity alignment is one of the important techniques for carrying out linguistic research and constructing a multilingual knowledge graph. At present, there is a lack of research content for Chinese and Korean bilingualism in the field of knowledge graphs. At the same time, mainstream entity alignment methods are susceptible to the impact of data set size and graph structure heterogeneity. This paper proposes a cross-language entity alignment model that combines GAT and TransH, which can alleviate the negative impact of the model on the number of data sets and the heterogeneous graph structure. Firstly, this article uses crawlers to collect and sort out a high-quality Chinese-Korean aligned bilingual data set for training the alignment model; secondly, using the model proposed in this article, through learning the structure and relationship characteristics of the bilingual knowledge graph, automatically discover the existence of cross-language entities with the same semantics. Experiments show that this method has a high accuracy rate. When tested on the Chinese-Korean alignment data set, Hits@1, Hits@5 and Hits@10 reached 49.62%, 80.89% and 91.76%, respectively.

Keywords: Cross-language entity alignment · Knowledge graph · Knowledge representation learning · Graph neural network

1 Introduction

The concept of knowledge graph was first proposed by Google in 2012 to optimize the query quality of search engines and the efficiency of user queries. With the rapid development of the field of intelligent information in recent years, technologies related to knowledge graphs have also been widely used in question answering [1], search [2], recommendation [3] and other systems. With the rise of multilingual online encyclopedias, more and more large-scale multilingual knowledge graphs have appeared, such as DBpedia [4], YAGO [5] and Xlore2 [6]. The multilingual knowledge graphs constructs a large number of entities, attributes and relationships described in different languages in the objective world into a huge knowledge network. In artificial intelligence applications oriented to multilingual scenarios, the multilingual knowledge graph can build a rich knowledge base, provide prior knowledge for artificial intelligence applications, and optimize its cognition and understanding capabilities. In the process of constructing a

© Springer Nature Singapore Pte Ltd. 2021
B. Qin et al. (Eds.): CCKS 2021, CCIS 1466, pp. 134–144, 2021.
https://doi.org/10.1007/978-981-16-6471-7_10

multilingual knowledge graph, sometimes it is necessary to process massive amounts of data in dozens of hundreds of languages, which consumes huge manpower and material resources. Therefore, how to make the machine efficiently and automatically process the above work has important research significance and value.

At present, cross-language entity alignment methods based on machine learning are widely used to automatically construct a multilingual knowledge graph. Given the knowledge graphs of two different languages, the entity alignment task aims to discover the entities with the same meaning contained in them. Entity alignment can also be reduced to a graph matching problem, that is, finding the node matching that makes the two knowledge graph structures the most similar.

Currently, cross-language entity alignment mainly adopts the method based on knowledge embedding and the method based on graph embedding. The method of entity alignment based on knowledge embedding is currently the main strategy adopted, but it requires very high data volume and it is difficult to achieve high accuracy when the data volume is low. Although the method based on graph embedding requires lower data volume in comparison, it has the problems of high memory usage, slow training speed and susceptibility to heterogeneous graph structure.

This paper proposes a method of combining knowledge embedding and graph embedding to achieve cross-language entity alignment. This method can simultaneously learn the relationship information between entities and the structural information of the knowledge graph, effectively alleviating the lack of data and the heterogeneous structure of the knowledge graph. The negative impact caused by the model effectively improves the accuracy of the alignment model. At present, the commonly used public data sets include FB15K [7] and DBP15K [8], etc., mainly in Chinese-English, Japanese-English, French-English and other languages. However, there are few researches in the field of Chinese and Korean bilingual knowledge graphs, and there is also a lack of data sets of entities aligned with Chinese-Korean. Therefore, this article uses crawler technology to collect and organize more than 80,000 high-quality Chinese-Korean bilingual aligned structured data sets from the Internet, which provides important basic data for the development of upstream and downstream research on the relevant knowledge graph.

2   Related Work

Cross-language entity alignment methods based on embedding usually use low-dimensional real-valued vectors to represent the relationships and entities in the knowledge graph, so that the entities that have relationships satisfy a certain functional relationship with each other. Finally, the vector representation of the entities is used to train the model, which automatically aligns potential cross-language entities with the same semantics. Cross-language entity alignment methods are mainly divided into two types: the method based on knowledge embedding and the method based on graph embedding.

The TransE model [7] is a classic model in the field of knowledge representation learning. TransE is analogous to the representation of word vectors and uses a distributed representation method to represent the entities and relationships in the knowledge graph. Inspired by the TransE model, Chen M et al. were the first to introduce an embedding-based representation learning method to cross-language entity alignment, and proposed the MTransE model [9]. Zhu H et al. improved on the basis of MTransE and proposed the ITransE model, which improves the accuracy of the model by selecting seeds that are added to the training set during iterations, and at the same time assigns different weights to different seeds according to their confidence to ease the spread the wrong question [10]. Sun Z proposed BootEA on the basis of ITransE, which can alleviate the problem of the accumulation of error seeds in the ITransE model by adding tags to the newly aligned entities and at the same time editing or deleting the new entities that have been added to the training set [11].

Graph neural network is very suitable for processing structured data, can effectively extract the structural features and node features of the graph, and has been widely used in many fields. Wang Z et al. proposed GCN-Align, an alignment model based on GCN, which uses GCNs to embed entities into a unified vector space, and implements entity alignment based on the distance between entities in the embedded space [12]. Yang H W et al. combined BERT and GCN, and used pre-training models to reduce the differences between different languages. GCN was used to extract entity topology, relationship and attribute information [13]. Mao X et al. directly model the cross-language entity embedding by paying attention to the node's input and output neighbors and the meta-semantics of their connection relations, and then use a two-way iterative strategy to perform entity alignment [14].

At present, many alignment models based on embedded representation learning models have high requirements on the scale of the data set, and the scale of the data set directly affects the accuracy of the alignment model. The use of graph convolutional neural network or graph convolutional neural network and the alignment model of the representation learning model, when facing the situation of heterogeneous knowledge graph of different languages, it is easy to cause the problem of underfitting of graph convolutional neural network, which will lead to a decrease in the accuracy of the model. In order to solve the above problems, this paper proposes a cross-language entity alignment method combining graph embedding and knowledge embedding.

3 Cross-Language Entity Alignment Model Combining Graph Embedding and Knowledge Embedding

The cross-language entity alignment model structure adopted in this paper is shown in Fig. 1, which mainly includes two parts, a graph embedding layer and a knowledge embedding layer. The input of the model is the knowledge graph of the two languages of Chinese and Korean and pre-aligned entity pairs.

The graph embedding layer and the knowledge embedding layer respectively extract the structural feature information of the knowledge graph and the relationship feature information between entities. After iterative training, the model maps the entities to the corresponding vector space. In the vector space, equivalent entities with the same semantics are close to each other, and the alignment degree between entities can be calculated by the L_2 norm. The cross-language alignment model is updated through multiple iterations, so that potential alignment entities with the same semantics are given a conflict-free, one-to-one relationship to form a new alignment entity pair. At the same time, new and appropriate vector representations are given to entities and relationships.

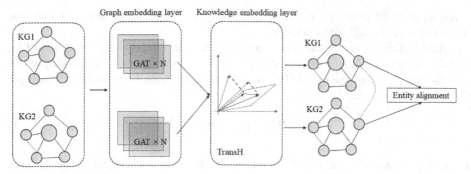

Fig. 1. Cross-language entity alignment model combining graph embedding and knowledge embedding.

Finally, according to the vector representation of the entities, the possible alignment relationships among all entities are calculated.

3.1 Graph Embedding Layer

Graph Attention Network (GAT), Graph Neural Network (GNN) and Graph Convolutional Network (GCN) are all graph machine learning models, which are mainly used to deal with non-Euclidean graph problems that traditional neural networks cannot handle. It is a problem related to the graph formed by several nodes and the edges connecting the nodes. GCN has a good effect when dealing with transductive learning tasks, that is, when both the training phase and the test phase have the same graph structure. However, GCN cannot handle inductive learning tasks, that is, the problem of heterogeneous graph structures in the training phase and the test phase. At the same time, when dealing with directed graphs, it is difficult to assign different weights to different neighbor nodes. GAT is a graph embedding model based on GCN combined with attention mechanism.

The goal of the graph embedding model is to embed the aligned entities in the bilingual KG into a unified vector space using the structural features of the KG. GAT has the following advantages:

(1) Adjacent node pairs can be calculated in parallel to improve model calculation speed;
(2) It is possible to assign weights of any size to edges with multiple connections to nodes;
(3) It can adapt to data with a different structure from the training set;
(4) After introducing the attention mechanism, the model only pays attention to neighbor nodes, so there is no need to obtain complete graph information at one time. It can reduce the memory usage and increase the model calculation speed.

Compared with other graph machine learning models, GAT has better robustness, faster calculation rate and lower algorithm complexity. In this paper, GAT is used as an encoder to obtain KG graph structure information, and different neighbor nodes are given different attention to ignore some relatively low importance nodes to reduce the impact of different KG heterogeneity.

The set of all entity neighbor nodes is used to represent the network structure of KG, and the entity embedding matrix $\mathbf{X} \in \mathbb{R}^{r \times d}$ is used as the input of the encoder, where d is the dimension of the entity. The encoder is implemented by overlaying multiple Graph Attention Layers:

$$\mathbf{H}^{(l+1)} = \sigma\left(\mathbf{A}^{(l)}\mathbf{H}^{(l)}\mathbf{W}^{(l)}\right) \tag{1}$$

$\mathbf{H}^l \in \mathbb{R}^{n \times d^{(l)}}$ and $\mathbf{W}^{(l)} \in \mathbb{R}^{d^l \times d^{(l+1)}}$ are the hidden state and weight of the l_{th} layer respectively, $\mathbf{H}^{(0)} = \mathbf{X}$; $\sigma(\cdot)$ is the non-linear activation function $ReLU(\cdot) = \max(0, \cdot)$; $\mathbf{A}^{(l)} \in \mathbb{R}^{n \times n}$ is a connected matrix obtained by calculating the graph input to the model using the self-attention mechanism. Let one element in the connected matrix $\mathbf{A}^{(l)}$ be $a_{ij}^{(l)}$, used to represent the weight of entities $\mathbf{e}_i$ to $\mathbf{e}_j$. $a_{ij}^{(l)}$ can be calculated by the self-attention mechanism:

$$a_{ij}^{(l)} = softmax\left(c_{ij}^{(l)}\right) = \frac{\exp\left(c_{ij}^l\right)}{\sum_{\mathbf{e}_k \in N_{\mathbf{e}_i} \cup \{\mathbf{e}_i\}} \exp\left(c_{ij}^{(l)}\right)} \tag{2}$$

where $N_{\mathbf{e}_i} \cup \{\mathbf{e}_i\}$ is the set of neighbor nodes of the self-loop edge of $\mathbf{e}_i$, c_{ij}^l is the attention coefficient of entities e_i to e_j. The calculation formula of the attention coefficient $c_{ij}^{(l)}$ is as follows:

$$c_{ij}^l = LeakyReLU\left(\mathbf{q}^T\left[\mathbf{W}^{(l)}\mathbf{h}_i^{(l)} \oplus \mathbf{W}^{(l)}\mathbf{h}_j^{(l)}\right]\right) \tag{3}$$

LeakyReLU is a nonlinear activation function, the formula is as follows:

$$LeakyReLU(x) = \begin{cases} x, & x \geq 0 \\ ax, & x < 0 \end{cases} \tag{4}$$

a is a fixed parameter in the interval $(0, +\infty)$, in this paper $a = 0.2$.

In Eq. (3), $\mathbf{h}_i^{(l)}$, $\mathbf{h}_j^{(l)} \in \mathbf{H}^{(l)}$ represent the hidden states of entities $\mathbf{e}_i$ and $\mathbf{e}_j$, respectivly, $\mathbf{q} \in \mathbb{R}^{2d^{(l)}}$ is a learnable parameter used to represent the weight between the conneced layers in the neural network. $\cdot^T$ represents the transposition of the matrix, and $\oplus$ represents the splicing operation of two vectors. The model is updated iteratively for $t + 1$ times to integrate the features of the entity and its neighbor nodes. Finally, the i_{th} row of the hidden state $\mathbf{H}^{(t+1)}$ represents the embedding vector of the entity $\mathbf{e}_i$ after the attention mechanism is updated. The loss function of the graph embedding layer is as follows:

$$O_G = \sum_{(\mathbf{e}_i, \mathbf{e}_j) \in S} \sum_{\left(\mathbf{e}_i', \mathbf{e}_j'\right) \in S'} \left[dist(\mathbf{e}_i, \mathbf{e}_j) + \gamma_1 - dist\left(\mathbf{e}_i', \mathbf{e}_j'\right)\right]_+ \tag{5}$$

where $dist(\mathbf{e}_i, \mathbf{e}_j) = \|\mathbf{e}_i - \mathbf{e}_j\|$ is the L_2 norm between two aligned entity pairs $(\mathbf{e}_i, \mathbf{e}_j)$, and S' represents the negative generated from the sample set S through nearest neighbor sampling. The set of sample pairs, $\gamma_1 > 0$ is a boundary hyperparameter.

3.2 Knowledge Embedding Layer

By mining the entity and relationship information in the text, knowledge can be organized into a structured knowledge network. Using these rich structured information will help us to better complete specific tasks in various scenarios under the knowledge-driven. Using traditional feature extraction methods to process knowledge graphs will have problems such as low computational efficiency and sparse structure, which will limit the deployment and application of knowledge graphs in specific task scenarios. In order to better apply the rich structured information in the knowledge graph to downstream tasks, the entities and relationships in the knowledge graph need to be represented as low-dimensional dense vectors.

At present, there are already many representation learning models that can handle knowledge graphs, among which the TransE model, a translation-based model, is a representative classic model. TransE maps entities and relationships to the same low-dimensional vector space, and expresses the relationship between entities and entities as translation operations between entity vectors, which can achieve better results with a simple structure. However, the TransE model has some shortcomings in dealing with reflexive relationships, one-to-many, many-to-one, and many-to-many relationships. It assigns completely different entities to very similar vectors in the vector space, resulting in poor vector representation. The TransH model is modified on the basis of the TransE model. The specific idea is to abstract the relationship in the triple into a hyperplane in a vector space, and map the head entity or tail entity to this hyperplane, then calculate the difference between the head and tail entities by the translation vector on the hyperplane. TransH defines a hyperplane W_r and a relationship vector r for the relationship in each triplet. Project the head entity e_h and the tail entity e_t on the hyperplane W_r to obtain the projections $e_{h\perp}$ and $e_{t\perp}$. In this way, the same entity has different meanings in different relationships, while the same entity has the same meaning in the same relationship.

The goal of the knowledge embedding model is to model the entities and relationships in KG, and use appropriate vectors to represent different entities. We use the TransH model as the knowledge embedding model. For a given triple (e_h, r, e_t), project on W_r to obtain $e_{h\perp}, e_{t\perp}$, so that the triple satisfies $e_{h\perp} + r \approx e_{t\perp}$. Model training uses the margin ranking loss function as the training target of the knowledge embedding model.

For a triple in KG, map the head entity e_h and tail entity e_t to the hyperplane to obtain the mapping vectors $e_{h\perp}$ and $e_{t\perp}$, the formulas are as follows:

$$\begin{cases} \mathbf{e}_{h\perp} = \mathbf{e}_h - \mathbf{e}_{hw_r} = \mathbf{e}_h - w^T \mathbf{e}_h w \\ \mathbf{e}_{t\perp} = \mathbf{e}_t - \mathbf{e}_{tw_r} = \mathbf{e}_t - w_r^T \mathbf{e}_t w_r \end{cases} \tag{6}$$

where:

$$\begin{cases} \mathbf{e}_{hw_r} = w^T \mathbf{e}_h w \\ \mathbf{e}_{tw_r} = w^T \mathbf{e}_t w \end{cases} \tag{7}$$

in $w^T \mathbf{e}_h w$, $w^T \mathbf{e}_h = |w||h|cos\theta$ represents the projection length of $\mathbf{e}_h$ in the direction of w_r, and then multiply it by the projection of e_h on w_r to get the projection on the hyperplane. Then according to Eq. (8) to find the score of the triplet:

$$f_r(\mathbf{e}_h, \mathbf{e}_t) = \|h - w_r^T \mathbf{e}_h w_r + r - \mathbf{e}_t + w_r^T \mathbf{e}_t w_r\|$$

If the triple relationship is correct, the result of the above formula is smaller. If the triple relationship is wrong, the result of the above formula is larger.

The objective function of the graph embedding model is shown in Eq. (9):

$$O_K = \sum_{(e_h,r,e_t)\in T} \sum_{\left(e_h',r',e_t'\right)\in T'} \left[\gamma_2 + f_r(\mathbf{e}_h, \mathbf{e}_t) - f_r\left(\mathbf{e}_h', \mathbf{e}_t'\right)\right]_+ \tag{9}$$

let $\mathbf{w}_r$ be the normal vector of plane W_r, w_r satisfies the constraint condition $\|w_r\|_2^2 = 1$. In the above formula $[\cdot]_+ = \max\{0, \cdot\}$, the embedded entities $\mathbf{e}_h, \mathbf{e}_r$ come from the embedding matrix $\mathbf{H}^{(L)}$ after the graph embedding model is updated, and the relationship r comes from the relationship that needs to be learned the matrix $\mathbf{R} \in \mathbb{R}^{|R|\times d}$, T' is a set of negative sample triples, and $\gamma_2 > 0$ is a boundary hyperparameter.

3.3 Model Construction and Training

Obtaining the structural features of KG only through the graph embedding model will ignore the relationship information existing between entities, and if only the knowledge representation learning model is used for entity alignment tasks, on the one hand, the structural information in KG will be ignored, and on the other hand, it will be easily affected by the amount of data. Therefore, the cross-language entity alignment model in this paper combines the graph embedding model and the knowledge embedding model. The overall objective function is defined as follows:

$$O = O_G + O_K \tag{10}$$

where O_G and O_K are given by Eq. (5) and Eq. (9). When performing entity alignment reasoning, we find a new entity with the same semantics by calculating the L_2 distance of two entities in the vector space. The L_2 distance between entities and relationships with the same semantics should be small. The AdaGrad is used to optimize the total objective function O shown in Eq. (10).

4 Experiment and Analysis

4.1 Data Set

This paper uses crawler technology to crawl a total of 702,645 triples in 13 languages from BabelNet [15]. Since this research only focuses on Chinese and Korean, only the data in these two languages are processed and cleaned. The details of the Chinese-Korean alignment data set are shown in Table 1. In the experiment, 70% of the data set is used as the training set and 30% as the test set.

As with most entity alignment tasks, this article uses Hits@k to evaluate the performance of the model. Hits@k in the cross-language entity alignment task means that when all entities in the current language are automatically aligned to the entities, the semantic similarity of the entities in another language is sorted, and the top k entities contain the target entity. Probability. This article records the results of Hits@1, Hits@5,

Table 1. Chinese-Korean alignment data set statistics

Data set	Quantity
Triples	86934
Entities	54795
Relation	1196
Train set	14289
Test set	33344

and Hits@10. In the experiment, the case where Chinese aligns with Korean entities and Korean entities aligns with Chinese entities is considered at the same time.

In order to verify the feasibility of the cross-language entity alignment model described in Sect. 2, we conducted corresponding experiments. The operating system of the server used in the experiment was Ubuntu 20.04 LTS, the memory was 128 GB, the GPU was NVIDIA Quadro RTX 5000, and the CPU was Intel Xeon® Gold 6128. The program is mainly written in python 3.7, and the industry's mainstream deep learning framework Pytorch 1.6 is used to implement data loading and model training.

When training in this paper, the learning rate of AdaGrade is $\lambda = 0.01$, the parameters of interval sorting loss are $\gamma_1 = 3$, $\gamma_2 = 3$, and the number of iterations is 1000. This paper uses the Chinese-Korean aligned data set to experiment on the method of combining the GAT model with multiple knowledge representation learning models. The experimental results are shown in Table 2.

Table 2. Chinese-Korean entity alignment experiment results based on the combination of graph embedding and knowledge embedding

Model	KO → ZH			ZH → KO		
	Hits@1	Hits@5	Hits@10	Hits@1	Hits@5	Hits@10
MTransE	0.4294	0.6833	0.7622	0.4282	0.675	0.7533
GCN-Align	0.4359	0.7089	0.8029	0.4387	0.7077	0.7981
GAT + TransR	0.4594	0.7643	0.8925	0.4551	0.7587	0.891
GAT + TransD	0.4632	0.7709	0.8942	0.457	0.7618	0.894
GAT + TransE	0.4443	0.7424	0.8719	0.4411	0.7393	0.8678
GAT + RotatE	0.477	0.7531	0.835	0.4758	0.744	0.8226
GAT + TransH	**0.4962**	**0.8089**	**0.9176**	**0.4978**	**0.8074**	**0.9167**

Observing the results in Table 2 can be found:

(1) The overall performance of the alignment model based on the combination of graph embedding and knowledge embedding is compared with the cross-language entity alignment models MTransR and GCN-Align based on knowledge embedding and graph embedding. The accuracy rate is generally significantly improved, and Hits@1 is improved by 2%–7%, Hits@5 increased by 6%–12%, Hits@10 increased by 7%–15%. Therefore, we can think that the cross-language alignment model based on the combination of knowledge embedding and graph embedding can effectively improve the accuracy of the model, and it is effective and reasonable to adopt this method.

(2) In the performance of all models, the performance of Korean aligning to Chinese exceeds the accuracy of Chinese aligning to Korean entities, which is generally about 1% higher. The analysis is due to the fact that Korean entities are more recognizable in the text representation of entities than Chinese entities. Different entities have fewer repetitions in Korean, and there are more entities with the same Chinese entity text but different meanings. Therefore, the model has certain difficulties when representing Chinese entities in the vector space, and the accuracy is slightly reduced.

(3) TransR, TransD, TransH, and TransE all use a linear relationship-based constraint method to model the entity, while RotatE uses a rotation-based constraint relationship to model the entity. According to Table 2, we can find that when GAT is combined with a model based on linear constraints, the accuracy of the model is higher than that of the model combined with RotatE. Therefore, we can think that the combination of the linear constraint-based knowledge embedding model and the GAT model can better represent the vector space when dealing with cross-language entity alignment tasks, and the alignment model has a higher accuracy rate.

(4) TransR, TransD, TransH, and TransE all use a constraint method based on linear relationship to model the entity, while RotatE uses a rotation-based constraint relationship to model the entity. According to Table 2, we can find that when GAT is combined with a model based on linear constraints, the accuracy of the model is higher than that of the model combined with RotatE. Therefore, we can think that the knowledge embedding model based on linear constraints combined with the GAT model can better represent the vector space when processing cross-language entity alignment tasks, and the alignment model has a higher accuracy rate.

The accuracy of the GAT+TransH combination is higher than other combinations, so this method is selected as the final cross-language entity alignment scheme in this paper. The loss curve of the model is shown in Fig. 2.

As the number of iterations increases, the Loss curve of the graph embedding model decreases uniformly and smoothly, without large oscillations, which can indicate that the graph embedding model is well trained and there is no over-fitting or under-fitting. The Loss curve of the knowledge embedding model converges quickly, and it rises to a certain extent in the later stage of training, and there is an over-fitting phenomenon. In the next work, further revisions to the knowledge embedding model are considered.

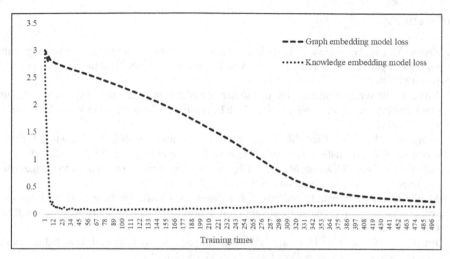

Fig. 2. Model loss curve

But in general, the training process of the model is relatively stable, the convergence is normal, and the trained model can satisfy the use of the alignment task.

5 Conclusion and Future Work

At present, the lack of data sets in the field of Chinese-Korean knowledge graphs has greatly restricted the development of research in related fields. This article uses crawler technology to crawl the Chinese and Korean aligned entity data from the Internet, and process the data to obtain a high-quality Chinese Korean bilingual aligned structured data set. At the same time, this paper analyzes the characteristics and defects of the current mainstream algorithm design for cross-language entity alignment based on knowledge representation learning, and proposes a cross-language entity alignment algorithm that combines GAT and TransH. The algorithm uses GAT to capture the network structure characteristics of the knowledge graph, and uses the TransH model to constrain the relationship between entities and update the vector, alleviating the problem that mainstream alignment algorithms are susceptible to the number of data sets and the model is affected by the heterogeneity of the knowledge graph. Experiments show that the combination of TransH and GAT is better than all other solutions that combine knowledge representation learning models and graph attention.

In further work, we will further analyze the factors that affect the effects of graph neural networks and knowledge representation models. At the same time, experiments are carried out on more different graph embedding and knowledge embedding combined methods to explore better cross-language entity alignment strategies.

Acknowledgements. This work was supported by the National Language Commission Scientific Research Project (YB135–76); Yanbian University Foreign Language and Literature First-Class Subject Construction Project (18YLPY13).

References

1. Zhang, Y., Dai, H., Kozareva, Z., et al.: Variational reasoning for question answering with knowledge graph. In: Proceedings of the AAAI Conference on Artificial Intelligence, vol. 32, issue1 (2018)
2. Xiong, C., Power, R., Callan, J.: Explicit semantic ranking for academic search via knowledge graph embedding. In: Proceedings of the 26th International Conference On World Wide Web, pp. 1271–1279 (2017)
3. Wang, H., Zhang, F., Zhao, M., et al.: Multi-task feature learning for knowledge graph enhanced recommendation. The World Wide Web Conference, pp. 2000–2010 (2019)
4. Lehmann, J., Isele, R., Jakob, M., et al.: Dbpedia–a large-scale, multilingual knowledge base extracted from wikipedia. Semantic web 6(2), 167–195 (2015)
5. Mahdisoltani, F., Biega, J., Suchanek, F.: Yago3: A knowledge base from multilingual wikipedias. In: 7th Biennial Conference on Innovative Data Systems Research. CIDR Conference (2014)
6. Jin, H., Li, C., Zhang, J., et al.: XLORE2: large-scale cross-lingual knowledge graph construction and application. Data Intell. 1(1), 77–98 (2019)
7. Bordes, A., Usunier, N., Garcia-Duran, A., et al.: Translating Embeddings for Modeling Multi-relational Data. Neural Information Processing Systems (NIPS), pp. 2787–2795. South Lake Tahoe, USA (2013)
8. Sun, Z., Hu, W., Li, C.: Cross-lingual entity alignment via joint attribute-preserving embedding. In: d'Amato, C., et al. (eds.) ISWC 2017. LNCS, vol. 10587, pp. 628–644. Springer, Cham (2017). https://doi.org/10.1007/978-3-319-68288-4_37
9. Chen, M., Tian, Y., Yang, M., et al.: Multilingual Knowledge Graph Embeddings for Cross-Lingual Knowledge Alignment. In: Proceedings of the 26th International Joint Conference on Artificial Intelligence, pp. 1511–1517. Melbourne, Australia (2017)
10. Zhu, H., Xie, R., Liu, Z., et al.: Iterative Entity Alignment via Joint Knowledge Embeddings, pp. 4258–4264. IJCAI. Melbourne, Australia (2017)
11. Sun, Z., Hu, W., Zhang, Q., et al.: Bootstrapping entity alignment with knowledge graph embedding. IJCAI. 18, 4396–4402 (2018)
12. Wang, Z., Lv, Q., Lan, X., et al.: Cross-lingual knowledge graph alignment via graph convolutional networks. In: Proceedings of the 2018 Conference on Empirical Methods in Natural Language Processing, pp. 349–357 (2018)
13. Yang, H.W., Zou, Y., Shi, P., et al.: Aligning Cross-Lingual Entities with Multi-Aspect Information. In: Proceedings of the 2019 Conference on Empirical Methods in Natural Language Processing and the 9th International Joint Conference on Natural Language Processing (EMNLP-IJCNLP), pp. 4422–4432 (2019)
14. Mao, X., Wang, W., Xu, H., et al.: MRAEA: an efficient and robust entity alignment approach for cross-lingual knowledge graph. In: Proceedings of the 13th International Conference on Web Search and Data Mining, pp. 420–428 (2020)
15. Navigli, R., Ponzetto, P., BabelNet, S.: Building a Very Large Multilingual Semantic Network. Proceedings of the 48th annual meeting of the associationfor computational linguistics. Uppsala, Sweden, pp. 216–225 (2010).

Incorporating Complete Syntactical Knowledge for Spoken Language Understanding

Shimin Tao[1(✉)], Ying Qin[1], Yimeng Chen[1], Chunning Du[1,2], Haifeng Sun[2],
Weibin Meng[1,3], Yanghua Xiao[4], Jiaxin Guo[1], Chang Su[1], Minghan Wang[1],
Min Zhang[1], Yuxia Wang[1], and Hao Yang[1]

[1] HUAWEI, Shenzhen, China
taoshimin@huawei.com
[2] Beijing University of Posts and Telecommunications, Beijing, China
[3] Tsinghua University, Beijing, China
[4] Fudan University, Shanghai, China

Abstract. Spoken Language Processing (SLU) is important in task-oriented dialog systems. Intent detection and slot filling are two significant tasks of SLU. State-of-the-art methods for SLU jointly solve these two tasks in an end-to-end fashion using pre-trained language models like BERT. However, existing methods ignore the syntax knowledge and long-range word dependencies, which are essential supplements for semantic models. In this paper, we utilize the Graph Convolutional Networks (GCNs) and dependency tree to incorporate the syntactical knowledge. Meanwhile, we propose a novel gate mechanism to model the label of the dependency arcs. Therefore, the labels and geometric connection of dependency tree are both encoded. The proposed method can adaptively attach a weight on each dependency arc based on dependency types and word contexts, which avoids encoding redundant features. Extensive experimental results show that our model outperforms strong baselines.

Keywords: Spoken language understanding · GCNs · Syntax

1 Introduction

In recent years, various intelligent voice assistants have been widely deployed and achieved great success. Typical systems include Google Home, Apple Siri, MI AI, etc. They extract the uses' intents from the goal-oriented dialogues and help users accomplish their tasks through voice interactions. The core component of the intelligent voice assistants is Spoken Language Understanding (SLU) [24]. Typically, SLU consists of two subtasks: intent detection and slot filling [20]. For example, given an utterance "Please play happy birthday", the intent detection aims to find the purpose of the user (i.e., PlayMusic in this example), and it can be modeled as a classification task. The other subtask is slot filling, which is usually modeled as a sequence labeling task to dig out the slots (i.e., happy birthday in this example).

© Springer Nature Singapore Pte Ltd. 2021
B. Qin et al. (Eds.): CCKS 2021, CCIS 1466, pp. 145–156, 2021.
https://doi.org/10.1007/978-981-16-6471-7_11

Deep learning methods achieve promising performances in SLU. Intent detection subtask is a sentence-level classification task, which is widely studied field. CNN [26], RNN [12] and pre-trained models like BERT [3] are widely used to solve the classification task. Slot filling is usually solved by the RNN and pre-trained models BERT as a sequence labeling task. Meanwhile, other efforts jointly solve the intent detection and slot filling problems in end-to-end fashion by exploiting and modeling the dependencies between the two subtasks to improve the performance of the two subtasks simultaneously [5,6,9,13]. Recently, pre-training methods have remarkably improved many natural language processing tasks, which motivates some preliminary studies to investigate the application of pre-trained model in SLU [1].

Despite the success, these methods lack a mechanism to realize relevant syntactical constrains and long-range word dependencies. Ideally, the SLU models should not only utilize the semantic features but also consider linguistic knowledge such as the syntax, which offers grammar information about texts. Parsing sentences to dependency trees is a widely used technique to represent the instructive grammatical structure of texts. It plays an essential role in SLU especially in slot filling subtasks. An example of a sentence and its corresponding dependency tree and slot filling labels are shown in Fig. 1. We can see that the dependency arcs offer important clues for the estimation of the slots: the dependency tree links the separate words to form a joint phrase "New York". And it also links with the core word "flight". For the situation where one sentence expresses multiple intents and has corresponding multiple slots, the dependency parse tree can effectively separate different intents and corresponding slots. Therefore, introducing the structure information about texts in SLU models is valuable.

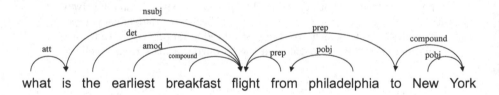

Fig. 1. Illustration of dependency tree.

Here, the core problem we should solve is how to integrate the syntactical structure information in existing SLU models, especially in pre-training methods. Recently, graph convolutional networks (GCNs) have become a popular approaches to integrate structure inductive biases into NLP models [18]. For example, some recent studies have successfully employed GCN to improve the performances in natural language processing tasks such as sentiment analysis [10,19], relation extraction [7,27], and semantic role labeling [14]. In general, these GCNs methods can enhance the performance of SLU tasks. However, the labels of the dependency arcs, which reflect the type of specific relations between words, are ignored in vanilla GCNs. GCNs methods only consider whether two words are connected in dependency tree and establish the adjacency matrix A, ignoring how these words are connected (the labels of dependency arcs). Generally, an ideal syntax-aware spoken language understanding models depends on not only the syntactically connection information but also the corresponding connection labels. Meanwhile, the GCNs syntax-aware models rely on a precise dependency parser. However, in some cases, the parser produces imperfect results.

To tackle these issues, we propose a new variant of GCNs, which can utilize the external dependency tree to enhance the widely-used pre-training models like BERT [3]. Specifically, our model uses BERT as backbone and stacks GCNs on the output of BERT for each token. The dependency tree servers as the external knowledge and can be interpreted to an adjacency matrix. Based on the adjacency matrix, the graph convolutional operation can effectively handle the syntactically related words and exploit long-range relations. Moreover, to take into account the types of dependency arcs, we further design a gate mechanism to dynamically select some specific dependency arcs based on the labels of dependency arcs and the word contexts. Accumulating evidences have shown that encoding the full dependency tree will bring redundant information and degrade the performance of models [7,27]. Therefore, we encode the labels of the dependency arcs and dynamically attach a weight on each word dependency arc according to the labels' embeddings and word contexts. We conducted extensive experiments on the joint SLU task, including intent classification and slot filling. Experimental results show that our model outperforms strong baselines in both tasks.

The contributions can be summarized as follows:

- We propose to integrate the external syntactical knowledge of text in spoken language understanding models. Graph convolutional networks are introduced, utilizing the adjacent matrix obtained from the dependency trees.
- We design a novel gate mechanism to prune the redundant dependency arcs based on the types (labels) of the dependency arcs and corresponding word contexts.
- Experiments on the joint SLU task show that our model effectively incorporates the external syntactical knowledge in BERT and outperforms strong baselines.

2 Related Work

2.1 Spoken Language Understanding

The SLU task includes intent detection and slot filling subtasks. In this section, we review methods that focus on intent detection and slot filling separately and joint methods.

For the intent detection, it is formulated as a text classification problem. Various classification methods such as support vector machine (SVM) [8] and RNN [17], and capsule network [16] were proposed. For the slot filling subtask, it can be regarded as a sequence tagging problem. Approaches for slot filling include CNN, LSTM, RNN-EM [15], and attention mechanism [28]. Despite of the success, existing SLU models ignore the syntactical information and thus have a limitation performance.

The joint models solve the intent detection and slot filling in a single model, which considers the high correlation between the two sub-tasks. Among them some methods were proposed to utilize the intent information to guild the slot filling [5,25]. There are also some studies to consider the cross-impact between the slot and intents [4,23]. Meanwhile, joint models based on large-scale pre-trained models BERT achieve the state-of-the-art performances [1].

2.2 Syntax-Aware Models

The dependency tree contains rich information about syntax of texts and integrating them in the NLP models has become a promising direction. Early studies include recursive neural network operating on dependency trees [6]. Recently, graph neural networks have become gradually popular to integrate structural inductive bias into NLP models. Meanwhile, some studies also reveal that encoding the full dependency tree will bring redundant information and degrade the ability of models [7,18,27]. Pruning graph methods were also an important research area.

3 Model

In this section, we introduce our model. We first briefly introduce BERT and GCNs. Then we elaborate our model: a gate mechanism and the training strategies used in our model. The architecture of our model is shown in Fig. 2.

3.1 Backgrounds of BERT

We adopt Bidirectional Encoder Representations from Transformers (BERT) as the basis of our model. The model architecture of BERT is a multi-layer of Transformers encoder [22]. The input of BERT includes token embeddings, position embeddings, and segment embeddings. Compared to the traditional Word2Vec and GloVe word embeddings, which only provide single context-independent representations for each token, BERT can generate deep contextualized representations. Moreover, BERT is pre-trained by the self-supervised tasks. The large-scale pre-training gives BERT powerful ability in language processing. However, the architecture and the pre-training strategy of BERT has no specific mechanism to employ the syntactical knowledge of texts.

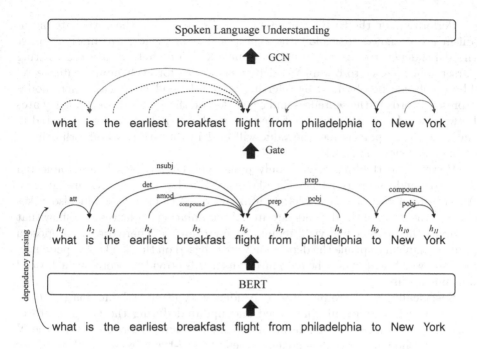

Fig. 2. Model architecture.

3.2   Backgrounds of GCN

Graph Convolutional Networks (GCNs) can be regarded as an adaptation of a conventional CNN. GCNs are unique in directly operating on graph structures [11]. A graph with k nodes can be represented by an adjacency matrix $A \in \mathbb{R}^{k \times k}$. The l-layer of GCN can be formulated as:

$$h_i^l = \sigma(\sum_{j=1}^{k} A_{ij} \boldsymbol{W}^l h_j^{l-1} + \boldsymbol{b}^l), \tag{1}$$

where h_i^0 represents the initial representation of node i. And h_i^l is the output of the l-th GCN layer for node i. $\boldsymbol{W}^l$ and $\boldsymbol{b}^l$ are trainable parameters of l-layer of GCNs. σ is a nonlinear function. Through graph convolution operation, each node can gather and summarize information from its neighboring nodes in the graph.

3.3   The Proposed Model

Our proposed model for SLU consists of BERT embeddings, GCNs and a novel gate mechanism. Specifically, given the input token: $X = \{r_1, r_2, r_3, \ldots, r_T\}$ of length T, BERT is adopted to get the deep contextualized representations: $H = \{h_1, h_2, h_3, \ldots, h_T\}$.

Performing on the BERT embeddings, GCN layers and the novel gate mechanism are designed to exploit the syntactical dependency. We firstly convert the dependency tree to the adjacency matrix. Note that we use the existing parser tool SpaCy[1] to obtain the dependency tree for each input sentence X. The graph can be obtained by referring to the dependency trees, where nodes represent words in the sentences and edges denote the syntactic dependency arcs between words. In other words, if the i-th word and j-th word are connected in the dependency parser tree, the value will be 1 in the i-th row and j-th column of the adjacency matrix A.

However, in this manner, we only judge whether two words are connected in the dependency parser trees. The labels of the dependency arcs are ignored. As argued in Sect. 1, a complete syntax-aware SLU model should also takes account the types of the dependency arcs. Accumulating evidences also show that encoding the full tree will yield redundant information. Therefore, we propose a gate mechanism to prune the dependency trees based on the labels of dependency arcs and words contexts. The gate mechanism will provide a score on each word dependency arc.

Specifically, we introduce the dependency type embedding matrix $P \in \mathbb{R}^{d_t \times N}$, which is randomly initialized and updated during the training procedure. The d_t denotes the dimension of dependency type embeddings, and the N represents the number of dependency types. If the i-the word and j-th word are connected in the dependency tree, the corresponding dependency type embedding t_{ij} is obtained by looking up the embedding matrix P using the type id. Moreover, the gate mechanism also considers the word contexts corresponding to the dependency arc when calculating the gate score:

$$a_{ij} = \sigma([h_i; h_j]^T \boldsymbol{W}_t t_{ij} + \boldsymbol{b}_t), \tag{2}$$

where h_i and h_j are word representations from BERT. And $[;]$ denotes the concatenation operation. $\boldsymbol{W}_t$ and $\boldsymbol{b}_t$ are trainable parameters. σ denotes the *sigmoid* activation function. After getting the gate score a_{ij}, the adjacency matrix A will be modified by it: If there is a dependency arc between word i and j: $A_{ij}^t = a_{ij}$. Then we conduct the graph convolution operation on the updated syntax-type-aware graph:

$$h_i^l = \sigma(\sum_{j=1}^n \frac{A_{ij}^t \boldsymbol{W}^l h_j^{l-1}}{d_i} + \boldsymbol{b}^l), \tag{3}$$

where $\boldsymbol{W}^l$ and $\boldsymbol{b}^l$ are the trainable parameters. d_i is the degree of the node i: $d_i = \sum_{j=1}^n A_{ij}^t$, which aims to normalize the activations of nodes at different degress. The input for the graph convolution network is initialized from BERT's output: $h_i^0 = h_i$. The output of h_i^l of the top GCN layer is the enhanced representation, which simultaneously encodes the syntactical geometric connection and syntactical types.

---

[1] https://spacy.io/.

3.4   Model Training

Our gate mechanism can adaptively adjust the contributions from different dependency arcs. Important dependency arcs will be retained and redundant arcs will be discarded. Meanwhile, The proposed gate mechanism and GCNs are both differentiable. Therefore, our model can be trained in an end-to-end fashion. In this subsection, we will describe how our model is applied to intent detection and slot filling tasks.

The intent detection is a sentence-level classification task. It demands to classify the input sentence to a pre-defined category $I = \{I_1, I_2, ..., I_n\}$. We firstly average the syntax-aware word representations and then predict the label using a softmax layer:

$$H_I = \frac{1}{T} \sum_{i=1}^{T} h_i^l, \tag{4}$$

$$y^I = \text{softmax}(W_I H_I + b_I), \tag{5}$$

where W_I and b_I are trainable parameters.

The slot filling is a sequence tagging problem. Each token has a label indicating the slots boundary using 'BIO' tagging scheme. We feed the syntax-aware word representations h_i^l to a softmax layer to predict the slot labels:

$$y_i^s = \text{softmax}(W_s h_i^l + b_s), \tag{6}$$

where W_s and b_s are trainable parameters.

To jointly model intent classification and slot filling, the objective is formulated as:

$$p(y^I, y_i^s | X) = p(y^I | X) \prod_{i=1}^{T} p(y_i^s | X). \tag{7}$$

The learning objective is to maximize the conditional probability $p(y^I, y_i^s | X)$. The model is fine-tuned end-to-end via minimizing the cross-entropy loss.

4   Experiments

In this seciton, we evaluate our proposed model on two public SLU benchmark datasets: ATIS and Snips.

4.1   Datasets

The ATIS [21] and Snips [2] are two widely used SLU datasets. The ATIS consists of audio recordings of people making flight reservations. The Snips dataset was collected from the Snips personal voice assistant. We follow the same data division as [1,5]. The statistics of these two datasets are listed in Table 1.

Table 1. Summary statistics of the datasets. Slot and Intent denote the number of slot labels and intent types, respectively.

Dataset	Train	Dev	Test	Slot	Intent
ATIS	4478	500	893	120	21
Snips	13084	700	700	72	7

4.2   Training Details

We adopted BERT$_{base}$ (uncased) as the basis, which has 12 layers and 12 heads. The dimension of the word embeddings is 768. The hidden dimensions of GCNs is 256, and the hidden dimension of the labels of word dependency is 24. During the training procedure, the batch size for ATIS and Snips are both 32. We use Adam optimizer with learning rate 2e–5, $\beta_1 = 0.9$, $\beta_2 = 0.999$, L2 weight decay of 0.01.

4.3   Baseline Methods

We compare our model with the following methods:

- **RNN-LSTM** [9]: A joint multi-domain model enabling intent detection and slot filling where data from each domain reinforces each other. The model takes an RNN-LSTM architecture.
- **Atten-BiRNN** [13]: An attention-based neural network model for joint intent detection and slot filling. Learning from the attention mechanism in encoder-decoder model, this model introduces attention to the alignment-based RNN models.
- **SLot-Gated** [5]: A slot-gated model that focuses on learning the relationship between intent and slot attention vectors in order to obtain better semantic frame results by the global optimization.
- **Joint BERT** [1]: A joint intent classification and slot filling model based on BERT.
- **Joint BERT CRF** [1]: A joint intent classification and slot filling model based on BERT and Conditional Random Field (CRF). The CRF layer can improve the slot filling by depending on the predicted surrounding labels.

4.4   Main Results

Table 2 presents the results of our model and baselines on ATIS and Snips datasets. From the table, we have the following observations:

For the previous models, RNN-LSTM, Atten-BiRNN, and Slot-Gated are all based on static word embeddings such as GloVe and Word2Vec. Compared to BERT-based methods, the traditional word embeddings can only offer separate word-level representations, without the ability to model the semantic changes among different contexts. Moreover, they lack the large-scale pre-training procedure. Therefore, BERT-based models outperform these three models by a large margin.

For the BERT-based models, stacking CRF layer on BERT achieves a comparable result to vanilla BERT, probably due to the self-attention in Transformers may sufficiently model the label distributions. On the contrary, introducing our proposed gate mechanism and GCNs brings a remarkable improvements over BERT-based solution. It verifies the effectiveness our model, which effectively incorporate the syntax knowledge of texts.

Table 2. Experimental results. Results are measured by intent classification accuracy (Intent), slot filling F1 (Slot), and sentence-level semantic frame accuracy (Sent).

Model	Snips			ATIS		
	Intent	Slot	Sent	Intent	Slot	Sent
RNN-LSTM [9]	96.9	87.3	73.2	92.6	94.3	80.7
Atten-BiRNN [13]	96.7	87.8	74.1	91.1	94.2	78.9
Slot-Gated [5]	97.0	88.8	75.5	94.1	95.2	82.6
Joint BERT [1]	98.6	97.0	92.8	97.5	96.1	88.2
Joint BERT CRF [1]	98.4	96.7	92.6	97.9	96.0	88.6
Our model	**98.9**	**98.3**	**93.1**	**98.3**	**97.3**	**90.2**

4.5 Ablation Study

To analyze the effect of different components including the GCNs and the proposed gate mechanism, we report the results of variants of our model. From Table 3, we can find that, without the proposed gate mechanism, the accuracy of intent detection and slot filling both drop. Especially for the slot filling subtask, the F1 scores drop remarkably. Normally, sequence labeling task demands the ability to resolve the syntax. The results show that our proposed gate mechanism can effectively utilize the syntactical knowledge. The model without GCNs degrade to normal BERT.

Table 3. Ablation study. w/o means without.

Model	Snips			ATIS		
	Intent	Slot	Sent	Intent	Slot	Sent
FULL-MODEL	**98.9**	**98.3**	**93.1**	**98.3**	**97.3**	**90.2**
-w/o Gate	98.6	97.4	93.1	98.0	96.2	89.0
-w/o GCNs	98.6	97.0	92.8	97.5	96.1	88.2

4.6   Case Study

To intuitively show the effectiveness of our proposed model. We randomly select a sample in test set of ATIS, and show the gate scores of each dependency arcs in Fig. 3. For this example, 'philadelphia' and 'New York' are core words that should be tagged, and the intent of this sentence is 'book flight'. Figure 3 shows that our model pay more attention on the part of dependency arcs relevant to core words. It verifies the effectiveness of our model.

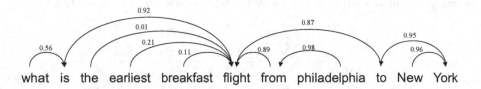

Fig. 3. Case study.

5   Conclusion

In this paper, we aim at incorporating the syntactic knowledge into the spoken language understanding model, which includes intent detection and slot filling tasks. We re-examine the drawbacks of existing GCNs models, and propose a gate mechanism. The gate mechanism effectively models the dependency types and attachs a weight to each dependency arc to avoid redundant features. Extensive experiments show that our model outperforms strong baselines.

References

1. Chen, Q., Zhuo, Z., Wang, W.: BERT for joint intent classification and slot filling. arXiv preprint arXiv:1902.10909 (2019)
2. Coucke, A., et al.: Snips voice platform: an embedded spoken language understanding system for private-by-design voice interfaces. arXiv preprint arXiv:1805.10190 (2018)
3. Devlin, J., Chang, M.W., Lee, K., Toutanova, K.: Bert: pre-training of deep bidirectional transformers for language understanding. arXiv preprint arXiv:1810.04805 (2018)
4. Niu, P., Chen, Z., Song, M.: A novel bi-directional interrelated model for joint intent detection and slot filling. In: Proceedings of the 57th Conference of the Association for Computational Linguistics, ACL 2019, Florence, Italy, Vol. 1, pp. 5467–5471 (2019)
5. Goo, C., et al.: Slot-gated modeling for joint slot filling and intent prediction. In: Proceedings of the 2018 Conference of the North American Chapter of the Association for Computational Linguistics: Human Language Technologies, NAACL-HLT, New Orleans, Louisiana, USA, Vol. 2, pp. 753–757 (2018)

6. Guo, D., Tür, G., Yih, W., Zweig, G.: Joint semantic utterance classification and slot filling with recursive neural networks. In: 2014 IEEE Spoken Language Technology Workshop, SLT 2014, South Lake Tahoe, NV, USA, pp. 554–559 (2014)
7. Guo, Z., Zhang, Y., Lu, W.: Attention guided graph convolutional networks for relation extraction. In: Proceedings of the 57th Conference of the Association for Computational Linguistics, ACL 2019, Florence, Italy, Vol. 1, pp. 241–251 (2019)
8. Haffner, P., Tür, G., Wright, J.H.: Optimizing svms for complex call classification. In: 2003 IEEE International Conference on Acoustics, Speech, and Signal Processing, ICASSP '03, Hong Kong, pp. 632–635 (2003)
9. Hakkani-Tür, D., et al.: Multi-domain joint semantic frame parsing using bidirectional RNN-LSTM. In: Interspeech 2016, 17th Annual Conference of the International Speech Communication Association, San Francisco, CA, USA, pp. 715–719 (2016)
10. Huang, B., Carley, K.M.: Syntax-aware aspect level sentiment classification with graph attention networks. In: EMNLP-IJCNLP 2019, Hong Kong, China, pp. 5468–5476 (2019)
11. Kipf, T.N., Welling, M.: Semi-supervised classification with graph convolutional networks. In: ICLR 2017, Toulon, France, Conference Track Proceedings (2017)
12. Lai, S., Xu, L., Liu, K., Zhao, J.: Recurrent convolutional neural networks for text classification. In: Proceedings of the Twenty-Ninth AAAI Conference on Artificial Intelligence, Austin, Texas, USA. pp. 2267–2273 (2015)
13. Liu, B., Lane, I.: Attention-based recurrent neural network models for joint intent detection and slot filling. In: Interspeech 2016, 17th Annual Conference of the International Speech Communication Association, San Francisco, CA, USA, pp. 685–689 (2016)
14. Marcheggiani, D., Titov, I.: Encoding sentences with graph convolutional networks for semantic role labeling. In: EMNLP 2017, Copenhagen, Denmark, pp. 1506–1515 (2017)
15. Peng, B., Yao, K., Li, J., Wong, K.: Recurrent neural networks with external memory for spoken language understanding. In: NLPCC 2015, Nanchang, China, pp. 25–35 (2015)
16. Riloff, E., Chiang, D., Hockenmaier, J., Tsujii, J. (eds.): Proceedings of the 2018 Conference on Empirical Methods in Natural Language Processing, Brussels, Belgium. Association for Computational Linguistics, (2018)
17. Sarikaya, R., Hinton, G.E., Ramabhadran, B.: Deep belief nets for natural language call-routing. In: ICASSP 2011, Prague Congress Center, Prague, Czech Republic. pp. 5680–5683 (2011)
18. Schlichtkrull, M.S., Cao, N.D., Titov, I.: Interpreting graph neural networks for NLP with differentiable edge masking. arXiv preprint arXiv:2010.00577 (2020)
19. Sun, K., Zhang, R., Mensah, S., Mao, Y., Liu, X.: Aspect-level sentiment analysis via convolution over dependency tree. In: EMNLP-IJCNLP 2019, Hong Kong, China, pp. 5678–5687 (2019)
20. Tur, G.: Spoken Language Understanding: Systems for Extracting Semantic Information from Speech. John Wiley & Sons (2011)
21. Tür, G., Hakkani-Tür, D., Heck, L.P.: What is left to be understood in atis? In: 2010 IEEE Spoken Language Technology Workshop, Berkeley, California, USA, pp. 19–24 (2010)
22. Vaswani, A., et al.: Attention is all you need. In: Advances in Neural Information Processing Systems 30: Annual Conference on Neural Information Processing Systems 2017, Long Beach, CA, USA. pp. 5998–6008 (2017)

23. Wang, Y., Shen, Y., Jin, H.: A bi-model based RNN semantic frame parsing model for intent detection and slot filling. In: NAACL-HLT, New Orleans, Louisiana, USA, Vol. 2, pp. 309–314 (2018)
24. Young, S.J., Gasic, M., Thomson, B., Williams, J.D.: Pomdp-based statistical spoken dialog systems: a review. Proc. IEEE **101**(5), 1160–1179 (2013)
25. Zhang, C., Li, Y., Du, N., Fan, W., Yu, P.S.: Joint slot filling and intent detection via capsule neural networks. In: Proceedings of the 57th Conference of the Association for Computational Linguistics, ACL 2019, Florence, Italy, Vol. 1, pp. 5259–5267 (2019)
26. Zhang, X., Zhao, J.J., LeCun, Y.: Character-level convolutional networks for text classification. In: Advances in Neural Information Processing Systems 28: Annual Conference on Neural Information Processing Systems 2015, Montreal, Quebec, Canada. pp. 649–657 (2015)
27. Zhang, Y., Qi, P., Manning, C.D.: Graph convolution over pruned dependency trees improves relation extraction. In: Proceedings of the 2018 Conference on Empirical Methods in Natural Language Processing, Brussels, Belgium, pp. 2205–2215 (2018). https://doi.org/10.18653/v1/d18-1244
28. Zhao, L., Feng, Z.: Improving slot filling in spoken language understanding with joint pointer and attention. In: Proceedings of the 56th Annual Meeting of the Association for Computational Linguistics, ACL 2018, Melbourne, Australia, Vol. 2, pp. 426–431 (2018)

NSRL: Named Entity Recognition with Noisy Labels via Selective Review Learning

Xiusheng Huang[1,2]([✉]), Yubo Chen[1,2], Kang Liu[1,2], Yuantao Xie[3], Weijian Sun[3], and Jun Zhao[1,2]

[1] National Laboratory of Pattern Recognition, Institute of Automation, CAS, Beijing, China
huangxiusheng2020@ia.ac.cn, {yubo.chen,kliu,jzhao}@nlpr.ia.ac.cn
[2] School of Artificial Intelligence, University of Chinese Academy of Sciences, Beijing, China
[3] Huawei Technologies Co., Ltd., Shenzhen, China
{xieyuantao2,sunweijian}@huawei.com

Abstract. Named entity recognition (NER) is a task of identifying both types and spans in the sentences. Previous works always assume that the NER datasets are correctly annotated. However, not all samples help with generalization. There are many noisy samples from a variety of sources (e.g., weak, pseudo, or distant annotations). Meanwhile existing methods are prone to cause error propagation in self-training process because of ignoring the overfitting, and becomes particularly challenging. In this paper, we propose a robust Selective Review Learning (NSRL) framework for NER task with noisy labels. Specifically, we design a Status Loss Function (SLF) which helps the model review the previous knowledge continuously when learning new knowledge, and prevents model from overfitting noisy samples in self-training process. In addition, we propose a novel Confidence Estimate Mechanism (CEM), which utilizes the difference between logit values to identify positive samples. Experiments on four distant supervision datasets and two real-world datasets show that the NSRL significantly outperforms previous methods.

Keywords: Overfitting · Self-training · Status loss function · Confidence estimation

1 Introduction

Named entity recognition (NER) is a fundamental task in natural language processing, and we treat the NER task as a sequence-labeling problem in this paper. For example, the NER aims to identify both spans and types of the entity *Manchester Unite* in the Table 1. As neural networks have significantly improved NER performances, the potential improvement of the existing frameworks is mainly limited by the data quality [18]. However, not all samples are

B. Qin et al. (Eds.): CCKS 2021, CCIS 1466, pp. 157–170, 2021.
https://doi.org/10.1007/978-981-16-6471-7_12

completely correct in the NER datasets. For example, in the Table 1, the entity *Manchester Unite* is an organization (ORG), but it is incorrectly annotated as location (LOC). Meanwhile, manually annotated datasets are considered to have a high quality. But in classical CoNLL03 dataset [21], more than 1300 samples are found mislabeled, which accounts for 3.7% of the entire dataset. Moreover, the quality of other datasets are worse [9,16] (e.g., weak supervision datasets [4,17], or distant supervision datasets [28]).

To relieve the negative impact of noisy samples, the most of existing methods correct labels with self-training strategy [11,12]. However, the model output is unreliable due to the influence of noisy samples in self-training process. As the increase of training iteration epochs, the model will overfit noisy samples and then generate incorrect tags [18]. This will hinder the generalization of model and cause error propagation. Thus how to prevent model from overfitting noisy samples in self-training process is a challenging problem.

Table 1. A noisy label example in CoNLL03 dataset. The *Manchester Unite* is a football club in Manchester England, but which is incorrectly annotated as class LOC.

	Keane	Signs	four-years	contract	with	Manchester	Unite
Gold labels	B-PER	O	O	O	O	B-ORG	I-ORG
Noisy labels	B-PER	O	O	O	O	B-LOC	I-LOC

In addition, although existing methods can utilize self-training strategies to correct labels, these labels are not absolutely correct, thus previous works mostly use the loss values to identify positive samples [8]. Meanwhile we observe that the loss values are obtained by the logit values through softmax layer and loss function. However, the softmax layer is a normalized exponential function, which will nonlinearly increase the weight of maximum value in the logit matrix and bring unfairness for identifying positive samples. Thus how to identify positive samples accurately is another challenging problem.

In this paper, we propose a novel framework named Selective Review Learning (NSRL) to solve these issues. To prevent model from overfitting noisy samples, we design a Status Loss Function (SLE). With exploiting the early-learning phenomenon [1,31], the SLF helps model constantly review what it learned in the previous stage when the model learns new knowledge. To identify positive samples accurately, we establish a novel Confidence Estimate Mechanism (CEM), which directly utilizes the difference between logit values for estimating samples instead of loss values. Experiments on four distant supervision datasets and two real-world datasets show that NSRL outperforms previous methods.

In summary, our major contributions are the following:

– We proposed the Selective Review Learning (NSRL) framework which can effectively alleviate the negative effect of noisy samples and improve the performance of the NER model.

- To prevent the model from overfitting noisy samples, we design a Status Loss Function (SLF) which helps model constantly review what it learned in the previous stage. To identify positive samples accurately, we propose a novel Confidence Estimate Mechanism (CEM) which utilizes the difference between logit values.
- We conduct extensive experiments on two real-world datasets and four distant supervision datasets, and our proposed NSRL significantly outperforms the previous methods. We will release our code soon.

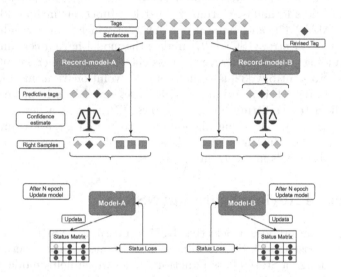

Fig. 1. An overview of our proposed NSRL framework. The NSRL employs dual-model architecture in self-training process. Meanwhile NSRL utilizes the Record-model to generate tags, and the predictive noisy tags are filtered by confidence estimate mechanism. Then predictive correct tags are delivered to another model. The Model-A will generate a Status Matrix for forming status loss. In addition, the Record-model-A is initialized by Model-A with the early stopping strategy.

2   Related Work

There are various approaches have been proposed for noisy tags, and we define them as two categories: 1) Label correction method, 2) Robust loss functions.

2.1   Labels Correction Methods

Label correction method is to improve the quality of original labels. Most of the previous methods used self-training to correct labels [11]. New labels equal to the probabilities estimated by the model (known as soft labels) or to one-hot vectors representing the model predictions (hard labels) [25,29]. In addition, another option is to set the new labels to equal a convex combination of the noisy labels and the soft or hard labels [20]. NLNCE [12] proposes strategies for estimating confidence scores based on local and global independence assumptions, that partially marginalize out labels of low confidence with a CRF model. However, these methods require the support from extra clean data or an inefficient detection process to estimate samples.

2.2   Robust Loss Functions

Robust loss functions improve the robustness of the model for noisy samples by designing new loss formulas. Current robust loss functions include Mean Absolute Error (MAE) [6], and Improved MAE [26] which is a reweighted MAE. Symmetric Cross Entropy(SCE) [27] makes the model have a certain noisy tolerance by adding a symmetric reverse cross entropy after the cross entropy. And Generalized cross entropy [32] is actually a new evolutionary form of MAE. Regularization (LSR) [24] is a technique which uses soft labels in place of one-hot labels to alleviate overfitting for noisy labels. ELR [13] is a new method that makes full use of early learning phenomenon to keep a large learning gradient for correct samples. But these methods can not effectively prevent model from overfitting noisy labels in self-training process.

3   Selective Review Learning(NSRL)

In this paper, we propose a Selective Review Learning (NSRL) framework for obtaining a robust model. Specifically, to prevent model from overfitting noisy samples, we design a Status Loss Function (SLE) which helps model constantly review what it learned in the previous stage when the model learns new knowledge. To identify positive samples accurately, we establish a novel Confidence Estimate Mechanism (CEM), which directly utilizes the difference between logit values for estimating labels instead of previous loss values.

3.1   Status Loss Function (SLF)

Previous loss functions always absolutely believed in tags (e.g., Cross Entropy), even if the tags are mislabeled. Meanwhile the predictive tags are unworthy full trust in self-training process [12]. In addition, deep neural networks will first fit the samples with correct tags during an "early learning" phase [1,31]. With exploiting the early-learning phenomenon, we proposed a new status loss function, which prevents model from overfitting noisy samples by utilizing tags information, previous status and current status. Specifically, the SLF obtains previous status by status matrix in the Fig. 1, and the status matrix is constantly updated by the current model.

KL-Divergence. We define two distributions matrix p and q, and the relationship between entropy, cross entropy and KL-divergence is as follows:

$$KL(q||p) \ = \ H(q,p) - H(q) \tag{1}$$

In self-training process, where $q = q(k|x)$ is the distribution of the predictive tags in sample x, and $p = p(k|x)$ is the predictive distribution of the model for sample x. KL-divergence optimizes the difference between entropy and cross entropy to make the model learn in the specified direction.

Definition SLF. Predictive tags $q = q(k|x)$ are unworthy full trust, and classical loss functions(e.g., Cross Entropy) is not suitable for self-training process. We design the SLF as follows:

$$L_{SLF} = L_{CE}(\Theta) + L_T = L_{CE}(\Theta) + \frac{\lambda}{n} \sum_{i=1}^{n} p \log t \tag{2}$$

where p is the predictive tags and λ is a hyperparameter. The t is a target function which combines tags information $q^{[i]}(j-1)$, previous status $t^{[i]}(j-1)$ and current status $p^{[i]}(j)$. The SLF enables the model to be continuously influenced by tags and previous status in the self-training process. That can effectively prevent model from overfitting original noisy tags and predictive negative tags. The target function t as follows:

$$t^{[i]}(j) = \beta t^{[i]}(j-1) + (1-\beta)(\alpha p^{[i]}(j) + (1-\alpha)q^{[i]}(j-1)) \tag{3}$$

Where $q^{[i]}(j-1)$ is the predictive tag of sample i in the j epoch, which is initialized by the original tag, and then is replaced by the predictive tag $q^{[i]}(j)$. The α and β are two hyperparameters, and $0 < \alpha < 1$, $0 < \beta < 1$.

By iterating the target function $t^{[i]}$, model is constantly affected by the previous status when it fits the new samples. That can effectively prevent the model from overfitting.

SLF Robustness Analysis. In order to simplify the calculation, we set λ as 1 and derive the gradient of L_{SLF}. The gradient of the L_{SLF} with respect to the logits Z_j can be derived as:

$$\frac{\partial L_{SLF}}{\partial Z_j} = \frac{\partial L_{CE}(\Theta)}{\partial Z_j} + \frac{\lambda}{n} \sum_{i=1}^{n} \left(\frac{\partial p_k}{\partial Z_j} \log t + \frac{p \cdot \frac{\partial t}{\partial Z_j}}{t} \right) \tag{4}$$

where $\frac{\partial p_k}{\partial Z_j}$ can be further derived based on whether $k = j$:

$$\begin{cases} \frac{\partial p_k}{\partial Z_j} = p_k(1-p_k) & k = j \\ \frac{\partial p_k}{\partial Z_j} = -p_k p_j & k \neq j \end{cases} \tag{5}$$

Particularly, we can obtain:

$$\frac{\partial p_k}{\partial Z_j} = \frac{\partial \left(\frac{Z_k}{\sum_{j=1}^{K} e^{Z_j}} \right)}{\partial Z_j} = \frac{\frac{\partial e^{Z_k}}{\partial Z_j} (\sum_{j=1}^{K}) - e^{Z_k} \frac{\partial (\sum_{j=1}^{K} e^{Z_j})}{\partial Z_j}}{(\sum_{j=1}^{K} e^{Z_j})^2} \tag{6}$$

When $k=j$:

$$\frac{\partial p_k}{\partial Z_j} = \frac{\partial p_k}{\partial Z_k} = \frac{e^{Z_k}}{\sum_{k=1}^{k} e^{Z_k}} - (\frac{e^{Z_k}}{\sum_{k=1}^{K} e^{Z_k}})^2 \tag{7}$$
$$= p_k - p_k{}^2 = p_k(1-p_k)$$

When $k \neq j$:

$$\frac{\partial p_k}{\partial Z_j} = \frac{0 \cdot (\sum_{k=1}^{K} e^{Z_j}) - e^{Z_k} e^{Z_j}}{(\sum_{k=1}^{K} e^{Z_j})(\sum_{k=1}^{K} e^{Z_j})} \tag{8}$$
$$= -\frac{e^{Z_k}}{\sum_{k=1}^{K} e^{Z_j}}\frac{e^{Z_j}}{\sum_{k=1}^{K} e^{Z_j}} = -p_k p_j$$

Simplifying the intermediate steps, and we set $\alpha = 0.5$ and $b_k = p_k \log t_k + \frac{p_k^2}{2t_k}$.

$$L' = \frac{\partial L_{SLF}}{\partial Z_j} = \frac{\partial L_{CE}(\Theta)}{\partial Z_j} + p_j \sum_{k=1}^{K} b_k - b_j \tag{9}$$

We set $L'_T = p_j \sum_{k=1}^{K} b_k - b_j$, then we can obtain:

$$L''_T = \sum_{k=1}^{K} b_k + p_j (\sum_{k=1}^{K} b_k)' - \frac{\partial b_j}{\partial p_j} = \sum_{k=1}^{K} b_k + (p_j - 1)\frac{\partial b_j}{\partial p_j} \tag{10}$$

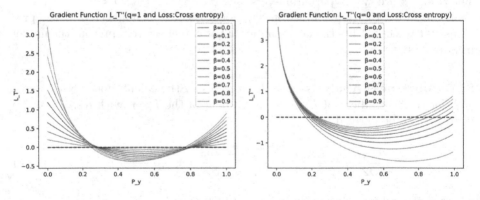

Fig. 2. A demonstration of the second derivative L''_T. The left: $q = 1$ means that the P_y values corresponding to tags. The right: $q = 0$ means that the P_y values corresponding to others. The x-axis refers to the model output p_y, and the y-axis refers to the L''_T function.

In the Fig. 2, we can observe that the L''_T will provide a large acceleration when p_y approaches to 0. On the contrary, the L''_T will provide a small acceleration when p_y approaches to 1. This will help model reduce the dependence for p_y, meanwhile improve the influence of tags information and previous status on the model. In addition, when p_y approaches to 0.5, it means that the model output is not self-confident, the L_T'' will provide a minimum value. This will help model reduce learning in imprecise directions. Meanwhile, the gradient of the L_{SLF} with respect to the logits Z_j equals:

$$\frac{\partial L_{SLF}}{\partial Z_j} = \nabla N_x^{[i]}(\Theta)((p^{[i]} - y^{[i]}) + \lambda(t^{[i]} - p^{[i]})) \tag{11}$$

Where $\lambda(t^{[i]} - p^{[i]})$ is an adaptive acceleration term. That makes the model output $p^{[i]}$ closes to the target function $t^{[i]}$. Due to the $t^{[i]}$ combines tags information, previous status and current status, the model will learn in these three directions. Meanwhile, the $(p^{[i]} - y^{[i]})$ term is helpful to improve the convergence ability of model.

3.2 Confidence Estimate Mechanism (CEM)

Preliminary. In self-training process, previous works always default the predictive samples as samples for next training. However, if the predictive samples are incorrect and are used as input to the model, this will cause error propagation. Establishing a confidence estimate mechanism for the predictive tags is necessary, which will help model identify high confidence samples for learning.

Utilizing Logit Value. Previous methods mostly use loss values to distinguish positive samples and negative samples [8,30]. But the predictive tags is not entirely trustworthy because of the model for predicting tags is not exactly correct. Meanwhile, neural networks will output a logit matrix in training process, which goes through the softmax layer and then gets into loss function. The softmax layer is a normalized exponential function, which will nonlinearly increase the weight of maximum value in the logit matrix and bring unfairness for identifying positive samples. Our proposed CEM directly utilizes logit matrix to identify positive samples instead of loss values. And the CEM will help model utilize the confidence scores of samples to identify positive samples.

Observing Logit Value. Given a sentence $x = [x_1, x_2, ..., x_n]$ and its tag sequence $y = [y_1, y_2, ...y_n]$, n is the sentence length. Every token x_i will obtain a corresponding logit matrix $Z = [z_1^i, z_2^i, ..., z_m^i]$, m denotes total number of tags. The Z_y is the logit value corresponding to predictive tag y. In the Fig. 3, we can observe that the logit value Z_y of positive samples is evidently higher than other values in the logit matrix. Meanwhile, the Z_y of negative samples is evidently smaller than other values in the early stage. Because of the predictive tags are replaced ceaselessly, Z_y gradually becomes the maximum in logit matrix.

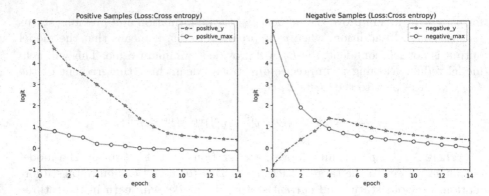

Fig. 3. The comparison of the positive and negative samples in logit value. The blue line: the logit value Z_y corresponding to the predictive tag y in logit matrix; the orange line: the maximal logit value $Z_{max}^{(t)}$ in logit matrix except Z_y. The x-axis refers to the number of training epochs, and the y-axis refers to the logit value.

Defining CEM. We propose a new confidence estimate mechanism, which utilizes the difference between the Z_y and the maximum Z_{max} in logit matrix. The CEM score be defined as:

$$CEM(x, y) \ = \ min(Z_y - \max_{i \neq y} Z_i) \tag{12}$$

Meanwhile, every token will obtain a CEM score. Because of a sentence is the minimum unit of the input in the NER tasks, if the CEM score corresponding to single token is small enough in the sentence, we can consider that the sentence is negative. By utilizing the difference between Z_y and Z_{max}, the CEM can effectively reflect the definite degree of model for predictive tags.

4   Experiments

4.1   Dataset and Baseline

Distant Supervision Datasets. We evaluate our method on four datasets including CoNLL03 [22], Webpage [19], Tweet [7], Wikigold [2]. In this setting, the distantly supervised tags are generated by the dictionary following BOND [11].

Real-World Datasets. We consider two real-world NER datasets. WUT-17 [4] is a user generated English dataset. Weibo NER [17] is a Chinese dataset in social domain. Since the WUT-17 has no development set, we randomly select 10% samples from the training set as the development sct. Compared with other real-world datasets, the noise ratio of these two datasets is larger.

Table 2. The result in four distant supervision datasets.

Method	Distant supervision			
	CoNLL03	Webpage	Tweet	Wikigold
AutoNER [23]	67.0	51.4	26.1	47.5
LRNT [3]	69.7	47.7	23.8	46.2
NLNCE [12]	79.9	61.9	47.3	57.7
BOND [11]	81.5	65.7	48.0	60.1
NSRL(our)	**82.2**	**66.5**	**49.2**	**60.8**

Baseline. For distant supervision datasets, we compare NSRL with four recently proposed robust methods: (1) AutoNER [23]; (2) LRNT [3]; (3) NLNCE [12]; (4) BOND [11]; For real-world datasets, we compare NSRL with the previous methods as well as the standard Cross Entropy(CE): (1) SCE [27]: symmetric cross entropy loss; (2) DSC [10]: dice loss function; (3) ELR [13]: early regularization.

Evaluation. Our primary evaluation metric is F1 score on the test set to compare the results of different methods.

Pre-trained Language Model. The **BERT** [5] employs a Transformer encoder to learn a BiLM from large unlabeled text corpora and sub-word units to represent textual tokens. We use the $BERT_{base}$ model in our experiments of real-world datasets. The **RoBERTa-base** [14] is a replication study of BERT pre-training that measures the impact of many key hyperparameters and training data size. We use the $RoBERTa_{base}$ model in our experiments of distant supervision datasets.

4.2 Implementation Details

In our experiments, we set the initial learning rate to $lr = 1e{-}5$ for all datasets. The loss is optimized by AdamW [15]. The parameters of our model is initialized by the Xavier initializer. Since the scale of each dataset varies, we set different training batch size for different datasets. Specifically, we set the batch sizes of four distant supervision datasets, WUT-17 and Weibo NER as 16, 40, 40. We stop the training when we find the best result in the development set.

4.3 Overall Performance

The Experiments on Distant Supervision Datasets. Table 2 shows the results of the baseline methods and our proposed NSRL. Compared with other methods, the NSRL shows obvious advantage in the four distant supervision datasets. Meanwhile NSRL has achieved new state-of-the-art (SOTA) with the

F1 score reached to **82.2**, **66.5**, **49.2**, and **60.8%** on CoNLL03, Webpage, Tweet and Wikigold datasets separately. Specifically, Our method outperforms previous SOTA model BOND by 1.2% in F1 score on Tweet dataset. This shows that our NSRL method can effectively prevent the model from overfitting noisy samples on distant supervision datasets.

The Experiments on Real-World Datasets. Table 3 presents the experiment results in two real-world datasets. Our NSRL outperforms other methods by 0.57%, and 0.35% in F1 score on WUT-17 and Weibo NER datasets separately. Compared with other methods, NSRL can accurately filter noise samples in real-wold datasets. This shows that our method is still effective on real-wold datasets with large noise ratio.

Table 3. The result in two real-world datasets.

Dataset	CE	SCE [27]	DSC [10]	ELR [13]	NSRL(our)
WUT-17	54.77	53.55	54.48	53.54	**55.34**
Weibo NER	59.71	60.97	60.78	57.41	**61.32**

4.4   Ablation Study

The Ablation Experiment of NSRL. The NSRL framework contains SLF loss function and CEM mechanism. Table 4 presents the results of ablation experiment. We can easily observe that without the SLF or CEM, the F1 scores appear significant decrease. Specifically, the F1 scores reach the minimum without SLF and CEM. Meanwhile the F1 scores dropped by 1.4, 1.1, 1.6, and 1.5% on CoNLL03, Webpage, Tweet and Wikigold datasets separately. This demonstrates that SLF and CEM are essential to improve the performance of the model.

Table 4. The result of ablation experiment on four distant supervision datasets. The "w/o SLF" means removing the SLF loss function from the complete model.

Method	Distant supervision			
	CoNLL03	Webpage	Tweet	Wikigold
w/o SLF	81.1	66.1	48.2	60.1
w/o CEM	81.6	65.9	48.4	60.3
w/o SLF & CEM	80.8	65.4	47.6	59.3
w/ SLF & CEM	**82.2**	**66.5**	**49.2**	**60.8**

Table 5. The ablation experiment for Tags information(T), Previous status(P) and Current status(C) in SLF loss function.

Method	Distant supervision			
	CoNLL03	Webpage	Tweet	Wikigold
w/o T	80.8	64.9	47.6	58.8
w/o P	81.4	65.4	48.2	59.3
w/o C	78.2	63.4	46.9	57.7
w/o T & P	78.9	64.0	47.8	59.7
w/o T & C	77.9	63.3	46.7	57.4
w/o P & C	74.8	61.4	44.6	55.9
NSRL(our)	**82.2**	**66.5**	**49.2**	**60.8**

The Ablation Experiment of SLF. The SLF loss function combines tags information, previous status and current status. Table 5 shows the ablation experiment of the three factors. We can clearly observe that removing any factor causes obvious performance degradation, but the importance of each factor is different. Specifically, the F1 score will decrease significantly without current status, which shows that the current status has a great influence on the model. In conclusion, from ablation experiments, we can find that each factor can be implemented independent of the other, but together they can achieve the best result, showing that all these three factors are essential to our model.

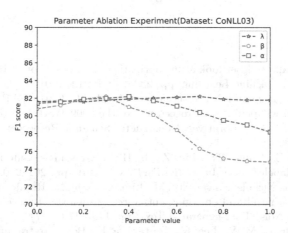

Fig. 4. Parameter analysis on λ, β, α in SLF loss function: (a) Tuning λ (fix β=0.3 and α=0.4); (b) Tuning β (fix λ=0.7 and α=0.4); (c) Tuning α (fix λ=0.7 and β=0.3). The x-axis refers to the parameter value, and the y-axis refers to the F1 score.

The Parameter Analysis of SLF. Figure 4 shows the parameter analysis for SLF on CoNLL03 dataset. We test λ, β, α in $[0.0,1.0]$. When $\lambda=0.7,\beta=0.3$ and $\alpha=0.4$, the model achieves the best performance. Specifically, when $\beta=1.0$, the F1 score reaches the minimum 74.8%. In addition, when β approaches to 1, we observe that the influence of previous status and current status on the model is weakened from the Eq. 3. This weakens the generalization of the model and makes model overfit noisy samples.

5   Conclusions

In this paper, we propose a robust Selective Review Learning (NSRL) framework for NER task with noisy labels. The NSRL contains SLF loss function and CEM mechanism. Specifically, the SLF can help model reduce the dependence for model output p_y. Meanwhile the SLF improves the robustness of model and prevents model from overfitting noisy samples. In addition, the CEM mechanism can help the model identify positive samples accurately for next training. Experiments on four distant supervision datasets and two real-world datasets show that NSRL outperforms the previous methods, and which obtained the state-of-the-art (SOTA) results on the four distant supervision datasets.

Acknowledge. This work is supported by the National Natural Science Foundation of China (No.61976211, No.61806201). This work is supported by Beijing Academy of Artificial Intelligence (BAAI2019QN0301) and the Key Research Program of the Chinese Academy of Sciences (Grant NO. ZDBS-SSW-JSC006). This work is also supported by a grant from Huawei Technologies Co., Ltd.

References

1. Arpit, D., et al.: A closer look at memorization in deep networks. In: International Conference on Machine Learning. pp. 233–242. PMLR (2017)
2. Balasuriya, D., Ringland, N., Nothman, J., Murphy, T., Curran, J.R.: Named entity recognition in wikipedia. In: Proceedings of the 2009 Workshop on The People's Web Meets NLP: Collaboratively Constructed Semantic Resources (People's Web). pp. 10–18 (2009)
3. Cao, Y., Hu, Z., Chua, T.S., Liu, Z., Ji, H.: Low-resource name tagging learned with weakly labeled data. In: EMNLP-IJCNLP 2019. pp. 261–270. ACL (2020)
4. Derczynski, L., Nichols, E., van Erp, M., Limsopatham, N.: Results of the wnut2017 shared task on novel and emerging entity recognition. In: Proceedings of the 3rd Workshop on Noisy User-generated Text, pp. 140–147 (2017)
5. Devlin, J., Chang, M.W., Lee, K., Toutanova, K.: Bert: pre-training of deep bidirectional transformers for language understanding. In: Proceedings of the 2019 Conference of the North American Chapter of the Association for Computational Linguistics: Human Language Technologies, Vol. 1, pp. 4171–4186 (2019)
6. Ghosh, A., Kumar, H., Sastry, P.: Robust loss functions under label noise for deep neural networks. In: Proceedings of the Thirty-First AAAI Conference on Artificial Intelligence, pp. 1919–1925 (2017)

7. Godin, F., Vandersmissen, B., De Neve, W., Van de Walle, R.: Multimedia lab@ acl wnut ner shared task: Named entity recognition for twitter microposts using distributed word representations. In: Proceedings of the Workshop on Noisy User-Generated Text, pp. 146–153 (2015)

8. Han, B., et al.: Co-teaching: robust training of deep neural networks with extremely noisy labels. In: Proceedings of the 32nd International Conference on Neural Information Processing Systems, pp. 8536–8546 (2018)

9. Lange, L., Hedderich, M.A., Klakow, D.: Feature-dependent confusion matrices for low-resource ner labeling with noisy labels. In: EMNLP-IJCNLP-2019, pp. 3545–3550 (2019)

10. Li, X., Sun, X., Meng, Y., Liang, J., Wu, F., Li, J.: Dice loss for data-imbalanced nlp tasks. In: Proceedings of the 58th Annual Meeting of the Association for Computational Linguistics, pp. 465–476 (2020)

11. Liang, C., et al.: Bond: Bert-assisted open-domain named entity recognition with distant supervision. In: Proceedings of the 26th ACM SIGKDD International Conference on Knowledge Discovery & Data Mining, pp. 1054–1064 (2020)

12. Liu, K., et al.: Noisy-labeled ner with confidence estimation. arXiv preprint arXiv:2104.04318 (2021)

13. Liu, S., Niles-Weed, J., Razavian, N., Fernandez-Granda, C.: Early-learning regularization prevents memorization of noisy labels. Adv. Neural Inf. Process. Syst. **33** (2020)

14. Liu, Y., et al.: Roberta: a robustly optimized bert pretraining approach. arXiv preprint arXiv:1907.11692 (2019)

15. Loshchilov, I., Hutter, F.: Fixing weight decay regularization in adam (2018)

16. Nooralahzadeh, F., Lønning, J.T., Øvrelid, L.: Reinforcement-based denoising of distantly supervised ner with partial annotation. In: Proceedings of the 2nd Workshop on Deep Learning Approaches for Low-Resource NLP (DeepLo 2019), pp. 225–233 (2019)

17. Peng, N., Dredze, M.: Named entity recognition for Chinese social media with jointly trained embeddings. In: Proceedings of the 2015 Conference on Empirical Methods in Natural Language Processing, pp. 548–554 (2015)

18. Pleiss, G., Zhang, T., Elenberg, E.R., Weinberger, K.Q.: Identifying mislabeled data using the area under the margin ranking. arXiv preprint arXiv:2001.10528 (2020)

19. Ratinov, L., Roth, D.: Design challenges and misconceptions in named entity recognition. In: Proceedings of the Thirteenth Conference on Computational Natural Language Learning (CoNLL-2009), pp. 147–155 (2009)

20. Reed, S.E., Lee, H., Anguelov, D., Szegedy, C., Erhan, D., Rabinovich, A.: Training deep neural networks on noisy labels with bootstrapping. In: ICLR (Workshop) (2015)

21. Reiss, F., Xu, H., Cutler, B., Muthuraman, K., Eichenberger, Z.: Identifying incorrect labels in the CoNLL-2003 corpus. In: Proceedings of the 24th Conference on Computational Natural Language Learning, pp. 215–226 (2020)

22. Sang, E.F., De Meulder, F.: Introduction to the CoNLL-2003 shared task: Language-independent named entity recognition. arXiv preprint cs/0306050 (2003)

23. Shang, J., Liu, L., Ren, X., Gu, X., Ren, T., Han, J.: Learning named entity tagger using domain specific dictionary. arXiv preprint arXiv:1800.03600 (2018)

24. Szegedy, C., Vanhoucke, V., Ioffe, S., Shlens, J., Wojna, Z.: Rethinking the inception architecture for computer vision. In: Proceedings of the IEEE Conference on Computer Vision and Pattern Recognition, pp. 2818–2826 (2016)

25. Tanaka, D., Ikami, D., Yamasaki, T., Aizawa, K.: Joint optimization framework for learning with noisy labels. In: Proceedings of the IEEE Conference on Computer Vision and Pattern Recognition, pp. 5552–5560 (2018)
26. Wang, X., Hua, Y., Kodirov, E., Robertson, N.M.: Imae for noise-robust learning: mean absolute error does not treat examples equally and gradient magnitude's variance matters. arXiv preprint arXiv:1903.12141 (2019)
27. Wang, Z., Shang, J., Liu, L., Lu, L., Liu, J., Han, J.: Crossweigh: training named entity tagger from imperfect annotations. In: EMNLP-IJCNLP, pp. 5157–5166 (2019)
28. Yang, Y., Chen, W., Li, Z., He, Z., Zhang, M.: Distantly supervised NER with partial annotation learning and reinforcement learning. In: Proceedings of the 27th International Conference on Computational Linguistics, pp. 2159–2169 (2018)
29. Yi, K., Wu, J.: Probabilistic end-to-end noise correction for learning with noisy labels. In: Proceedings of the IEEE/CVF Conference on Computer Vision and Pattern Recognition, pp. 7017–7025 (2019)
30. Yu, X., Han, B., Yao, J., Niu, G., Tsang, I., Sugiyama, M.: How does disagreement help generalization against label corruption? In: International Conference on Machine Learning, pp. 7164–7173. PMLR (2019)
31. Zhang, C., Bengio, S., Hardt, M., Recht, B., Vinyals, O.: Understanding deep learning requires rethinking generalization. arXiv preprint arXiv:1611.03530 (2016)
32. Zhang, Z., Sabuncu, M.R.: Generalized cross entropy loss for training deep neural networks with noisy labels. In: Proceedings of the 32nd International Conference on Neural Information Processing Systems, pp. 8792–8802 (2018)

Knowledge Enhanced Target-Aware Stance Detection on Tweets

Xin Zhang[1], Jianhua Yuan[1], Yanyan Zhao[1], and Bing Qin[1,2(✉)]

[1] Research Center for Social Computing and Information Retrieval,
Faculty of Computing, Harbin Institute of Technology, Harbin, China
{xzhang,jhyuan,yyzhao,qinb}@ir.hit.edu.cn
[2] Peng Cheng Laboratory, Shenzhen, China

Abstract. Stance detection aims to determine the stance of a text towards a given target. Different from aspect-level sentiment classification, the target may not appear in the text. While existing models have achieved great success in this task using deep neural networks, their performances still drop sharply on cases where targets are not directly mentioned in texts, even with 'target-aware' structures. We argue that the nonalignment between targets and potentially opinioned terms in texts causes such failure and this could be remedied with external knowledge as a bridge. To this end, we propose RelNet, which leverages multiple external knowledge bases as bridges to explicitly link potentially opinioned terms in texts to targets of interest. Experiments on the well-adopted SemEval 2016 task 6 dataset demonstrate the effectiveness of the proposed model, especially on the subset where targets do not appear in texts.

Keywords: Stance detection · External knowledge · Target-awareness

1 Introduction

Stance detection is the task of classifying attitude of a text towards a given target entity or claim [12]. Its techniques can be utilized for various downstream applications, such as online debates analysis [14,17,20], opinion groups identification [1,15], election results prediction [10], and fake news detection [7]. Thus, it has attracted extensive attention from the Natural Language Processing (NLP) community.

Unlike conventional aspect-based sentiment classification tasks, the main challenge of stance classification is that a target may not be explicitly mentioned in the text. In other words, many texts express their stance laterally by expressing their views on other entities. As shown in Table 1, more than 73% of tweets don't have direct mentions of corresponding targets in the dataset of SemEval 2016 task 6. Furthermore, the relation between the targets and opinioned terms in texts may not be easily inferred from the training corpus. As shown in Fig. 1, the tweet refers to Matthew from the Bible while the target is Atheism. If the model cannot capture the relation between Atheism and Matthew

© Springer Nature Singapore Pte Ltd. 2021
B. Qin et al. (Eds.): CCKS 2021, CCIS 1466, pp. 171–184, 2021.
https://doi.org/10.1007/978-981-16-6471-7_13

Table 1. The statistics of whether a target is clearly mentioned in a sentence in SemEval 2016 task 6 subtask A. "In" means that the keyword of a target appears in the tweet. For example, for target "Climate change is a real concern", when "climate" or "change" appears in a sentence, we treat the sentence as "In".

	All	In		Out	
	Num.	Num.	Proportion	Num.	Proportion
Training set	1249	335	26.83%	914	73.17%
Test set	2914	760	26.09%	2154	73.91%

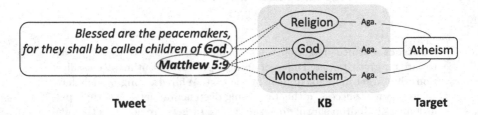

Fig. 1. An example of stance classification. Note that the target *Atheism* is not directly mentioned in the tweet.

5:9 (Christianity), it is most possible that a model will falsely classify the stance as None.

Various models have been proposed for stance detection. Some of them use feature engineering to manually extract features [13], and some use classical neural network-based models such as Recurrent Neural Networks (RNNs) [2] and Convolutional Neural Networks (CNNs) [22]. [2,6] learn target-aware tweet representations with target feature initialization and target embedding attention respectively. However, all these methods learn the relation between opinioned terms and targets implicitly. And text features alone usually fail to provide adequate context to capture such relations.

Thus, many turn to external knowledge for enriching the context. CKEMN [5] exploits ConceptNet to expand the concept words for entities in the text and further models the relation between targets and those concepts with attention. AEKFW [8] obtain relevant concept words for targets and learns relation representations between those concepts and each word in the text. However, these methods do not directly model the relation between target and opinioned terms, which is sub-optimal for effectively fusing knowledge.

To remedy this, we propose RelNet to directly model relations between targets and opinioned terms by utilizing multiple external knowledge bases. RelNet begins with expanding targets with knowledge from multiple sources. Then it chooses opinioned terms in the text by measuring the similarity to expanded entities from knowledge. Afterwards, the relations between opinioned terms and targets can be induced according to the similarity between expanded knowledge and opinioned terms. RelNet employs an auxiliary relation classification task to enforce connections between targets and tweets. The learned relation representations between targets and opinioned terms are further combined with textual features for final stance classification in a multi-task learning manner.

We conduct experiments on the SemEval 2016 task 6 dataset [12]. RelNet outperforms BERT by 1.55% in terms of macro-f1. Case studies reveal that RelNet is better at handling cases in which targets are not explicitly mentioned. In summary, the contributions of this work are two-fold:

- We propose a novel method named RelNet, which explicitly connects targets and potential opinioned terms in the text.
- We provide a simple way to fuse knowledge from multiple external knowledge bases and outperform the BERT model by a large margin, especially on cases where targets are not explicitly mentioned in texts.

2 Related Work

2.1 Stance Detection

Previous works mostly focus on identifying stance in debates [9,18], QA pairs [24]. Recently, there is a growing interest in performing stance detection on microblogs. [13] use SVM based model, [22] use a CNN based model, and [2] use a bidirectional LSTM based model. However, all these methods ignore the target information in the stance detection task. [6] proposed a target-specific attention network, which made full use of the target information in the stance detection. [4] proposed a two-phase LSTM-based model with attention and [21] proposed an end-to-end neural memory model realizing the interaction between target and tweet. [16] proposed a neural ensemble method that combines the attention-based densely connected Bi-LSTM and nested LSTMs models with the multi-kernel convolution in a unified architecture. Adequate context information cannot be extracted only from the text, so later work introduce commonsense knowledge to enhance context information. [5] models the relation between targets and concepts in ConceptNet with attention. [8] propose a model that attends to related concepts and events when encoding the given text to incorporate Wikipedia into stance detection.

2.2 Incorporating External Knowledge

Commonsense Knowledge bases have been widely used in various NLP tasks, such as open-domain conversation generation, visual question answering, sentiment classification, and stance detection. For conversation generation, there are several end-to-end conversation models [23] leveraging CKB to improve the relevance and diversity of generated responses in open-domain conversations. For visual question answering, Su et al. [19] proposed the visual knowledge memory network to leverage self-built CKB for supporting visual question answering. For sentiment analysis, Ma et al. [11] integrated external CKB into RNN cell to improve the performance on aspect-level sentiment classification. In recent years,

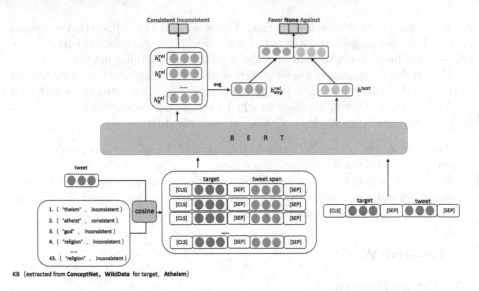

Fig. 2. An overview of the proposed model.

adding external knowledge to stance detection has attracted increasing attention. Ka et al. [8] built a new stance detection dataset consisting of 6,701 tweets on seven topics with associated Wikipedia articles and confirmed the necessity of external knowledge for this task through this dataset. Du et al. [5] incorporated commonsense knowledge into memory network for stance detection.

3   Model

The task of stance detection can be formalized as follows: given a tweet text composed of a number of words $s = \{w_1, w_2, \ldots, w_N\}$ and a target t, the model computes the probability distribution $y \in R^3$ over the three stance labels, Favor, Against, and None.

The overall architecture of our model is shown in Fig. 2. It consists of four main components: 1) Target Related Knowledge Filtration Module, acquiring knowledge triples from multiple knowledge bases, filtering knowledge, and integrating knowledge relations into two categories: consistent and inconsistent; 2) Textual Encoding Module, getting target-aware textual representation h^{text} of target and tweet; 3) Opinioned Terms Encoder Module, selecting K opinioned terms in tweets using target related knowledge, and getting the opinioned terms feature referring to target $h_i^{rel}, i \in \{1, 2, \ldots, K\}$; 4) Classification Module, for introducing the relation between target and opinioned terms, we use a multi-task framework to combine the relation classification task and the stance detection task. We describe the details of these four components in the following subsections.

Table 2. The search terms of targets in SemEval2016 Task 6 SubtaskA. "*" means that the search term is a summary of the target. "-" means that the target cannot be found in the knowledge base.

Target	ConceptNet	WikiData
Atheism	Atheism	Atheism
Climate change is a real concern	Climate change*	Climate change*
Feminist movement	-	Feminist movement
Hillary clinton	-	Hillary clinton
Legalization of abortion	Abortion*	Abortion*

3.1 Textual Encoding Module

The textual encoding module is shown in the right part in Fig. 2. To obtain the target-aware textual representation h^{text}, we concatenate the tweet and target into a single input sequence by using special tokens ([CLS]) and ([SEP]): [CLS]$target$[SEP]$tweet$[SEP]. The concatenated sequence is tokenized by the WordPiece tokenizer. The representation of the [CLS] token h^{text} is treated as the target-aware textual representation.

3.2 Target Related Knowledge Filtration Module

The goal of the target related knowledge filtration module is to get the target related knowledge triples from the external knowledge base, and organize the relation of knowledge into a suitable format (consistent, inconsistent) for stance detection. This module is composed of three parts: knowledge acquisition, knowledge filtering, and relation integration.

Knowledge Acquisition. Specifically, we take two structured knowledge bases, ConceptNet and WikiData. ConceptNet mainly contains commonsense knowledge, WikiData mainly contains social knowledge. They complement each other and supply rich target related knowledge for target understanding. Each of the targets in the dataset is treated as a key for searching for the most related commonsense knowledge from them. Table 2 shows the search terms and results for targets.

In this way, we build a knowledge base KB D of triples $(t_{kb}, r, c) \in D$ for this dataset , each consisting of a target t_{kb}, a relation r, and a knowledge concept in the WikiData or ConceptNet c, e.g. *(atheism, different from, religion)*.

Knowledge Filtering. To remove less relevant ones in built knowledge base D, we use training data to filter knowledge base as follows: for each pair of the target t and tweet s in training data, we take knowledge-word in D which t_{kb} is the same as target t and take tweet-spans from the tweet. The similarity between tweet-spans and knowledge-word is calculated by Cosine Similarity, and

Table 3. Target relational knowledge base construction results.

Target	Consistent	Inconsistent	Total
Atheism	36	7	43
Climate change	26	0	26
Feminist movement	6	2	8
Hillary clinton	77	0	77
Abortion	68	6	74

the top five pairs ($knowledge\ word, tweet\ spans$) with the highest similarity are selected. Delete triples whose knowledge words do not appear in all selected pairs ($knowledge\ word, tweet\ spans$). Through the above process, we filter out irrelevant knowledge in the knowledge base D.

Relation Integration. Although WikiData and ConceptNet both are structured knowledge bases, the types and numbers of relations are quite different. For WikiData, relations and query terms are highly correlated, the relations for different query terms are different, and there are many types of relations. For ConceptNet, the relation is relatively fixed, but the number of concept words expanded by a relation is large. Meanwhile, previous work [8] has shown that extracting consistent and inconsistent relations from commonsense knowledge is a promising approach to enhancing stance detection performance, therefore we unified the relations in ConceptNet and WikiData, and integrated the originally complicated relation into a consistent (1) and inconsistent (-1), e.g.from ($atheism, different from, religion$) to ($atheism, -1, religion$) .

Built knowledge base D is obtained by searching targets, so the relation between knowledge concept c and the target is consistent. Based on the above premise, we use some rules to label inconsistent relations. 1) Integrating the relations with opposite meanings into inconsistent, such as "antonym" and "different from". 2) For a triples, if its relation is "relatedto" or "derivedfrom", and its knowledge concept starts with antonym prefixes such as "anti", replace the relation with inconsistent. The final results are shown in Table 3.

3.3   Opinioned Terms Encoder Module

The opinioned terms encoder module, shown in the left part in Fig. 2, aims to get the representations of vital opinioned terms in the tweet based on the target related knowledge in KB. The content in the tweet is obtained by sliding a window of size w across the sentence, producing a sequence of word fragments of length w , $w = \{1, 2, 3, 4, 5\}$, as the tweet-spans of the tweet. For a pair of the target t and tweet s , we take knowledge-word c in $(t_{kb}, r, c) \in D$ which t_{kb} is the same as target t and take tweet-spans from the tweet. The similarity between tweet-spans and knowledge-word is calculated by cosine similarity, and the top K ($knowledge\ word, tweet\ spans$) pairs with the highest similarity scores are selected. Every selected tweet-span and its corresponding target are fused into a single input sequence by using a special classification token ([CLS]) and a separator token ([SEP]): [CLS]target[SEP]tweet span[SEP]. The input sequences are tokenized using the WordPiece tokenizer. The final hidden state representation corresponding to the [CLS] token is used as $h_i^{rel} \in R^H, i \in \{1, 2, \ldots, K\}$.

3.4 Classification Module

Relation Classifier. Opinioned terms and knowledge concepts are semantically siermsmilar, so the relation between knowledge concepts and targets is equivalent to the relation between opinioned terms and targets. Based on this, every opinioned term representation is fed into the relation classifier for relation prediction using the relation r in the knowledge base D as the gold label.

$$y_{rel} = W_r h_i^{rel} + b_r \tag{1}$$

Stance Classifier. The opinioned terms representation h_{avg}^{rel} of a tweet is obtained by averaging $\{h_1^{rel}, h_2^{rel} \ldots h_K^{rel}\}$. To jointly leverage the tweet text, target, and relation information between opinioned terms in the tweet and the target, the opinioned terms presentations and the target-aware textual representation are concatenated and fed into the stance classifier.

$$y = W_p(h_{avg}^{rel} \oplus h^{text}) + b_p \tag{2}$$

Where $h_{avg}^{rel} \oplus h^{text} \in R^{2d_h}$ is the concatenated representation, d_h is the hidden dimension of Bert. W_p is the weight of stance classifier, b_p is a bias term, y is the predicted probability of stance.

3.5 Joint Learning

Parameters of the RelNet model are learned by optimizing the joint loss function:

$$loss = loss_{ce} + \lambda loss_{rel} \tag{3}$$

Where λ is the weight of $loss_{rel}$ in the total $loss$. $loss_{ce}$ and $loss_{rel}$ are cross-entropy loss for stance detection and relation classification respectively. With this joint loss function, we enforce the relation between the target and opinioned terms in a tweet.

4 Experimental Setup

For our experiments, we consider the uncased base version of BERT with 12 layers, 768 hidden sizes, and 12 attention heads. We fine-tune BERT$_{base}$ models using the Adam optimizer with learning rates $\{1, 3, 5\} \times 10^{-5}$ and training batch sizes $\{24, 28, 32\}$. We choose the best parameters based on the performance of the

Table 4. Statistics of SemEval2016 Task 6 SubtaskA.

Target	#total	% of instances in Train				% of instances in Test			
		#Train	Favor	Against	None	#Test	Favor	Against	None
Atheism	733	513	17.9	59.3	22.8	220	14.5	72.7	12.7
Climate change	564	395	53.7	3.8	42.5	169	72.8	6.5	20.7
Feminist movement	949	664	31.6	49.4	19.0	285	20.4	64.2	15.4
Hillary clinton	984	689	17.1	57.0	25.8	295	15.3	58.3	26.4
Abortion	933	653	18.5	54.4	27.1	280	16.4	67.5	16.1
All	4163	2914	25.8	47.9	26.3	1249	24.3	57.3	18.4

evaluation set. For the commonsense knowledge module, we set $K = 5$, $\lambda = 0.01$, and use stochastic dropout rate of 0.1. For measuring the performance, we use the macro-average of the F1-score for *Favor* and the F1-score for *Against* as the bottom-line evaluation metric, which is the official evaluation measure for SemEval-2016. Note that the official metric does not disregard the *None* class. By taking the average F-score for only the *Favor* and *Against* classes, the final metric treats *None* as a class that is not of interest.

4.1   Dataset and External Knowledge Bases

SemEval-2016 Task 6 released a dataset for stance detection on English tweets. In total, there are 4,163 tweets in this dataset, and the stance of each tweet is manually annotated towards one of five targets, which are *Atheism*, *Climate Change is a Real Concern (Climate Change)*, *Feminist Movement*, *Hillary Clinton*, and *Legalization of Abortion (Abortion)*. This dataset has two subtasks, including subtaskA supervised learning and subtaskB unsupervised learning. In this paper, we merely work on subtaskA, in which the targets provided in the test set can all be found in the training set. Table 4 shows the statistics of subtaskA.

ConceptNet and WikiData are used as the CKB in our proposed model. ConceptNet contains 1.5 million entities and 18.1 million relations. WikiData is a free knowledge base with 93,704,491 data items that anyone can edit. The knowledge in both commonsense knowledge bases is organized as entity-relation triples.

4.2   Baselines

We compare our model with the following methods:

- SVM-ngrams [13]: Five SVM classifiers (one per target) trained on the corresponding training set for the target using word n-grams (1-, 2-, and 3-g) and character n-grams (2-, 3-, 4-, and 5-g) features.
- MITRE [25]: The best system in SemEval-2016 subtaskA is MITRE. This model uses two RNNs: the first one is trained to predict the task-relevant hashtags on a very large unlabeled Twitter corpus. This network is used to initialize the second RNN classifier, which was trained with the provided subtask-A data.

Table 5. Results on the SemEval dataset.

Models	Atheism	Climate	Feminist	Hillary	Abortion	Overall
SVM-ngrams	65.19	42.35	57.46	58.63	66.42	68.98
MITRE	61.47	41.63	62.09	57.67	57.67	67.82
BiCond	61.47	41.63	48.94	57.67	57.28	67.82
TAN	59.33	53.59	55.77	65.38	68.79	68.79
Bert$_{base}$	68.67	44.14	61.66	62.34	58.60	69.51
CKEMN	62.69	53.52	61.25	64.19	64.19	**69.74**
Our	70.55	57.20	61.55	62.33	63.65	**71.06**

- BiCond [2]: Model employs conditional LSTM to learn a representation of the tweet considering the target.
- TAN [6]: A neural network-based model, which incorporates target-specific information into stance classification with an attention mechanism.
- CKEMN [5]: A commonsense knowledge enhanced memory network (CKEMN) for stance classification using LSTM as embedding.
- Bert$_{base}$ [3]: Bert model encodes the single input sequence of the pair of target and tweet and uses a linear layer to classify the stance.

5 Results and Analysis

5.1 Main Results

The experimental results of baselines and our proposed model on the SemEval dataset are reported in Table 5. For models that do not introduce commonsense knowledge, SVM-ngrams, MITRE, BiCond, and TAN, their macro-F1 values are lower than those of Bert$_{base}$, CKEMN, and RelNet that introduce external knowledge. Among them, the baseline SVM-ngrams of the SemEval task and the macro-F1 of TAN are higher because they train separate models for different targets, indicating that the feature gap between different targets is quite large, and training together will cause some target-specific features to be lost. BiCond and TAN both enhance interactions between tweets and targets, but their improvements of macro-f1 value are relatively weak, indicating that the task of stance detection requires external knowledge to help the model better understand the relation between target and opinioned terms in the tweet.

Since the BERT$_{base}$ model incorporates the knowledge acquired from massive external corpora, the performance improves compared to the previous models that did not introduce commonsense knowledge. CKEMN uses transE to introduce the entities and relations of ConceptNet into the stance detection task, and the performance is further improved. However, CKEMN did not integrate different relationships according to the requirements of stance detection task. Our model integrates the relationship into consistent and inconsistent, and establishes the relationship between target and opinioned terms in tweets.

Table 6. The comparison of whether the target is in the sentence or not. Accuracy(%) is adopted for evaluation.

	All	In	Out
Bert$_{base}$	69.33	77.31	66.41
Our	71.33	76.71	69.36

Table 7. The effect of different numbers of text-span

The number of text-span (K)	3	4	5	6	7
Macro-F1		70.27	70.33	**71.06**	69.72	69.94

Our model RelNet outperforms all these baselines and achieves a improvement of about 1.32 points in F1-score over the strong baseline CKEMN. This highlights the value of leveraging multiple external knowledge bases as the bridge to explicitly link opinioned terms in tweets to given targets.

5.2 Influence of Absence of Targets *In* Tweets

To further testify the capability of RelNet in connecting targets and opinioned terms in tweets, we divide the dataset (All) into two subsets, one with the target explicitly mentioned in the sentence (*In*) and the other one with the target not in the sentence (*Out*). If a tweet contains important words related to the target, it falls to (*In*) set, otherwise to (*Out*) set. For example, for the target "Climate Change Is A Real Concern", if "Climate" or "Change" appears in the tweet, the target is considered to be mentioned in the sentence. For Bert$_{base}$ and RelNet, we calculate the accuracy on the In and Out subsets, the experimental results are reported in Table 6.

It can be seen that on the *Out* dataset, the accuracy of RelNet improves over 2.9% compared with Bert$_{base}$. This proves that our model can effectively establish a relation between targets and opinioned terms for texts where targets are not mentioned, thus help to improve the performance of stance detection.

5.3 Effects of Text-Span

In order to verify whether the number of text-spans affects the results of stance detection, we compare the macro-f1 of RelNet with $k = 3, 4, 5, 6, 7$. k is the number of pairs (*knowledge word, tweet spans*) extracted from a tweet using the similarity function. The experimental results are reported in Table 7. It can be seen from Table 7 that when $k = 5$, the performance of the model is the best. When k gradually increases from 5, the performance gradually decreases, indicating that a certain amount of noise is introduced, which has a negative effect on stance detection. When k gradually decreases from 5, the performance also gradually declines. It may be that some common text-spans have high semantic similarity with knowledge. When k is gradually reduced, some target-related text-spans cannot be effectively selected.

Table 8. The comparison of whether the anti-knowledge is added into KB (D).

	Macro-F1
RelNet	71.06
+ anti-knowledge	70.27

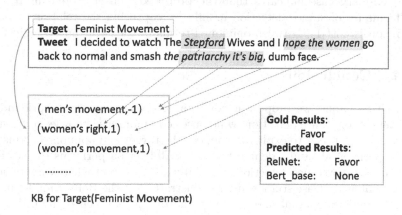

study.pdf   KB for Target(Feminist Movement)

Fig. 3. The example of case study.

5.4 Adding *Anti*-Knowledge

We use the target as the key to search the knowledge base, which causes that the anti-knowledge in KB D is very sparse. So we add anti-knowledge to KB by using the anti-*target* as the key to search the knowledge base and analyze the macro-F1 change of RelNet. Experimental results are reported in Table 8. For Climate Change and Feminist Movement, do not increase anti-knowledge. For target: Abortion, Atheism, and Hillary Clinton, search WikiData and ConceptNet for anti-abortion, theism, and Donald Trump respectively, and use strict rules to filter the expanded concept words. Finally, for Abortion, {pro-life} is added, and for Atheism {Belief in god} is added, for Hillary Clinton, we add {donald trump, make america great again, america first, realdonaldtrump}. It can be seen from Table 8 that after adding anti-knowledge into D, the performance of the model declines, which is contrary to our expectations. It is possible that manually selected anti words are not suitable for the dataset, resulting in a lot of noise.

6 Case Study

To make it more intuitive, we selected an example from the test set that is misclassified by other models while correctly classified by RelNet. We show the knowledge base established in Target Related Knowledge Filtration Module, and the text-span captured by Opinioned Terms Encoder Modul in Fig. 3. It can be seen that this tweet does not directly express one's stance against Feminist Movement (target), but through expressing opposition to *patriarchy* and support for *Stepford Wives* (A science-fiction comedy depicting the awakening of female consciousness) indirectly favor the target. Our model uses external knowledge as a bridge to capture the opinioned terms (Stepford, women, hope women, patriarchy, patriarchy big) in the tweet and their consistent relations with the target, helping to enhance the tweet's awareness of the target. This exemplary case indicates that the proposed RelNet model improves the performance of stance detection mainly by capturing the relations (consistent(1), inconsistent(-1)) between opinioned terms in the tweet and the given target.

7　Conclusion

In this paper, we propose a novel target-aware neural network model named RelNet for stance detection, which uses commonsense knowledge from WikiData and ConceptNet to explicitly connect the target and opinioned terms in a tweet. Experimental results demonstrate that RelNet outperforms the state-of-the-art methods for stance detection, especially on the subset where targets do not occur in sentences. Case studies further illustrate that the RelNet model benefits from introducing external knowledge as the bridge between opinioned terms in the tweet and the target.

Acknowledgements. This work was supported by the National Key R&D Program of China (Grant No. 2018YFB1005103), and the National Natural Science Foundation of China (Grant Nos. 61632011 and 61772153). We would particularly like to acknowledge Yanyue Lu, Yijian Tian, Hao Yang, and Yang Wu, for their kind help and useful discussion. Our deepest gratitude goes to the anonymous reviewers for their careful work and thoughtful suggestions that have helped improve this paper substantially.

References

1. Abu-Jbara, A., Dasigi, P., Diab, M., Radev, D.: Subgroup detection in ideological discussions. In: Proceedings of the 50th Annual Meeting of the Association for Computational Linguistics, Vol. 1, pp. 399–409 (2012)
2. Augenstein, I., Rocktäschel, T., Vlachos, A., Bontcheva, K.: Stance detection with bidirectional conditional encoding. arXiv preprint arXiv:1606.05464 (2016)
3. Devlin, J., Chang, M.W., Lee, K., Toutanova, K.: Bert: pre-training of deep bidirectional transformers for language understanding. arXiv preprint arXiv:1810.04805 (2018)
4. Dey, K., Shrivastava, R., Kaushik, S.: Topical stance detection for twitter: a two-phase LSTM model using attention. In: Pasi, G., Piwowarski, B., Azzopardi, L., Hanbury, A. (eds.) ECIR 2018. LNCS, vol. 10772, pp. 529–536. Springer, Cham (2018). https://doi.org/10.1007/978-3-319-76941-7_40
5. Du, J., Gui, L., Xu, R., Xia, Y., Wang, X.: Commonsense knowledge enhanced memory network for stance classification. IEEE Intell Syst **35**(4), 102–109 (2020)
6. Du, J., Xu, R., He, Y., Gui, L.: Stance classification with target-specific neural attention networks. In: International Joint Conferences on Artificial Intelligence (2017)

7. Ferreira, W., Vlachos, A.: Emergent: a novel data-set for stance classification. In: Proceedings of the 2016 conference of the North American chapter of the association for computational linguistics: Human language technologies, pp. 1163–1168 (2016)
8. Hanawa, K., Sasaki, A., Okazaki, N., Inui, K.: Stance detection attending external knowledge from wikipedia. J Inf Process **27**, 499–506 (2019)
9. Hasan, K.S., Ng, V.: Stance classification of ideological debates: Data, models, features, and constraints. In: Proceedings of the Sixth International Joint Conference on Natural Language Processing, pp. 1348–1356 (2013)
10. Kim, S.M., Hovy, E.: Crystal: analyzing predictive opinions on the web. In: Proceedings of the 2007 Joint Conference on Empirical Methods in Natural Language Processing and Computational Natural Language Learning (EMNLP-CoNLL), pp. 1056–1064 (2007)
11. Ma, Y., Peng, H., Cambria, E.: Targeted aspect-based sentiment analysis via embedding commonsense knowledge into an attentive LSTM. In: Proceedings of the AAAI Conference on Artificial Intelligence, vol. 32 (2018)
12. Mohammad, S., Kiritchenko, S., Sobhani, P., Zhu, X., Cherry, C.: Semeval-2016 task 6: detecting stance in tweets. In: Proceedings of the 10th International Workshop on Semantic Evaluation (SemEval-2016), pp. 31–41 (2016)
13. Mohammad, S.M., Sobhani, P., Kiritchenko, S.: Stance and sentiment in tweets. ACM Transac Internet Technol **17**(3), 1–23 (2017)
14. Murakami, A., Raymond, R.: Support or oppose? Classifying positions in online debates from reply activities and opinion expressions. In: Coling 2010, Posters, pp. 869–875 (2010)
15. Qiu, M., Yang, L., Jiang, J.: Modeling interaction features for debate side clustering. In: Proceedings of the 22nd ACM international conference on Information & Knowledge Management, pp. 873–878 (2013)
16. Siddiqua, U.A., Chy, A.N., Aono, M.: Tweet stance detection using an attention based neural ensemble model. In: Proceedings of the 2019 Conference of the North American Chapter of the Association for Computational Linguistics: Human Language Technologies, Vol 1, pp. 1868–1873 (2019)
17. Somasundaran, S., Wiebe, J.: Recognizing stances in ideological on-line debates. In: Proceedings of the NAACL HLT 2010 Workshop on Computational Approaches to Analysis and Generation of Emotion in Text, pp. 116–124 (2010)
18. Sridhar, D., Foulds, J., Huang, B., Getoor, L., Walker, M.: Joint models of disagreement and stance in online debate. In: Proceedings of the 53rd Annual Meeting of the Association for Computational Linguistics and the 7th International Joint Conference on Natural Language Processing, Vol. 1, pp. 116–125 (2015)
19. Su, Z., Zhu, C., Dong, Y., Cai, D., Chen, Y., Li, J.: Learning visual knowledge memory networks for visual question answering. In: Proceedings of the IEEE Conference on Computer Vision and Pattern Recognition, pp. 7736–7745 (2018)
20. Thomas, M., Pang, B., Lee, L.: Get out the vote: determining support or opposition from congressional floor-debate transcripts. arXiv:cs/0607062v3 (2006)
21. Wei, P., Mao, W., Zeng, D.: A target-guided neural memory model for stance detection in twitter. In: IJCNN, pp. 1–8. IEEE (2018)
22. Wei, W., Zhang, X., Liu, X., Chen, W., Wang, T.: pkudblab at semeval-2016 task 6: a specific convolutional neural network system for effective stance detection. In: SemEval-2016, pp. 384–388 (2016)
23. Young, T., Cambria, E., Chaturvedi, I., Zhou, H., Biswas, S., Huang, M.: Augmenting end-to-end dialogue systems with commonsense knowledge. In: Proceedings of the AAAI Conference on Artificial Intelligence. vol. 32 (2018)

24. Yuan, J., Zhao, Y., Xu, J., Qin, B.: Exploring answer stance detection with recurrent conditional attention. In: Proceedings of the AAAI Conference on Artificial Intelligence, vol. 33, pp. 7426–7433 (2019)
25. Zarrella, G., Marsh, A.: Mitre at semeval-2016 task 6: transfer learning for stance detection. arXiv preprint arXiv:1606.03784 (2016)

Towards Nested and Fine-Grained Open Information Extraction

Jiawei Wang[1], Xin Zheng[2], Qiang Yang[3], Jianfeng Qu[1], Jiajie Xu[1],
Zhigang Chen[2], and Zhixu Li[1(✉)]

[1] School of Computer Science and Technology, Soochow University, Suzhou, China
jwwang97@stu.suda.edu.cn, {jfqu,xujj,zhixuli}@suda.edu.cn
[2] iFLYTEK CO., LTD, Suzhou, China
{xinzheng3,zgchen}@iflytek.com
[3] King Abdullah University of Science and Technology, Jeddah, Saudi Arabia
qiang.yang@kaust.edu.sa

Abstract. Open Information Extraction is a crucial task in natural language processing with wide applications. Existing efforts only work on extracting simple flat triplets that are not minimized, which neglect triplets of other kinds and their nested combinations. As a result, they cannot provide comprehensive extraction results for its downstream tasks. In this paper, we define three more fine-grained types of triplets, and also pay attention to the nested combination of these triplets. Particular, we propose a novel end-to-end joint extraction model, which identifies the basic semantic elements, comprehensive types of triplets, as well as their nested combinations from plain texts jointly. In this way, information is shared more thoroughly in the whole parsing process, which also lets the model achieve more fine-grained knowledge extraction without relying on external NLP tools or resources. Our empirical study on datasets of two domains, Building Codes and Biomedicine, demonstrates the effectiveness of our model comparing to state-of-the-art approaches.

Keywords: Open information extraction · Algorithm · Triplet

1 Introduction

Open Information Extraction (OIE) aims to extract structured information in the triplet form of (`arg1`, `rel`, `arg2`) from natural language texts without predefined schema. Considering the sentence *"The building is equipped with chairs"*, an OIE system aims to extract the triplet (`building`, `is equipped with`, `chairs`). Since it plays a significant role in various semantic tasks including question answering [15], text comprehension [22] and man-machine dialogue system [10], the task has received considerable interests [3,5,8].

Conventional OIE methods extract triplets that consist of two arguments connected by their relational phrases. Unfortunately, these models heavily rely on rule-based syntax or semantic parser and thus inevitably suffer from error propagation. To break this limitation, neural OIE methods are proposed, which can be

© Springer Nature Singapore Pte Ltd. 2021
B. Qin et al. (Eds.): CCKS 2021, CCIS 1466, pp. 185–197, 2021.
https://doi.org/10.1007/978-981-16-6471-7_14

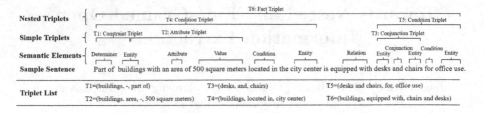

Fig. 1. An example sentence for nested and fine-grained OIE

classified into two categories: sequence generation [5,17] that employs a Seq2Seq
model and span selection [25] that uses a span selection model. However, this shallow triplet form reflected by binary relation is too simple and brings two challenges. Firstly, these methods tend to extract tediously long rather than minimized
triplets, which can be further decomposed into more fine-grained meaningful ones.
For example, the triplet (`Beckett, is sending, 100 old buses refitted with
desks and chairs to serve as temporary classrooms`) is extracted from the
given sentence *"Beckett is sending 100 old buses refitted with desks and chairs to
serve as temporary classrooms"*. Note that `arg2` contains more than one kind of
semantic information that is concatenated together. To some extent, it greatly
damages the effect of downstream semantic applications. Secondly, triplets at the
same level cannot express higher level relations well. Only when each related triplet
works together, semantic information could be fully covered. In the above example, if we have minimized the extracted triplet, several independent triplets can be
obtained. Nevertheless, a triplet representing a phrase such as *"serve as temporary
classrooms"* has an implicit meaning for other related triplets.

Therefore, many researches focus on complex relations that go beyond binary
relations. One way is to consider other types of semantic contexts on the basis of
factual relations. In real scenario, these facts are often conditioned by these contexts. OLLIE [21] adds extra fields to make factual extractions valid. MinIE [9]
provides four semantic annotations for each extraction including polarity, modality, attribution, and quantities. MIMO [14] explores condition triplets, which
usually describes a condition on which a fact holds. Apart from these types of
contexts, there are some fine-grained structured information which receives few
attentions yet; Another is to consider nested representations. Due to the complex
nature of human language, nested relations are frequently expressed, especially in
some domain-specific texts. The nested form is capable of combining triplets and
connecting their interdependent relationships. Although there are some attempts
of nested forms [4,19] to represent sentence semantics, these traditional methods heavily rely on grammar structures or manual templates. Besides, while the
simple triplets are identified based on elements, the nested triplets are identified
based on simple triplets. This cascading hierarchical extraction pattern is prone
to cumulative errors, that is, when an error occurs in any extraction step, the
subsequent extractions of higher levels are all erroneous.

To address the first challenge, we further nominate three more types of fine-
grained triplets, including *attribute triplet*, *constraint triplet* and *conjunction*

triplet, as a complementary to fact triplet and condition triplet. Formal definitions for these five types of triplet will be given in Sect. 3. The second challenge is handled by our new nested triplets. Unlike aforementioned methods, we not only work on identifying *simple triplets* consisting of basic semantic elements, but also *nested triplets* consisting of both semantic elements and other triplets. As the example shown in Fig. 1, we could identify three simple fine-grained triplets T_1, T_2, T_3, as well as three nested fine-grained triplets T_4, T_5, T_6. According to our sampling statistics on the Building Codes dataset employed in our experiments, the three newly-defined fine-grained types of triplets make up for 29.5% of all the samples, while the nested ones form 44.8% of all the samples. Similarly, the sampling statistics on the Biomedicine dataset also show that the three newly-defined fine-grained types of triplets make up for 45.5% of all the samples, while the nested ones form 52.5% of all the samples.

Towards this, we propose an iterative two-step parsing model, which recursively extracts elements and triplets jointly in a bottom-up fashion. At each iteration, the spans of elements or triplets are detected with a joint sequence labeling model in the first step, and then each triplet span will be parsed to obtain its components in the second step. More specifically, different from the existing efforts that identify elements and triplets separately, we take the triplet as a special kind of element, such that we could identify spans for elements and triplets with one model. In this way, we could better utilize the correlations between elements and triplets. Eventually, our model is able to reach state-of-the-art triplet extraction without relying on any NLP tools.

2 Related Work

Open Information Extraction. Open Information Extraction(OIE) is a task that extracts triplets without the limitation of specific relations. Several OIE systems [1,23,24] focus on the extraction of fact triplets. Under realistic scenario, complex relations that go beyond binary relations are frequently expressed in plain texts. Binary relational facts tend to suffer from significant information loss, which affects the completeness and correctness of extractions.

In order to handle this problem, a line of works have studied extracting n-ary relations. OLLIE [21] extends a triplet with an extra field to cover contextual information. ClausIE [6] identifies a coherent piece of information for each contributing entity to generate an extraction. MinIE [9] built on top of ClausIE aims to minimize extractions by using three minimization modes. Bast and Haussmann [3] point out that additional information supporting for factual triples is likely to be contained in long argument phrases of extractions, which will influence the downstream tasks that rely on this information. MIMO [14] generates relational phrases for fact and condition triplets by sequence labeling method and then distinguish them. However, these methods fail to deal with nested relations. To combine triplets, CSD-IE [2] allows each fact triplet to contain references of unique ids to its related triplets. NestIE [4] is the first to study nested representation of complex relations for OIE. StuffIE [19] uses grammatical

Table 1. All types of elements, and their corresponding symbols and examples

Element type	Symbol	Examples
Entity	E	USA, Mike, room, window, coronavirus
Relation	R	marry, equip_with, increase, use, lead_to
Attribute	A	birthdate, population, ventilation_area, growth_rate
(Attribute) value	V	9.02 million, lower than MU7.5, 93 ng/ml, higher than 85°C
Condition	C	(tourists) *on the top of* (the mountain), (a boy) *named* (John), (classrooms) *next to each other*
Determiner	D	*any* (chapter), *each* (residential building), *all* (schools), *one of* (these rooms)
Conjunction	J	(desks) *and* (chairs), (teachers) *as well as* (students), (the first floor) *or* (the second floor)

dependencies to capture the links between factual and contextual information to obtain nested triplets. In this paper, we pay more attention to fine-grained information, and nested relationships extracted in a hierarchical manner.

Span-based Models. Span-based models dealing with span-level tasks have been applied to many research fields, such as syntactic analysis [27], semantic parsing [11], and information extraction [25]. For OIE, Zhan et al. [25] first convert this task into a span selection task for n-ary extractions and propose a span model SpanOIE to fully exploit span-level features. The differences between our method and SpanOIE lies in that: (1) SpanOIE targeting at n-ary extractions, in which elements are all at the same level, cannot express nested relationships; (2) SpanOIE is a two-stage pipeline approach i.e., first identify predicate spans and then find argument spans for predicate spans. The characteristic of our method is that we can jointly extract nested representations of text in an iterative way. In addition, elements (argument spans and predicate spans) are identified based on triplets spans and then triplets are determined.

Sequence Labeling Models. Sequence labeling methods is very prevalent and effective in many tasks. Traditional sequence labeling models, such as Support Vector Machine (SVM) and Conditional Random Fields (CRF) [18], heavily rely on hand-crafted or language-specific features. After that, various researches explore neural networks to learn textual features. Huang et al. [13] first apply BiLSTM-CRF model to sequence labeling. Zheng et al. [26] transform the joint extraction task into a sequence labeling problem, and uses LSTM-based model to extract entities and relations in an end-to-end way. MIMO [14] is a sequence labeling model designed for fact and condition triplets. Compared with these methods, we pay more attention to comprehensive semantic information.

3   Definitions and Task Formulation

In this subsection, we give definitions to all the concerned semantic elements and fine-grained triplets and then formally define our task.

Table 2. All types of triplets, and their corresponding sketches and examples

Triplet type	Triplet sketch	Examples
Fact triplet	$T = (E_1, R, E_2) : T$	$T = $ (Gates, founded, Microsoft) : T
Condition triplet	$T = (E_1, C, E_2) : E_1$	$T = $ (tourists, on the top of, mountain): tourists
Attribute triplet	$T = (E.A, -, V) : T \mid E$	$T = $ (room. area, -, less than 5 m): $T \mid$ room
Constraint triplet	$T = (E, -, D) : E$	$T = $ (rooms, -, one of) : rooms
Conjunction triplet	$T = (E_1, J, E_2) : E_3$	$T = $ (desks, and, chairs) : E_3

Definition 1 (Semantic Elements). *The semantic elements (or elements for short) of a sentence are the basic semantic units in expressing the semantic meaning of the sentence, which includes seven different types as listed in Table 1.*

- **Entity.** A nominal phrase that refers to an object or a concept.
- **Relation.** A phrase describing the semantic relationship betweentwo entities.
- **Attribute.** A phrase referring to a particular property of an entity.
- **(Attribute) Value.** A phrase expressing the value of an attribute of an entity.
- **Condition.** A phrase that used to qualifying an entity with some other entity (or entities).
- **Determiner.** An indefinite determinative phrase that constraints an entity.
- **Conjunction.** A conjunctive phrase describing a parallel or selection relationship between two entities.

Based on the seven types of semantic elements obtained from a sentence, we could then identify the semantic triplets with the sentence.

Definition 2. *A semantic triplet (or a triplet for short) consists of three semantic elements, denoted by $T = (s, p, o) : rv$, where s, p, o correspond to the subject, predicate and object of the triplet respectively, and rv is the return value of the triplet, which refers to the key information of the triplet. We are concerned with five kinds of triplets as listed in Table 2.*

- **Fact Triplet.** A triplet describing a relation R between two entities E_1 and E_2, which takes the whole triplet as the return value.
- **Condition Triplet.** A triplet qualifying an entity E_1 with some other entity (or entities) E_2, which takes E_1 as the return value.
- **Attribute Triplet.** A triplet presenting the value V of an attribute A of an entity E, which takes either E or the whole triplet as the return value.
- **Constraint Triplet.** A triplet that constraints an entity E with a determiner D, which takes E as the return value.
- **Conjunction Triplet.** A triplet that combines two entities E_1 and E_2 with a conjunction, which takes the combined entity E_3, referring to $E_1 \bowtie J \bowtie E_2$ where $\bowtie$ denotes the conjunction operation between entity and relation to get the combined entity, as the return value.

Definition 3 (Simple Triplet and Nested Triplet). *We say a triplet $T = (s, p, o) : rv$ is a simple triplet if both s and o are basic semantic elements, otherwise, we call T as a nested triplet if at least one of the components s and o is the return value of some triplet.*

Given a sentence in plain text, this task aims to identify all the five kinds of triplets, including both simple ones and nested ones. More formally:

Definition 4 (Nested and Fine-Grained OIE). *Given a sentence $X = \{x_1, x_2, ..., x_l\}$ where l is the length of X, the task is to map X into a set of triplets $\mathbb{T} = \{T_1, T_2, ..., T_{|\mathbb{T}|}\}$, where $\mathbb{T}$ contains all the fine-grained triplets defined in Table 2, including both simple ones and nested ones, from X.*

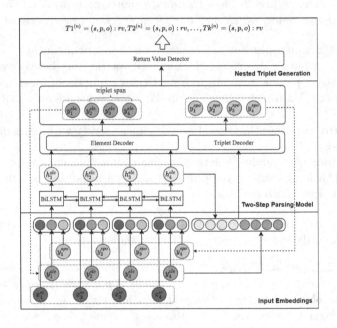

Fig. 2. The architecture of the two-step parsing model

4   Methodology

In this paper, we propose to tackle the nested and fine-grained OIE task with an iterative two-step parsing approach. In the first step, we take triplet as a special kind of element, such that we could identify spans for elements and triplets with one model. Then in the second step, each triplet span would be parsed to obtain its three components and its type. Each iteration takes the results of the current layer as the input for the next layer. For example, elements such as *"desks"*, *"and"* and *"chairs"* are identified at the second layer. The triplet spans such as *"desks and chairs"* at the third layer are identified based on the second layer. Given that we have triplets of different layers, this two-step parsing model needs to be applied layer-by-layer iteratively until all the triplets are identified.

4.1   Inputs Embedding

We encode a sentence X into a word-level vector representation $\mathbf{x}^w$ by applying an average pooling function to the outputs of the pre-trained BERT model [7]. Given an element label sequence $Y^{ele} = \{y_1^{ele}, ..., y_i^{ele}, ..., y_l^{ele}\}$ and a triplet label sequence $Y^{spo} = \{y_1^{spo}, ..., y_i^{spo}, ..., y_l^{spo}\}$, where $y_i^{(\cdot)}$ denotes the i-th label in the element label sequence or triplet label sequence. In the initial layer, the model has not yet output the element label sequence and triplet label sequence, so they are denoted as $\{O_1, O_2, ..., O_l\}$, where O refers to a non-element token. In other words, elements are not identified in this round. Each sequence is mapped into a vector via $\mathbf{y}_i^{(\cdot)} = \mathcal{W}_{es}^{(\cdot)} y_i^{(\cdot)}$, where $\mathcal{W}_{es}^{(\cdot)}$ is an embedding matrix for the element label sequence or triplet label sequence and i is the i-th token. Then the word embedding representation $\mathbf{x}^w$, the element label vector $\mathbf{y}^{ele}$ and the triplet label vector $\mathbf{y}^{spo}$ are concatenated to represent the input hidden state $\mathbf{x}$:

$$\mathbf{x} = [\mathbf{x}^w; \mathbf{y}^{ele}; \mathbf{y}^{spo}] \tag{1}$$

4.2   Iterative Encoder with Two-Step Parsing

In this section, we first introduce the two-step parsing model, and then present how to generate the nested triplet iteratively.

Two-Step Parsing Model. Different from traditional methods, we propose an end-to-end parsing model, which first recognizes the element spans and then extracts the triplets, as shown in Fig. 2.

Step 1: Element Span Recognition Module. The goal of this module is to recognize the element spans at each iteration. Different from previous work, we treat the triplet as a specific element that participates in the model training phase along with other semantic elements. We employ a BiLSTM neural network [12] to get the H^{ele}, where $H^{ele} = \{h_1^{ele}, ..., h_i^{ele}, ...h_l^{ele}\}$. We project the obtained H^{ele} into the element label space with a two-layer fully connected layer as following:

$$\mathcal{U} = \tanh(\mathcal{W}_{12} \tanh(\mathcal{W}_{11} H^{ele})) \tag{2}$$

The bias parameters are ignored here. Then the probability score for the i-th word in a sentence corresponds to the j-th label is calculated. We then pass it into to the CRF layer with a transition matrix $\mathcal{T}$ to predict the labels, i.e., from label i to label j, represented in Eq. 3.

$$p(Y^{ele}|X) \propto \exp\left(\sum_{i=0}^{l} \mathcal{T}_{y_i^{ele}, y_{i+1}^{ele}} + \sum_{i=1}^{l} \mathcal{U}_{i, y_i^{ele}}\right) \tag{3}$$

After this step, we have acquired all element spans in an element label sequence at the current iteration. The sequence with triplet spans will be used for the next layer of iteration (indicated by dotted arrow in Fig. 2).

Step 2: Triplet Span Parsing Module. The goal of this module is to parse out the three components i.e. *subject*, *predicate* and *object* in a triplet. In this module, all the possible relation categories are recognized in all triplet spans at the current layer via the encoded vector representation H^{ele}, and the representations $\mathbf{y}^{ele}$ of the element label sequence. In details, we first concentrate the above two vectors to obtain the context-sensitive triplet representation $H^{spo} = \{h_1^{spo}, ..., h_i^{spo}, ...h_l^{spo}\}$ and then project them into the triplet label space. Similar to Eq. 3, the probability of $p(Y^{spo}|X)$ is computed as following:

$$p(Y^{spo}|X) \propto \exp\left(\sum_{i=0}^{l} \mathcal{T}_{y_i^{spo}, y_{i+1}^{spo}} + \sum_{i=1}^{l} \mathcal{U}_{i, y_i^{spo}}\right) \tag{4}$$

Nested Triplet Generation. Different from existing methods that only focus on the coarse-grained triplets and simple triplets, we employ an iterative way to generate nested triplets using the previous iteration results.

Since triplets from different iterative layers have their own semantics that need to be focused on, the self-attention mechanism is applied to generate an embedding feature vector for updating the triplet representation. In details, we follow [16] to mask the triplets that have been visited in the current iteration to focus on different triplets.

$$h_i^{spo(n+1)} = \mathcal{W}_h h_i^{spo(n)} + \mathcal{W}_a A_i^{(n)} \tag{5}$$

$$A_i^{(n)} = \alpha_i^{(n)} \mathcal{W}_{mask} \tag{6}$$

$$\alpha_i^{(n)} = \frac{q^T h_i^{spo(n)}}{\sum q^T h_j^{spo(n)}} \tag{7}$$

where $\mathcal{W}_h, \mathcal{W}_a, \mathcal{W}_{mask}, q$ are model parameters and n denotes the n-th iteration.

4.3 Optimization

During training, we optimize the model by using the maximum likelihood estimation. The total loss consists of two parts, i.e., the element span recognition prediction loss ($\mathcal{L}_{ele}$) and the triplet span parsing prediction loss ($\mathcal{L}_{spo}$), which can be defined as:

$$\mathcal{L}_{total} = \gamma_{ele}\mathcal{L}_{ele} + \gamma_{spo}\mathcal{L}_{spo} \tag{8}$$

where γ_{ele} and γ_{spo} are two hyper-parameters to balance the two parts.

Given a training sample $\{(X_m, Y_m^{ele})\}$ where $m = \{1, 2, ..., |D|\}$ and $|D|$ is the size of the training samples. The loss function for element span recognition is given by:

$$\mathcal{L}_{ele} = \sum_{m=1}^{|D|} log(p(Y_m^{ele}|X_m)) \tag{9}$$

Likewise, the loss function for triplet span parsing is used to maximize the probability of the output triplet sequence at each decoding process.

5 Experimentation

5.1 Experimental Settings

Datasets. To evaluate our model, we experimented with two publicly available datasets: Building Codes (BuildCod for short), and Biomedicine (BioMed). The characteristics of these datasets are summarized in Table 3.

Building Codes (BuildCod). It is annotated from *National Building Codes* that contains architectural design plain texts issued by government authorized agencies. It consists of 7,400 sentences (6,858 for training and 542 for testing). We randomly select 400 sentences as the development set.

Biomedicine (BioMed). We refer to the annotated dataset from [14], where the train and development set are from Biomedical Conditional Fact Extraction that is manually annotated from 31 biomedical paper abstracts, and the test set is from Cancer Genetics task dataset of BioNLP Shared Task 2013 ([20]). We annotate the dataset for the nested and semantic parsing task which consists of 194 training sentences, 142 development sentences and 100 testing sentences.

Since source code of NestedIE [4] seems to be unavailable online, it is not used as our compared method. Several compared methods are listed as follows:

Table 3. Statistics for BuildCod and BioMed

Datasets		Element							Triplet	
		Entity	Relation	Attribute	Value	Condition	Determiner	Conjunction	Newly-defined triplet	Nested triplet
BuildCod	Train	25,812	10,334	6,955	3,101	1,800	690	3,485	7,276	11,034
	Test	2,331	813	623	349	182	46	287	682	1,070
BioMed	Train	1,118	332	321	79	391	56	247	624	727
	Test	403	156	122	27	127	29	70	223	249

Table 4. Performance comparison (in %) on BuildCod and BioMed

Mehods	BuildCod						BioMed					
	Element prediction			Triplet prediction			Element prediction			Triplet prediction		
	Prec.	Rec.	F1	Prec.	Rec.	F1	Prec.	Rec.	F1	Prec.	Rec.	F1
OpenIE5	–	–	–	–	–	–	–	–	–	44.8	42.0	43.0
StuffIE	–	–	–	–	–	–	–	–	–	47.7	51.6	49.8
CRF	56.7	54.3	56.8	53.4	49.7	51.2	56.1	50.2	52.6	53.5	47.3	50.7
BiLSTM-LSTMd	75.7	76.3	75.9	69.0	69.9	69.4	70.9	65.9	70.7	64.2	61.4	63.3
MIMO	81.1	82.7	79.6	72.6	71.5	75.3	75.7	71.6	74.7	78.5	79.6	78.0
Our model	**86.6**	**86.9**	**86.7**	**86.9**	**80.3**	**84.7**	**81.8**	**80.4**	**80.7**	**83.9**	**81.2**	**82.4**

OpenIE5[1] is a popular used OIE system. Since this system is only available for English language, we only employ it to conduct this experiment on BioMed dataset. **StuffIE** [19] is an OIE system for fine-grained nested relations using Stanford NLP and lexical databases. For the same reason above, the experiment is only carried out on BioMed dataset. **CRF** [18] is a probability graph model used to evaluate the conditional probability of a label sequence. **BiLSTM-LSTMd** [26] is a LSTM-based sequence labeling model to extract entities and relations jointly. **MIMO** [14] is a multi-input multi-output sequence labeling model for fact triplets and condition triplets. We use BERT as its encoder.

Evaluation Protocols and Parameter Settings. We adopt early stopping to determine the number of epochs. The hyper-parameters are determined on the development set. The label embedding dimension is set to 50. The max length of the input sentence is set to 200. The batch size is 32. During training, we set the LSTM dropout rate to 0.5. We use BiLSTM model with 128 units. Our model is trained via using Adam optimizer with initial learning rate of 10^{-5}. All the models are trained on a single GTX 1080Ti GPU. The pre-trained BERT model we used is [BERT-Base, Chinese] for BuildCod and [BERT-Base, Cased] for BioMed. To evaluate element prediction, we use the standard micro Precision (Prec.), Recall (Rec.) and F1 score. To evaluate triplet extraction, we regard a triplet as correct when it is mathched with the gold triplet.

5.2 Performance Comparison

Main Results. Table 4 presents our comparisons of element prediction and triplet prediction with other methods on BuildCod and BioMed datasets. From this table, we have the following observations: (1) On element prediction, we

Table 5. Human evaluation on the testing set of BuildCod and BioMed

Methods	BuildCod			BioMed		
	Coverage(0–5)	Minimality(%)	Nestedness(%)	Coverage(0–5)	Minimality(%)	Nestedness(%)
StuffIE	–	–	–	4.15	36.2	39.0
MIMO	3.79	68.5	59.7	4.10	54.1	68.2
Our model	**4.81**	**86.9**	**87.3**	**4.53**	**85.6**	**81.2**

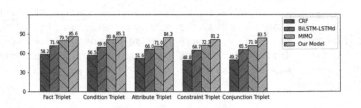

Fig. 3. Comparing F1 score of triplets extraction on Buildcod

[1] https://github.com/dair-iitd/OpenIE-standalone.

see that our model performs the best, which demonstrates that our model has significant advantages in identifying multiple types of elements in plain texts; (2) On triplet prediction, the results show that our method has a significant F1 score improvement over all compared methods. It indicates that our model is more capable of handling the situation that there are fine-grained triplets and nested triplets in texts.

Detailed Results on Fine-Grained Triplets. To further study the ability of our model on fine-grained triplet prediction, we also illustrate the F1 score of our model and compared methods on different types of triplets from BuildCod dataset in Fig. 3. Compared with the newly-defined triplets, fact triplets and condition triplets are easier for previous methods to extract. It is worth noting that these methods can not pay attention to different triplet types, but our model can deal with these cases well.

Human Evaluation. We sample 100 testing sentences from BuildCod dataset and choose all 100 BioMed tesing sentences for human evaluation to compare our model with StuffIE and MIMO. The evaluation is carried out with regard to three aspects: (1) Coverage. The semantic proportion covered by all triplets from a sentence is labeled of 0–5 (0 for bad and 5 for good), referring to the method for informativeness [4]; (2) Minimality. If components of a triplet extracted by the model do not contain phrases that can be further decomposed, this triplet is considered to be minimized. We calculate the proportion of minimized triplets among correctly extracted triplets; (3) Nestedness. A triplet is a nested triplet, when a component of this triplet contains a component of another triplet extracted from the same sentence. The proportion of nested triplets identified by the model is calculated. Table 5 shows the results of the three aspects. Overall, our model performs better with regard to minimized and nested information.

6   Conclusion and Future Work

This paper works towards nested and fine-grained open information extraction, and proposes a novel end-to-end neural model. Without relying on any NLP tools, it could identify basic semantic elements and all kinds of triplets iteratively. Extensive empirical study verifies the effectiveness of our model. Our future work will consider to incorporate knowledge bases into our model to provide knowledge for further improvement.

Acknowledgment. This research is partially supported by National Key R&D Program of China (No. 2018AAA0101900), National Natural Science Foundation of China (Grant No. 62072323, 61632016), Natural Science Foundation of Jiangsu Province (No. BK20191420), the Priority Academic Program Development of Jiangsu Higher Education Institutions, and the Collaborative Innovation Center of Novel Software Technology and Industrialization.

References

1. Angeli, G., Premkumar, M.J.J., Manning, C.D.: Leveraging linguistic structure for open domain information extraction. In: Proceedings of the 53rd Annual Meeting of the Association for Computational Linguistics and the 7th International Joint Conference on Natural Language Processing (Volume 1: Long Papers), pp. 344–354 (2015)
2. Bast, H., Haussmann, E.: Open information extraction via contextual sentence decomposition. In: 2013 IEEE Seventh International Conference on Semantic Computing, pp. 154–159. IEEE (2013)
3. Bast, H., Haussmann, E.: More informative open information extraction via simple inference. In: de Rijke, M., et al. (eds.) ECIR 2014. LNCS, vol. 8416, pp. 585–590. Springer, Cham (2014). https://doi.org/10.1007/978-3-319-06028-6_61
4. Bhutani, N., Jagadish, H., Radev, D.: Nested propositions in open information extraction. In: Proceedings of the 2016 Conference on Empirical Methods in Natural Language Processing, pp. 55–64 (2016)
5. Cui, L., Wei, F., Zhou, M.: Neural open information extraction. arXiv preprint arXiv:1805.04270 (2018)
6. Del Corro, L., Gemulla, R.: Clausie: clause-based open information extraction. In: Proceedings of the 22nd International Conference on World Wide Web, pp. 355–366 (2013)
7. Devlin, J., Chang, M.W., Lee, K., Toutanova, K.: Bert: Pre-training of deep bidirectional transformers for language understanding. arXiv preprint arXiv:1810.04805 (2018)
8. Fader, A., Soderland, S., Etzioni, O.: Identifying relations for open information extraction. In: Proceedings of the 2011 Conference on Empirical Methods in Natural Language Processing, pp. 1535–1545 (2011)
9. Gashteovski, K., Gemulla, R., Corro, L.D.: Minie: Minimizing Facts in Open Information Extraction. Association for Computational Linguistics (2017)
10. Han, S., Bang, J., Ryu, S., Lee, G.G.: Exploiting knowledge base to generate responses for natural language dialog listening agents. In: Proceedings of the 16th Annual Meeting of the Special Interest Group on Discourse and Dialogue, pp. 129–133 (2015)
11. Herzig, J., Berant, J.: Span-based semantic parsing for compositional generalization. arXiv preprint arXiv:2009.06040 (2020)
12. Hochreiter, S., Schmidhuber, J.: Long short-term memory. Neural Comput. **9**(8), 1735–1780 (1997)
13. Huang, Z., Xu, W., Yu, K.: Bidirectional lstm-crf models for sequence tagging. arXiv preprint arXiv:1508.01991 (2015)
14. Jiang, T., Zhao, T., Qin, B., Liu, T., Chawla, N., Jiang, M.: Multi-input multi-output sequence labeling for joint extraction of fact and condition tuples from scientific text. In: Proceedings of the 2019 Conference on Empirical Methods in Natural Language Processing and the 9th International Joint Conference on Natural Language Processing (EMNLP-IJCNLP), pp. 302–312 (2019)
15. Khot, T., Sabharwal, A., Clark, P.: Answering complex questions using open information extraction. arXiv preprint arXiv:1704.05572 (2017)
16. Kim, W., Goyal, B., Chawla, K., Lee, J., Kwon, K.: Attention-based ensemble for deep metric learning. In: Proceedings of the European Conference on Computer Vision (ECCV), pp. 736–751 (2018)

17. Kolluru, K., Aggarwal, S., Rathore, V., Chakrabarti, S., et al.: Imojie: Iterative memory-based joint open information extraction. arXiv preprint arXiv:2005.08178 (2020)
18. Lafferty, J., McCallum, A., Pereira, F.C.: Conditional random fields: Probabilistic models for segmenting and labeling sequence data (2001)
19. Prasojo, R.E., Kacimi, M., Nutt, W.: Stuffie: semantic tagging of unlabeled facets using fine-grained information extraction. In: Proceedings of the 27th ACM International Conference on Information and Knowledge Management, pp. 467–476 (2018)
20. Pyysalo, S., Ohta, T., Ananiadou, S.: Overview of the cancer genetics (cg) task of bionlp shared task 2013. In: Proceedings of the BioNLP Shared Task 2013 Workshop, pp. 58–66 (2013)
21. Schmitz, M., Soderland, S., Bart, R., Etzioni, O., et al.: Open language learning for information extraction. In: Proceedings of the 2012 Joint Conference on Empirical Methods in Natural Language Processing and Computational Natural Language Learning, pp. 523–534 (2012)
22. Stanovsky, G., Dagan, I., et al.: Open IE as an intermediate structure for semantic tasks. In: Proceedings of the 53rd Annual Meeting of the Association for Computational Linguistics and the 7th International Joint Conference on Natural Language Processing (Volume 2: Short Papers), pp. 303–308 (2015)
23. Yahya, M., Whang, S., Gupta, R., Halevy, A.: Renoun: fact extraction for nominal attributes. In: Proceedings of the 2014 Conference on Empirical Methods in Natural Language Processing (EMNLP), pp. 325–335 (2014)
24. Yates, A., Banko, M., Broadhead, M., Cafarella, M.J., Etzioni, O., Soderland, S.: Textrunner: open information extraction on the web. In: Proceedings of Human Language Technologies: The Annual Conference of the North American Chapter of the Association for Computational Linguistics (NAACL-HLT), pp. 25–26 (2007)
25. Zhan, J., Zhao, H.: Span model for open information extraction on accurate corpus. In: Proceedings of the AAAI Conference on Artificial Intelligence, vol. 34, pp. 9523–9530 (2020)
26. Zheng, S., Wang, F., Bao, H., Hao, Y., Zhou, P., Xu, B.: Joint extraction of entities and relations based on a novel tagging scheme. arXiv preprint arXiv:1706.05075 (2017)
27. Zhou, J., Zhao, H.: Head-driven phrase structure grammar parsing on penn treebank. arXiv preprint arXiv:1907.02684 (2019)

Toward a Better Text Data Augmentation via Filtering and Transforming Augmented Instances

Fei Xia[1,2]([✉]), Shizhu He[1,2], Kang Liu[1,2], Shengping Liu[3], and Jun Zhao[1,2]

[1] National Laboratory of Pattern Recognition Institute of Automation, Chinese Academy Sciences, Beijing 100190, China
xiafei2020@ia.ac.cn, {shizhu.he,kliu,jzhao}@nlpr.ia.ac.cn
[2] School of Artificial Intelligence, University of Chinese Academy of Sciences, Beijing 100190, China
[3] Beijing Unisound Information Technology, Beijing 100028, China
liushengping@unisound.com

Abstract. Thanks to a large amount of high-quality labeled data (instances), deep learning offers significant performance benefits in a variety of tasks. However, instance construction is very time-consuming and laborious, and it is a big challenge for natural language processing (NLP) tasks in many fields. For example, the instances of the question matching dataset CHIP in the medical field are only 2.7% of the general field dataset LCQMC, and its performance is only 79.19% of the general field. Due to the scarcity of instances, people often use methods such as data augmentation, robust learning, and the pre-trained model to alleviate this problem. Text data augmentation and pre-trained models are two of the most commonly used methods to solve this problem in NLP. However, current experiments have shown that the use of general data augmentation techniques may have limited or even negative effects on the pre-trained model. In order to fully understand the reasons for this result, this paper uses three types of data quality assessment methods from two levels of label-independent and label-dependent, and then select, filter and transform the results of the three text data augmentation methods. Our experiments on both generic and specialized (medical) fields have shown that through analysis, selection/filtering, and transformation of augmented instances, the performance of intent understanding and question matching in the pre-trained model can be effectively improved.

Keywords: Text augmentation · Pre-trained model · Quality assessment

1 Introduction

In recent years, machine learning and deep learning have achieved high accuracy in many NLP tasks such as intent understanding [5] and question matching [7],

B. Qin et al. (Eds.): CCKS 2021, CCIS 1466, pp. 198–210, 2021.
https://doi.org/10.1007/978-981-16-6471-7_15

but high performance often depends on the size and quality of training data. Preparing a large annotated dataset is very time-consuming and it is a big challenge for NLP tasks in many fields. We can compare the medical field which usually has small instances with the general field which usually has big instances. The medical intention understanding dataset CMID [2] has 12,254 instances, which are 16.2% of the general field dataset THUCNews [7] (74,000), and its performance on the BERT model is only 76.33% of THUCNews (Acc: 67.63% vs. 88.6%). Similarly, the data size of the medical question matching dataset CHIP2019 is 2.7% (20,000 vs. 25,000) of the general field dataset LCQMC [7], and its performance is only 79.19% of LCQMC (Acc: 68.9% vs. 87%).

In order to reduce the impact of instances scarcity, we often use methods such as data augmentation [12], robust learning [18], and the pre-trained model [4]. Among them, text data augmentation and pre-trained models are two of the most commonly used methods to solve this problem in NLP due to its usability and wide applicability. People can usually observe the improvement of the effect brought by the text data augmentation when it is suitable for the task: such as back translation [14] of machine translation and easy data augmentation (EDA) [11] in text classification tasks.

Data augmentation has proven to be widely effective in computer vision, as well as on pre-trained models. In NLP, similar results are most common in the context of low-data modes, non-pretrained models. However, Shayne Longpre [8] conducted experiments on the effectiveness of the data augmentation method under the pre-trained model, and the results were not very satisfactory. They used two data enhancement methods, EDA and BT, and systematically checked their effects across 5 classification tasks, 6 data sets, and 3 variants of modern pre-trained transformers, including BERT, XLNET, and ROBERTA. Experimental results demonstrate that improvements brought by text data augmentation methods on pre-trained models are marginal for 5 of the 6 datasets, and the average improvement is not more than 1%. In the performance on the remaining data set, 2 of the 3 pre-trained models are mostly negative results.

In order to fully understand the reasons for this result, we must make a careful analysis of the augmented instances. **The most challenging** aspect is to analyze what the augmented text features are and how to make better use of them on the pre-trained model. This paper first analyzes the quality of augmented instances from two perspectives of label-independent and label-dependent. And we found that the augmented instances had text noise, low confidence, labeling errors, and other issues as shown in Fig. 1. Therefore, in order to get a better text data augmentation result, we propose to analyze, filter, and transform the augmented instances. In specially, We filter low-quality instances, select high-confidence and high-quality instances and transform labels for low-confidence and high-quality instances.

In the General Field, our experiments on the LCQMC dataset show that our method can effectively prevent the risk that text augmentation may reduce the fine-tuned performance. At the same time, our method can also improve the fine-tuning performance on the pre-trained model. Specifically, when using 30% augmented instances, our approach outperforms the fine-tuned model without

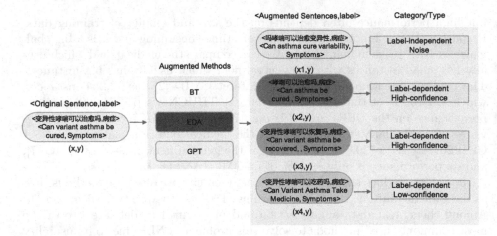

Fig. 1. Quality analysis and types of augmented text instances.

text augmentation by 0.42% and outperforms the fine-tuned model with all augmented instances by 1.21% on average. **In the specialized field (medicine)**, experiments on two public datasets (CMID and CHIP) show that our method can effectively improve the fine-tuned performances on pre-training models. On the entire intent understanding dataset CMID, our method is 1.76% better than the fine-tuned model with 100% augmented instances when using 30% augmented instances. On the entire question matching dataset CHIP, our method is 1.8% better than the fine-tuned model with 100% augmented instances when using 30% augmented instances. In the small sample scenario, the improvement brought by our method is even greater. On the CMID intention understanding dataset, the 30% augmented instances obtained by using our methods in a small sample scenario is 8.04% better than the fine-tuned model without text augmentation. On the CHIP question matching dataset, the 30% augmented instances obtained by using our methods in a small sample scenario is 12.33% better than the fine-tuned model without text augmentation. We also conducted experiments on two other common text augmentation methods (BT [14] and GPT [19]), and the experiments proved that our methodS are also very effective.

In a word, our main contributions are as follows:

- After observing the poor effect of general data augmentation methods under the pre-trained model, we analyzed the quality of text data augmented by general data augmentation methods and classified it into noise data, high-confidence data, and low-confidence data.
- We utilize three methods to evaluate the quality of augmented instances, and use different operations (select/filter and transform) according to different types of augmented instances.
- We demonstrate that our method shows hope to achieve significant results on question matching and intent understanding tasks on our benchmarks, especially in the small sample scenario.

2 Related Work

Text data augmentation technologies can be roughly divided into two categories: label-dependent and label-independent. The difference between the two is whether it needs to rely on label information during the augmentation process.

Label-Independent: It does not need to rely on label information in the enhancement process. The main methods are synonym substitution, back translation [14], mixup [16], and unconditional text generation. Methods based on synonym substitution include wordnet [17] and EDA [11]. EDA [11] is a recent data augmentation method containing a set of 4 text augmentation techniques, including synonym replacement, random insertion, random swap, and random deletion. Back translation [14] is a method of translating the original text into other languages and then translating back to obtain a new expression in the original language. Similar to the word replacement method, the augmented instances generated by back translation have the same semantics as the original data as much as possible. The well-known machine reading comprehension models QANet [15] and UDA [14] both use back translation technology for data augmentation. Mixup [16] is a recently proposed data augmentation method through linearly interpolating inputs and modeling targets of random samples. There are wordMixup and sendMixup to mix word vectors and sentence vectors in NLP. Generative augmentation methods utilize generative methods like GPT2 [10] to generate new data to achieve the purpose of data amplification.

Label-Dependent: The augmentation process requires label information, which can usually be divided into deep generative models and pre-trained models. When we need to introduce label information for data augmentation, we may think of CAVE [9] for the first time. CAVE is a common generative model. However, to generate high-quality enhanced data, a sufficient amount of annotation is often required which contradicts the premise of our small sample dilemma. CBERT [13] can generate new samples by using the class label and a few initial words as the prompt for the model. For example, we can use 3 initial words of each training text and generate one synthetic example for each point in the training data. lambda [1] uses GPT-2 [10] to splice the label information and the original text as training data for finetune, and also uses a discriminator to filter and reduce noise on the generated data.

3 The Proposed Method

3.1 Analysis of Augmented Instances

In response to the problems raised above, we analyzed and studied the augmented instances, and summarized it into 2 categories (label-dependent/label-independent) and 3 types of data. The three types are label-independent noise data, label-dependent high-confidence data, and low-confidence data. Noise data means that the augmented text does not completely conform to the text structure of a normal sentence. In this paper, we use two methods to determine whether it is noise data. High confidence means that the text of the augmented instances is semantically consistent with the label. That is to say, the label predicted based on the augmented text has a high degree of similarity to the original label without augmentation. Low-confidence data is the opposite of high-confidence data.

3.2 Data Quality Assessment Methods

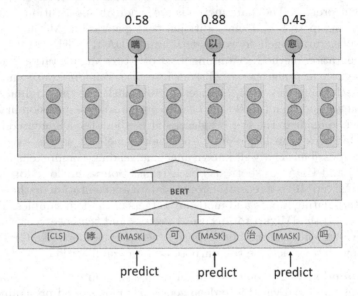

Fig. 2. Augmented text data quality assessment with BertLM method.

To better classify the augmented instances and quantify the quality of the enhanced data, we used the label-independent method and the label-dependent method to conduct experiments. The label-independent method is to evaluate the degree of noise data in the augmented instances. For example, we can use a language model to evaluate whether the input sentence is a complete sentence with smooth semantics. The label-dependent method is to evaluate whether the augmented text is consistent with the original label. For example, we can evaluate whether the augmented text is a high-confidence instance. The label-independent methods we use in this article are BertForMaskedLM (BertLM) and Probabilistic Context-Free Grammars (PCFG).

BertLM: The BertLM method picture is shown in Fig. 2. Equation (1) is the formula. When inputting the sentence S, we randomly mask t words W_i in positions i ($0 \leq i \leq n - 1$, n is the length of S). Then we can get the predicted words. By comparing the similarity $P(W_i)$ between the predicted word and the original word, we can know whether the original sentence is noise. As shown in Fig. 2, the score of the sentence is $Score_{bert} = \frac{1}{3} * (0.58 + 0.88 + 0.45) = 0.64$.

$$Score_{bert} = \frac{1}{t} * \sum_{i=0}^{n-1} P(W_i) \tag{1}$$

PCFG: The phrase structure analysis method based on probabilistic context-free grammar can be said to be the most successful grammar-driven statistical syntax analysis method at present, and it can be considered as a combination of the rule method and statistical method. Given a sentence, PCFG can estimate the probability of producing the sentence, so PCFG can be used as a language model. The method used in this article comes from the PCFG parser [6] of Standford. The formula 2 is to use PCFG to calculate the probability of the parse tree t, and r is the node of the tree.

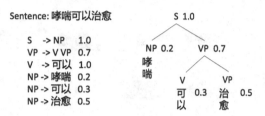

Fig. 3. Augmented text data quality assessment with PCFG method.

$$P(t) = \prod_{i=1...n} P(r_i) \tag{2}$$

Let us use a simple example to explain the process of the PCFG method. The PCFG rule set and probability is the left part in Fig. 3. The parse tree t and the corresponding probability values for the example sentence are the right part of Fig. 3. So the probability of t is $P(t) = 1.0 \times 0.2 \times 0.7 \times 0.3 \times 0.5 = 0.021$. If there are multiple parse trees for the same sentence, we can add up to get the final value. This value can be regarded as the result of syntactic analysis of the sentence.

Label-Dependent Confidence-Based Model: Because the method needs to use the label of the data, it is a method of the label-dependent class. Let (x_i, y_i^*) be the pair of the sample x_i and its true label y_i^*, and $D = \{(x_i, y_i^*)|1 \le i \le N\}$ be the training data set. D is the original training set without augmentation. Because D is a data set labeled by dedicated persons, we approximate it as a clean and noise-free data set. First, we use D as input of the Bert to train a model which is called the confidence model M_{conf}. We believe that a model trained with clean samples may have a certain ability to predict the true label of the data. Based on this, we use the M_{conf} model to predict the augmented instances' labels and get the confidence score, where the function of obtaining the score is f. Equation 3 is the formula of this method.

$$Score_{conf} = f(M_{conf}(x_{aug}, y_{aug}^*)) \tag{3}$$

3.3 Choice Strategy of Augmented Instances

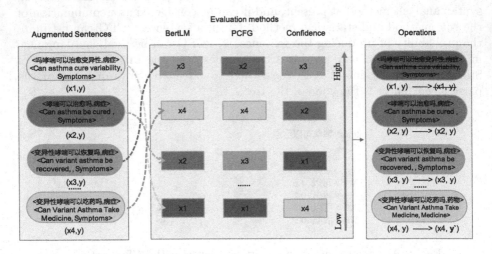

Fig. 4. The quality assessment process and choice strategy of augmented text instances.

For the augmented instances, we adopt the strategy of 'analyze, filter and transform' to better augment the data. After the previous analysis of the augmented text data, the augmented text data can be divided into noise data, low-confidence data, and high-confidence data. As shown in Fig. 4, we use label-dependent and label-independent methods to evaluate the quality of the augmented instances and rank them according to the score.

- **Analyse:** First, we evaluate the augmented text through BertLM and PCFG, which are label-independent methods. Through the score of the method, we can know whether it is a grammatically normal and complete sentence, that is, whether it is noise data.
- **Select/Filter:** Then we filter out the data with the lowest score (most likely to be noise data) and keep the data that is least likely to be noise data. For example, we drop the x_1 data where BertLM and PCFG scores are relatively low in Fig. 4. We keep the x_2, x_3, x_4 data which have higher scores. In order to know more precisely how much data is appropriate to keep and remove, we also did a series of experiments from the 10%, 30%, 50%, etc. of the augmented text data.
- **Transform:** At last, for non-noise augmented instances, we use the label-dependent method of confidence to evaluate how well the prediction result matches the original label (the label of the original data before augmentation). We keep the augmented instances with high confidence and change the label of the data with low confidence and non-noise. For example, x_4 in Fig. 4 has low confidence but non-noise, we change its label from y to y'.

4 Experiment

4.1 Dataset

CMID [2]: It is a dataset of Chinese medical intention understanding. It includes four types of user intentions and 36 sub-types of user intentions. CMID contains 12,254 Chinese medical questions. Each question is marked with intent.

CHIP: This data set comes from the CHIP-2019 Ping An Medical Technology Disease Quiz and Migration Learning Competition. The main goal of this evaluation task is to carry out migration learning between disease types based on Chinese disease question and answer data. Specifically, given question pairs from 5 different diseases, it is required to determine whether the semantics of the two sentences are the same or similar.

LCQMC [7]: LCQMC is a general Chinese question matching dataset published by HIT. The dataset has a total of 260068 pairs of annotation results, divided into three parts, 238766 training set, 8802 validation set, and 12500 test set.

Dataset Split: We divide the CMID data into the train set, validation set, and test set according to 8:1:1. The proportions of various labels are consistent with the original data set. We take the same way to divide CHIP and LCQMC.

4.2 Experimental Setup

We compared the results of different Chinese pre-trained models on different tasks and datasets. Based on the results of Table 1, we finally chose the pre-trained model BERT-wwm [3] as our baseline. We evaluate the performance of our methods by adding the choice strategy to the baseline.

Table 1. The results of different Chinese pre-trained models on different tasks and datasets.

Dataset	LCQMC		THUCNews	
Model	dev	test	dev	test
BERT	88.4	86.4	97.4	97.6
BERT-wwm	**89.2**	**86.8**	**97.6**	**97.6**
RoBERTa-wwm-ext	88.7	86.1	97.5	97.5

We choose EDA [11], BT [14] and GPT2-ML (GPT) [19] as the general augmentation methods to augment the text first. Based on the recommendation of EDA, we set the number of expanded sentences generated by each original sentence, $n_{aug} = 4$. In other words, we make each input text into five sentences. For BT, we use Baidu Translation API[1] and adopt the strategy of *Chinese* →

---

[1] http://api.fanyi.baidu.com/.

Table 2. The effect of different k values on the GPT text data augmentation experiment.

no-aug	aug		
67.63	K = 3	K = 7	All words
	67.5	68.37	68.61

English → *Chinese* for text data augmentation. For GPT, we use the first k words of the original data to generate text. As shown in Table 2, we have experimented with $k = \{3, 7, all\ words\}$ and found that the performance is best when all the original data is input. The reason for this result may be that most of our original texts are short text. Therefore, we directly use the original text as input for amplification. We use PCFG, BertLM, and confidence methods to score and sort the enhanced data. We select, filter, and transform the results of the three text data augmentation methods, and choose the fractions (%): {10, 30, 50, 100} to add into the original data as input of the model for comparison.

In order to verify whether our method is still effective under a small sample, we specially selected a small sample scene to conduct the same experiment.For CMID, we sample 10% of the training set as a small sample scene. For CHIP, we choose 10% of the samples in the training set whose disease category is AIDS as the small sample scenario.

4.3 Results and Discussion

First, we use the text data augmentation methods to augment all data of these datasets. Then we use PCFG, BertLM, and confidence methods to evaluate the quality of the augmented instances. After that, we adopt selection/filtering, and transform operations, and then select different proportions of processed data {0, 10%, 30%, 50%} to add to the input data for experiments.

Result on General Field Dataset LCQMC with all Instances Dataset: We have conducted experiments on the generic field and the special field (medical), and Table 3 shows the result on the generic LCQMC dataset. From Table 3 we can see that: 1) There is a risk that the text augmentation combined with the pre-trained model on the generic dataset may degrade the experimental accuracy 2) Our approach is not only effective in preventing this, but also in further improving the model. Specifically, when using 30% augmented instances, our approach has an average performance of 0.43% better than the fine-tuned model without text augmentation and outperforms the fine-tuned model with all augmented instances by 1.22%.

Table 3. Results of using augmented instances quality analysis and choice strategies after the EDA text augmentation method on LCQMC with all instances dataset.

Dataset	Methods	The ratio of augmented instances				
		0	10%	30%	50%	All
LCQMC	PCFG	87	87.2 (+0.2)	**87.36 (+0.36)**	87.02 (+0.02)	86.21 (−0.79)
	BERTLM		87.28 (+0.28)	**87.5 (+0.5)**	87.14 (+0.14)	
	Confidence		87.35 (+0.35)	**87.42 (+0.42)**	87.22 (+0.22)	

Result on Medical Field CMID and CHIP with all Instances Dataset:
From the Table 4, we can observe that: 1) The addition of all the augmented text data will make the accuracy improvement on the intent understanding task and question matching task very limited (CMID: +0.83) or even counterproductive (CHIP: −0.05). The above experimental results are consistent with the research of Shayne Longpre [8]. After our quality analysis of the augmented instances, we found that one of the most important reasons may be the uneven quality of the enhanced text data. 2) More high-quality but less total augmented instances are better than all augmented instances. On the entire intent understanding dataset CMID, our method is 0.43% and 0.74% better than the fine-tuned model with 100% augmented instances when using 10% and 30% augmented instances. On the entire question matching dataset CHIP, our method is 0.30% and 1.57% better than the fine-tuned model with 100% augmented instances when using 10% and 30% augmented instances. Our experiments on two public datasets show that through analysis, selection/filtering, and transformation augmented instances, the performance of intent understanding and question matching in pre-trained model can be effectively improved.

Table 4. Results of using augmented instances quality analysis and choice strategies after the EDA text augmentation method on CMID and CHIP with all instances dataset.

Dataset	Methods	The ratio of augmented instances				
		0	10%	30%	50%	All
CMID	PCFG	72.52	74.78 (+2.26)	**75.11 (+2.59)**	73.11 (+0.59)	**73.35 (+0.83)**
	BertLM		73.51 (+0.99)	**73.67 (+1.15)**	72.88 (+0.36)	
	Confidence		72.77 (+0.52)	**73.48 (+0.96)**	73.42 (+0.90)	
CHIP	PCFG	86.00	86.15 (+0.15)	**87.60 (+1.60)**	86.15 (+0.15)	85.59 (−0.05)
	BertLM		87.12 (+0.12)	**87.75 (+1.75)**	86.45 (+0.45)	
	Confidence		86.49 (+0.49)	**87.21 (+1.21)**	86.05 (+0.05)	

Result on Medical Field CMID and CHIP with Small Instances Dataset: Table 5, Table 6 and Table 7 are the results of small sample scenes we experimented with EDA, BT and GPT text data augment methods. The

Table 5. Results of using augmented instances quality analysis and choice strategies after the EDA text augmentation method on CMID and CHIP small instances dataset.

Dataset	Methods	The ratio of augmented instances					
		0	30%	50%	70%	90%	All
CMID	PCFG	67.63	72.54 (+4.91)	68.42 (+0.79)	72.54 (+4.91)	**75.89 (8.26)**	73.33 (+5.70)
	BertLM		**75.67 (+8.04)**	71.76 (+4.13)	74.33 (+6.70)	73.55 (+5.92)	
	Confidence		66.63 (−1.00)	70.76 (+3.13)	**74.11 (+6.48)**	72.77 (+5.14)	
CHIP	PCFG	68.90	79.14 (+10.24)	75.52 (+6.62)	**80.82 (+11.92)**	77.42 (+8.52)	80.59 (+11.69)
	BertLM		**81.23 (+12.33)**	76.52 (+7.62)	77.25 (+8.35)	74.52 (+8.52)	
	Confidence		80.45 (+11.55)	**81.56 (+12.66)**	75.52 (+6.62)	74.25 (+5.35)	

purpose of these experiments is to verify whether our method is still applicable to small sample scenarios and other data augmentation methods.

From the Table 5, we can observe that: 1) The methods proposed in this paper can improve the performance of intent understanding and question matching in pre-trained model. Specifically, On the intent understanding dataset CMID in a small sample scenario, the best accuracy of our methods is 2.56% better than the 100% data when using part of the augmented instances (90%). On the question matching dataset CMID, the best accuracy of our methods is 0.97% better than the 100% data when using part of the augmented instances (50%). 2) In the CMID and CHIP small sample scenarios, after using our methods to obtain 30% augmented instances and adding them to the training set, the performance on the fine-tuned pre-trained model is 8.04% and 12.33% better than before the augmentation. In the same situation, the improvement of all CMID and CHIP data is 1.57% and 1.52%, which is far less than in the small sample scenario. The reason may be that the pre-trained model lacks sufficient training data in small sample scenarios. The text data augmentation methods increase the amount of data, and our methods can further improve the quality of the augmented text.

To fully verify the effectiveness of our method, we also conducted the same experiment on BT and GPT, which are two other text data augmentation methods. From Table 6 and Table 7, we can observe that: Our methods are effective in both BT and GPT data augmentation methods. The BT effect is slightly worse than the EDA method, and the GPT effect is the least obvious. Specifically, GPT's best performance is 2.45% and 2.28% less than EDA and BT on the CMID dataset, and 5.33% and 5.29% on the CHIP dataset. The possible reason is that GPT for data augmentation is hard to control and may not produce the results we want, which is a challenge and should be given more attention in future work.

Table 6. Results of using augmented instances quality analysis and choice strategies after the BT text augmentation method on CMID and CHIP small instances dataset.

Dataset	Methods	The ratio of augmented instances					
		0	30%	50%	70%	90%	All
CMID	PCFG	67.63	**73.42 (+5.79)**	69.15 (+1.52)	72.15 (+4.52)	73.34 (5.71)	73.21 (+5.58)
	BertLM		71.24 (+3.61)	**75.17 (+7.54)**	73.35 (+5.72)	73.18 (+5.55)	
	Confidence		68.65 (+1.02)	71.58 (+3.95)	**73.52 (+5.89)**	72.48 (+4.85)	
CHIP	PCFG	68.90	78.16 (+9.26)	**81.02 (+12.3)**	77.45 (+8.55)	77.12 (+8.22)	80.51 (+11.61)
	BertLM		**80.85 (+11.95)**	77.51 (+8.61)	78.36 (+9.46)	75.15 (+6.25)	
	Confidence		80.52 (+11.62)	**81.52 (+12.62)**	76.33 (+7.34)	74.85 (+5.95)	

Table 7. Results of using augmented instances quality analysis and choice strategies after the GPT text augmentation method on CMID and CHIP small instances dataset.

Dataset	Methods	The ratio of augmented instances					
		0	30%	50%	70%	90%	All
CMID	PCFG	67.63	**72.75 (+5.12)**	71.05 (+3.42)	70.15 (+2.52)	70.89 (3.26)	71.54 (+3.91)
	BertLM		71.75 (+3.88)	**71.88 (+4.25)**	71.23 (+3.60)	70.95 (+3.32)	
	Confidence		72.65 (+5.02)	**73.44 (+5.81)**	72.48 (+4.85)	71.22 (+3.59)	
CHIP	PCFG	68.90	**76.14 (+7.24)**	74.82 (+5.92)	73.52 (+4.62)	74.42 (+5.52)	73.25 (+6.35)
	BertLM		**76.23 (+7.33)**	75.52 (+6.62)	74.25 (+5.35)	74.46 (+5.56)	
	Confidence		74.45 (+5.55)	75.26 (+6.36)	**76.16 (+7.26)**	75.03 (+6.13)	

5   Conclusion

In this paper, we have analyzed the reasons why data augmentation technologies are not obvious or even counterproductive for improving the fine-tuned model on NLP tasks (especially intent understanding and question matching tasks). We use three methods to evaluate the quality of the augmented text data. We have conducted separate experiments on the general and specialized (medical) fields. In the medical field, we conducted detailed experiments and set up a small sample scenario to further verify the effectiveness of our method. Our experimental results demonstrate that analyzing the augmented instances quality, making reasonable choices, and transform low-confidence high-quality instances can get better performance on the pre-trained model.

Acknowledgement. This work was supported by the Science and Technology Program of the Headquarters of State Grid Corporation of China, Research on Knowledge Discovery, Reasoning and Decision-making for Electric Power Operation and Maintenance Based on Graph Machine Learning and Its Applications, under Grant 5700-202012488A-0-0-00. This work was also supported by the independent research project of National Laboratory of Pattern Recognition, the Youth Innovation Promotion Association CAS and Beijing Academy of Artificial Intelligence (BAAI).

References

1. Anaby-Tavor, A., et al.: Not enough data? deep learning to the rescue! arXiv:1911.03118 (2019)
2. Chen, N., Su, X., Liu, T., Hao, Q., Wei, M.: A benchmark dataset and case study for Chinese medical question intent classification. BMC Med. Inform. Decis. Making **20**(3), 1–7 (2020)
3. Cui, Y., Che, W., Liu, T., Qin, B., Wang, S., Hu, G.: Revisiting pre-trained models for Chinese natural language processing. In: Proceedings of the 2020 Conference on Empirical Methods in Natural Language Processing: Findings, pp. 657–668. Association for Computational Linguistics, Online, November 2020. https://www.aclweb.org/anthology/2020.findings-emnlp.58
4. Devlin, J., Chang, M.W., Lee, K., Toutanova, K.: Bert: Pre-training of deep bidirectional transformers for language understanding. In: NAACL-HLT (2019)
5. Hu, J., Wang, G., Lochovsky, F., Sun, J.T., Chen, Z.: Understanding user's query intent with wikipedia. In: Proceedings of the 18th International Conference on World Wide Web, pp. 471–480 (2009)
6. Klein, D., Manning, C.D.: Accurate unlexicalized parsing. In: Proceedings of the 41st Annual Meeting of the Association for Computational Linguistics, pp. 423–430 (2003)
7. Liu, X., et al.: Lcqmc: A large-scale Chinese question matching corpus. In: Proceedings of the 27th International Conference on Computational Linguistics, pp. 1952–1962 (2018)
8. Longpre, S., Wang, Y., DuBois, C.: How effective is task-agnostic data augmentation for pretrained transformers? arXiv:2010.01764 (2020)
9. Malandrakis, N., Shen, M., Goyal, A., Gao, S., Sethi, A., Metallinou, A.: Controlled text generation for data augmentation in intelligent artificial agents. arXiv:1910.03487 (2019)
10. Radford, A., Wu, J., Child, R., Luan, D., Amodei, D., Sutskever, I., et al.: Language models are unsupervised multitask learners. OpenAI Blog **1**(8), 9 (2019)
11. Wei, J., Zou, K.: Eda: Easy data augmentation techniques for boosting performance on text classification tasks. arXiv:1901.11196 (2019)
12. Wong, S.C., Gatt, A., Stamatescu, V., McDonnell, M.D.: Understanding data augmentation for classification: when to warp? In: 2016 International Conference on Digital Image Computing: Techniques and Applications (DICTA), pp. 1–6. IEEE (2016)
13. Wu, X., Lv, S., Zang, L., Han, J., Hu, S.: Conditional bert contextual augmentation. arXiv:1812.06705 (2019)
14. Xie, Q., Dai, Z., Hovy, E., Luong, M.T., Le, Q.V.: Unsupervised data augmentation for consistency training. arXiv: Learning (2020)
15. Yu, A.W., et al.: Qanet: Combining local convolution with global self-attention for reading comprehension. arXiv:1804.09541 (2018)
16. Zhang, H., Cissé, M., Dauphin, Y., Lopez-Paz, D.: mixup: Beyond empirical risk minimization. arXiv:1710.09412 (2018)
17. Zhang, X., Zhao, J., LeCun, Y.: Character-level convolutional networks for text classification. In: NIPS (2015)
18. Zhang, X., Wu, X., Chen, F., Zhao, L., Lu, C.T.: Self-paced robust learning for leveraging clean labels in noisy data. In: AAAI (2020)
19. Zhang, Z.: Gpt2-ml: Gpt-2 for multiple languages. https://github.com/imcaspar/gpt2-ml (2019)

**Knowledge Graph Applications:
Semantic Search, Question Answering,
Dialogue, Decision Support,
and Recommendation**

A Visual Analysis Method of Knowledge Graph Based on the Elements and Structure

Qiying He[1,2]([✉]), Wenjun Hou[1,2], and Yujing Wang[1,2]

[1] School of Digital Media and Design Arts, BUPT, Beijing, China
[2] Beijing Key Laboratory of Network Systems and Network Culture, Beijing, China

Abstract. As Knowledge Graphs (KG) are widely used in various industries, it is crucial to improve KG's readability and acceptance by the public. To assist the public to understand KG more comprehensively, we proposed a visual analysis method of KG and verified, evaluated, and applied the method. First, we extracted the visualization analysis dimensions of KG from its essential elements. Secondly, supported by the Beijing cultural KG data, we verified the algorithm of each analysis dimension. Finally, we evaluated the usability of this method through user experiments and verified the feasibility of this method by implementing a visual analysis application of KG. The results show that the method can effectively assist users in understanding KG from multiple angles.

Keywords: Knowledge graphs · Visual analysis · Usability evaluation

1 Introduction

KG describes entities and their relationships in graphs, which has been widely applied to many intelligent fields [1, 2]. However, the current KG has problems such as the single form of expression, high barriers to understanding, and single interpretation dimension, which bring some challenges to the universality of KG.

To solve those problems, this paper mainly makes the following contributions. Firstly, we explore the regular composition and structure of KG from its essential elements. Secondly, we extracted and summarized the visualization analysis dimensions and method of KG. Finally, to verify the usability and feasibility of this method, we conducted user experiments and developed a visualization analysis system of KG. In general, this method aims to lower the understanding threshold of KG by expressing it in a better-understood visual form and allowing KG researchers to understand KG from multiple angles more comprehensively.

2 Related Research

2.1 Visual Representation of the Knowledge Graph

Davenport [3] first proposed to visualize knowledge relations through KG. At present, the visual expression of KG mainly includes five categories: space-filling, node-link

Foundation item: Social Science Foundation of Beijing (18ZDA08).

graph, heat map, adjacency matrix, and others. People usually use a node-link graph combined with a classical force-oriented layout algorithm to visualize KG [4].

Although a node-link diagram can intuitively display the network structure of entities and relations, it leads to a low utilization rate of the display space. When the amount of data is too large, the dense KG will cause visual pressure.

2.2 Visualization Analysis of Knowledge Graph

G Qiu [5] summarized the development and hotspot evolution in database construction relying on the visual analysis of bibliometrics and CiteSpace. Weng [6] used two crowdsourcing methods to generate a KG of the node-link graph. Liang [7] concluded that visualization could help understand data more intuitively and conveniently. Wang [8] designed a general visualization analysis system oriented to KG. However, it only expresses KG by the node-link graph. Yang [9] designed a visualization expression service system for exploratory analysis of KG data. However, it was limited to the visualization of the vertical KG and lacked standardized analysis of general one.

At present, most people interpret KG by focusing on the narrow concept [10]. They construct field KG [11] by self-organizing or using tools such as CiteSpace, VosViewer [12, 13], and methods like citation and co-word analysis [14]. Most KG research focuses on node-link graphs or time zone graphs. People are limited to analyzing KG form development trends, partnerships, and research hotspots, which is not conducive to discovering other analytical dimensions and interpreting KG in depth.

Visualization can effectively convey our understanding of data [15]. Therefore, we combine various information charts to construct a general visualization analysis method to improve the people's understanding of KG.

3 Method

3.1 Extraction of Analysis Dimension of the Knowledge Graph

To combine data visualization to understand KG better, we have to start from the structure of the data itself. According to the definition of KG, its two core elements are Entity and Relationship. Therefore, extracting the composition and structure between Entity and Relationship is necessary for analyzing and understanding KG. This section will explore the basic structure of KG and extract its analytical dimensions.

Let us call the collection describing the same type of entities as a dimension. Entity elements include one-dimensional Entity, two-dimensional Entity, three-dimensional Entity, and high-dimensional Entity. The composition form of Relationship elements will show different forms according to the different Entity dimensions. We will discuss in-depth in the subsequent analysis.

The core idea of permutation is to list all possible elements according to requirements [16]. To explore and enumerate the possible structures between Entity and Relationship, we extracted several feasible basic structures of KG through permutation (see Fig. 1). We will discuss the following three issues in detail: 1. The feasibility of these basic structures; 2. The representative visual expressions that these structures can produce; 3. The scope of application of these visual expressions.

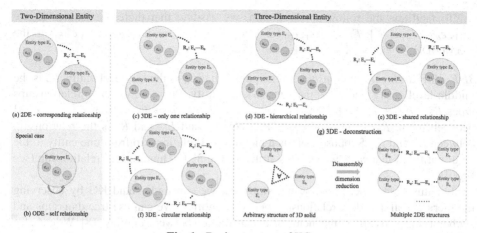

Fig. 1. Basic structure of KG

One-Dimensional Entity (ODE). There are only two possible structures for ODE: 1. There is no relationship within ODE. The KG of this structure will appear as a series of scattered points. This structure rarely appears in practical applications, so there is not much visual significance. Therefore, we will not repeat this situation in the subsequent discussion. 2. ODE has a relationship with itself (see Fig. 1b). We can think that the essence of this structure is the relationship structure of a two-dimensional Entity (2DE) with equivalent dimensions. Therefore, we regard this situation as a special case of 2DE relationship structure and analyze it through the subsequent 2DE method.

Two-Dimensional Entity (2DE). 2DE can only have a corresponding relationship $R_x : E_a \sim E_b = \{r_{x1}, r_{x2}, \ldots, r_{x|R|}\}$ between one-dimensional entity $E_a = \{e_{a1}, e_{a2}, \ldots, e_{a|E|}\}$ and the other-dimensional entity $E_b = \{e_{b1}, e_{b2}, \ldots, e_{b|E|}\}$ (see Fig. 1a). In the KG of this structure, we can use one-dimensional entities as dimensional variables and the other-dimensional entities as measurement variables. In this way, we can understand the measurement relationship of 2DE through traditional visualization measurement diagrams. For example, we can use line charts or histograms to analyze the fluctuations and proportions of measurement entities on dimensional entities [17].

Three-Dimensional Entity (3DE). 3DE has diverse relationship structures. After discussing through permutation, we can get 1. Only one relationship type cannot connect all 3DE (see Fig. 1c), which does not apply to 3DE, so we will not discuss it later; 2. Two relationship types connecting 3DE may have two relationship structures: (see Fig. 1d and 1e): hierarchical relationship and shared relationship; 3. Three relationship types connecting 3DE can only have one feasible relationship structure (see Fig. 1f): circular relationship; 4. We can interpret 3DE by disassembling and reducing the above-mentioned relational structure (see Fig. 1). We can conclude that 3DE has the following relationship structures: hierarchical relationship, shared relationship, circular relationship, and deconstruction.

To facilitate the follow-up discussion, we assume that there are entities $E_a = \{e_{a1}, e_{a2}, \ldots, e_{a|E|}\}$, $E_b = \{e_{b1}, e_{b2}, \ldots, e_{b|E|}\}$, and $E_c = \{e_{c1}, e_{c2}, \ldots, e_{c|E|}\}$ in the KG of 3DE:

3DE - Hierarchical Relationship. We stipulate that E_a is the top-level entity, E_b is the middle-level entity, and Ec is the bottom-level entity. There may be two relationships types $R_x : E_a \sim E_b = \{r_{x1}, r_{x2}, \ldots, r_{x|R|}\}$ and $R_y : E_b \sim E_c = \{r_{y1}, r_{y2}, \ldots, r_{y|R|}\}$ between them. R_x is the relationship between E_a and E_b, and R_y is the relationship between E_b and E_c. Suppose there is a hierarchical relationship from one entity to the other entity. In that case, we call this structure the 3DE - hierarchical relationship (see Fig. 1d).

For this hierarchical relationship structure, we can understand KG by observing the hierarchical and flow relationship between entities. We can use tree diagrams and sunburst charts to express the inclusion and affiliation between entities [18]. We can also use Sankey diagrams to reflect adjacent objects' relationship and data flow [19].

3DE - Shared Relationship. We stipulate E_a is a public entity, and E_b and E_c are edge entities. There may be two relationships types $R_x : E_a \sim E_b = \{r_{x1}, r_{x2}, \ldots, r_{x|R|}\}$ and $R_y : E_a \sim E_c = \{r_{y1}, r_{y2}, \ldots, r_{y|R|}\}$ between them. R_x and R_y are the relationship between E_a and E_b and between E_a and E_c, respectively. If two different entities have a relationship with the same entity, we call it the 3DE - shared relationship (see Fig. 1d).

We can use two edge entities as dimensional variables and the public entity as a measurement variable for this shared relationship structure. Then we can analyze and display the public entity from the edge entity dimension. This structure is very similar to the corresponding structure of 2DE, so we can still consider using traditional measurement diagrams for visual expression. The only difference is that we need to upgrade the traditional 2d coordinate structure to a 3d. For example, we can use 3d histograms or polar diagrams to observe the probability distribution of public entity [20].

3DE - Circular Relationship. We stipulate that there are 3DE E_a, E_b, E_c, and there may be relationships $R_x : E_a \sim E_b = \{r_{x1}, r_{x2}, \ldots, r_{x|R|}\}$, $R_y : E_b \sim E_c = \{r_{y1}, r_{y2}, \ldots, r_{y|R|}\}$, and $R_z : E_c \sim E_a = \{r_{y1}, r_{y2}, \ldots, r_{y|R|}\}$ between them. If there is a closed-loop relationship between 3DE, we call it the circular relationship structure of 3DE (see Fig. 1f).

For this circular relationship structure, it is difficult to find other perspectives to interpret the KG to ensure that all entities and relationships are displayed. Therefore, we can only convert a better-understood visualization form to show all entities and relationships. For example, we can use chord diagrams to show internal associations and transition differences [19] and then explore the structural relationships of KG.

3DE - Deconstruction. In addition, we can interpret 3de by disassembling and reducing the above-mentioned structures (see Fig. 1g). For example, we can reduce or dis-assemble the hierarchical relationship or sharing relationship of 3DE into correspond-ing relationships of 2DE. However, we need to combine different fields to determine whether it is necessary to deconstruct and the way of deconstructing.

In addition, 3DE may have relationships with themselves, or there are more than three relationship types, which are no longer the basic structures discussed above. Therefore, we can also understand complex structures in 3DE through deconstruction.

High-Dimensional Entity (HDE). When the entity exceeds three dimensions, more complex structures will appear. However, the above analysis of low-dimensional structure is also applicable to entity structures higher than three-dimensional. For example, high-dimensional entities may also have hierarchical, shared, and circular relationships, so we can still use the above methods to analyze high-dimensional KG. More complex KG is beyond the scope of the basic structure we are discussing. How-ever, we can still use deconstruction to convert them into multiple basic structures of 2DE and 3DE, then combine multiple small KG to understand complex network.

3.2 Visualization Analysis Methodology of Knowledge Graph

So far, we have extracted the basic structure of KG that can be composed of different dimensions. Each basic structure corresponds to an analysis dimension of KG. Therefore, we can summarize visual expression and analysis method of KG (see Table 1).

Table 1. Analysis and summary of the basic structure of KG

Basic structure	Visual expression	Analytical method
2DE - corresponding relationship [special case: ODE - self-relationship]	Traditional measurement graphs such as line graphs, bar graphs, and pie graphs	Use one type of entity as a dimensional variable and another type of entity as a measurement variable
3DE - hierarchical relationship	Tree diagram, sunburst diagram, Sankey diagram and other tree-like or flow layouts	Observe the containment or flow relationship between the top-level entity and the bottom-level entity
3DE - shared relationship	Polar diagram, 3d histograms, other 3d measurement graphs	Analyze and display the public entity from the edge entity dimension
3DE - circular relationship	Node relationship graphs such as chord diagrams	Analyze internal associations and transition differences
3DE - deconstruction	A variety of low-dimensional visual expressions	Compare multiple low-dimensional analysis angles
HDE	Low-dimensional combined visual expression	Analyze the combination structure of 2DE and 3DE

With the above basic structure, we should do as follows when we understand KG. First, we should determine the composition structure of KG, including its entity dimension, the relationship structure between entities, and the analysis method. Secondly, we can find the corresponding visual analysis expression in the above table, then determine the corresponding analysis method. From a visual point of view, the boring node-link expression form is enriched and diversified so that the KG users can understand the disordered KG more rationally and efficiently.

4 Algorithm and Verification

In the previous section, we extracted the analysis dimensions of KG by summarizing the entity-relationship structure and proposed a visual analysis method of KG. To further prove the feasibility of the visualization analysis of the KG and show the final visual analysis effect, we will propose and verify the data conversion algorithms of the above analysis dimensions.

Dataset is an important prerequisite for instance verification. Thanks to the support of the Beijing Ancient Capital Cultural KG Project, we have collected a large amount of data from Beijing's cultural fields such as Beijing opera, time-honored brands, architecture, traditional customs. We used these data to construct a KG about Beijing culture, which was finally stored in the neo4j database. In addition, since these data cover many different Beijing cultural fields, we can also use them to verify the universality of our proposed method in different fields. Next, we will verify the analysis dimensions summarized in Sect. 3.2 one by one:

4.1 2DE - Corresponding Relationship

From Sect. 3.1, we know that for the 2DE - corresponding relationship structure, we can use one entity type as a dimensional variable and another entity type as a measurement variable. Then we can use traditional measurement graphs such as line graphs and bar graphs to analyze KG. Therefore, we need to convert the graph data structure of KG into an array form conducive to drawing measurement graphs (see Fig. 2).

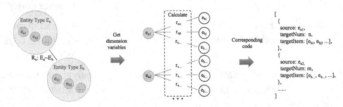

Fig. 2. Data conversion flow for 2DE - corresponding relationship structure

Processing Steps. First, obtain all entities of any entity type (such as E_a) in the 2DE from neo4j as dimensional variables. Secondly, perform measurement statistics on the correspondence between E_a and E_b, and the algorithm pseudocode is as follows (Table 2):

Table 2. Data conversion algorithm for 2DE - corresponding relationship structure

Algorithm 1. Two-Corres(entities)
For each entity in *entities*
Get R_x of the current entity through the cypher, and return all corresponding triples (*links*).
Use the current entity as the *source* and the number of *links* as the *targetNum* (the number of relationships related to the current entity).
For each triple in *links*
Store the tail entities of all triples (*targetItem*).
Store the *source*, *targetNum*, and *targetItem* in the array *Results*, which is the measurement relationship between the current and other entities.
Return *Results*

Since the 2DE - corresponding relationship structure is relatively basic, and we will use examples for verification in the subsequent discussion of 3DE - deconstruction.

4.2 3DE - Hierarchical Relationship

From Sect. 3.1, we know that for the 3DE - hierarchical relationship structure, we can use visual expressions such as tree diagrams and Sankey diagrams to observe KG from hierarchy and flow. So we need to transform the graph data structure (see Fig. 3).

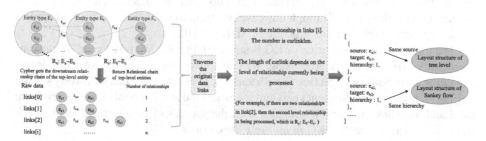

Fig. 3. Data conversion flow for 3DE - hierarchical relationship structure

Processing Steps. First, determine the top-level entity type E_a and the bottom-level entity type E_c. Secondly, determine the entire flow relationship from top to bottom according to the middle-level entities, and the algorithm pseudocode is as follows (Table 3):

Table 3. Data conversion algorithm for 3DE - hierarchical relationship structure

Algorithm 2. Three-Hier(topEntities)
For each entity in *topEntities*
Get all the downstream relationship chains (*links*) of the current top entity through cypher.
For each relationship chain in *links*
Process the *hierarchy*-level relationship (*hierarchy* is the number of relationships in the current relationship chain).
Use the tail entity of the *hierarchy*-1 relationship in the current relationship chain as the *source*, and the tail entity of the *hierarchy* relationship as the *target*.
Store the *source*, *target*, *hierarchy* in the array *Results*, which is the flow relationship set from the top to the bottom of the entity.
Return Results

After getting the flow relationship, we can put the same head node (*source*) items into an array. The *target* in the array is the next level entity of the head node. We can use these data to draw the visual expression of the tree layout, and we can also draw the visual expression of the Sankey flow layout based on the *hierarchy*.

Example Verification. We use data in the field of architecture in Beijing culture to verify. There are four-dimensional entities of "year", "architecture", "architectural elements" and "source of architectural elements", and there are three relations of "year - architecture", "architecture - architectural elements" and "architecture elements - source of architectural elements". Through this case, we want to show that this method applies to the hierarchical relationship structure of 3DE and HD.

The traditional representation of KG is shown in Fig. 4a. We take "year" as the top-level entity. After the above processing, we can convert the KG of this structure into a tree graph and a Sankey graph (see Fig. 4b and 4c). In this way, we can intuitively see the Beijing architectures of different years and see where the architectural elements originate from and their flow proportion structure with the sources of architectural elements. It can help readers to interpret KG in multiple dimensions.

4.3 3DE - Shared Relationship

From Sect. 3.1, we know that for the 3DE - shared relationship structure, we can use two edge entities as dimensional variables and public entities as measurement variables. Then we can analyze public entities from edge entities. Furthermore, we can understand the entire KG through a polar diagram or 3D histograms. Therefore, we need to rearrange the data structure based on edge entities (see Fig. 5).

Processing Steps. Firstly, obtain all entities of the two edge entities types E_b and E_c from database as two observation dimensions. Secondly, obtain all relationships in R_x called relX. Each relationship includes a head entity (public entity E_a) and a tail entity (edge entity E_b). Then, obtain all the relationships in R_y called relY. Each relationship

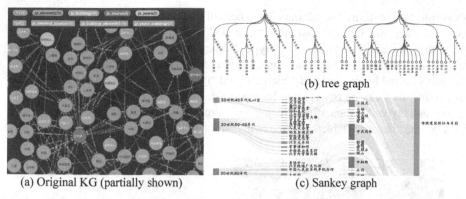

(a) Original KG (partially shown)   (b) tree graph

(c) Sankey graph

Fig. 4. Visual analysis results of 3D entity-hierarchy structure

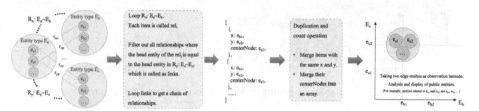

Fig. 5. Data conversion flow for 3DE - shared relationship structure

contains a head entity (public entity E_a) and a tail entity (edge entity E_c). Then associate edge entities through public entities, and the algorithm pseudocode is as follows (Table 4):

Table 4. Data conversion algorithm for 3DE - shared relationship structure

Algorithm 3. Three-Share(relX, relY)
For each relationship in *relX*
Filter out all relationships (*links*) where the head entity of the current relationship (*rel_i*) is equal to the head entity in *relY*.
For each relationship in *links*
x represents the tail entity of *rel_i*, y and *centerNode* represent the current relationship's tail entity and head entity, respectively.
Store x, y, *centerNode* in the array LinkGroup to connect edge entity with the public entity.
Return LinkGroup

Finally, the LinkGroup is deduplicated and counted to obtain the final data form [x, y, quantity]. For example, [e_{b1}, e_{c1}, 4] means that four public entities are related to the edge entities e_{b1} and e_{c1}, which are [$e_{ai}, e_{aj}, \ldots$].

Example Verification. We use data in the field of traditional customs in Beijing culture to verify. There are 3DE of "customs", "festivals" and "types", and two relationships of "customs - festivals" and "customs - types". They obviously satisfy the 3DE - shared relationship structure, whose traditional representation is shown in Fig. 6a. We regard "customs" as public entities, and "festivals" and "types" as edge entities. After the above processing, they can be represented by a polar diagram (see Fig. 6b). "festivals" is the radian, and "types" is the radius of concentric circles. We can clearly see the "delicious food" of the "lantern festival" is "rice ball", "dumplings" and so on.

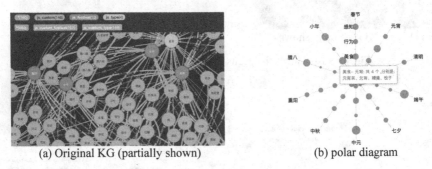

(a) Original KG (partially shown)   (b) polar diagram

Fig. 6. Visual analysis results of 3DE - shared relationship structure

4.4   3DE - Circular Relationship

From Sect. 3.1, to show all the entities and relationships in the circular structure, we can only interpret the KG by looking for better visualization, such as chord diagram. Therefore, we need to store entity and relationship data separately (see Fig. 7).

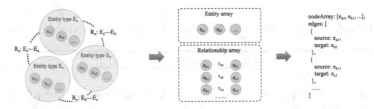

Fig. 7. Data conversion flow for 3DE - circular relationship structure

Processing Steps. Use the entity array to store all the entities, and use the relationship array to store all the relationships. Note that each item in the entity array must include "id", and each item in the relationship array must include source and target entities.

Example Verification. We use data of the time-honored brand in Beijing culture to verify. There are 3DE of "location", "brand", and "celebrity", and three relationships of "location - brand", "brand - celebrity", and "celebrity - location". They form a circular relationship structure (see Fig. 8a). After the above processing, we can get a chord diagram (see Fig. 8b). We also collected many plaques of the time-honored brand as nodes' background to facilitate readers' attention to the internal connection of KG.

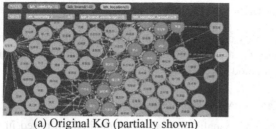
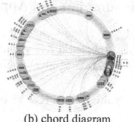

(a) Original KG (partially shown)   (b) chord diagram

Fig. 8. Visual analysis results of 3DE - circular relationship structure

4.5 3DE - Deconstruction

From Sect. 3.1, we can interpret the KG of the 3DE structure by disassembling and reducing the 3DE structure. However, this method is so flexible that we need to combine different fields to determine how to deconstruct, which means that it cannot perform unified data conversion. Therefore, we only use an example to verify this method.

Example Verification. In the field of Peking Opera, there are 3DE of "types", "actors", and "years", and two relationships of "types - actors" and "years - actors". The "types" has four entities: "Sheng", "Dan", "Jing", and "Chou". Based on the relationship of "types - actors" (see Fig. 9a), we can disassemble the KG of Peking Opera into four entity types: "Sheng actor", "Dan actor", "Jing actor" and "Chou actor" and their respective relationship with "years". In this way, this KG is disassembled into four KG of 2DE structure, so that the visualization expression of 2DE can be used for analysis. Finally, we use "year" as a public dimension to display these four KG on a 2d visualization expression (see Fig. 9b), then we can clearly observe the activity frequency and fluctuation of actors in various industries from the "years" dimension.

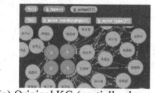
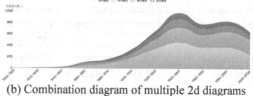

(a) Original KG (partially shown)   (b) Combination diagram of multiple 2d diagrams

Fig. 9. Visual analysis results of 3DE – deconstruction

4.6 HDE

From Sect. 3.1, HDE has more complex structures, and we can still disassemble them into a combination of 2DE and 3DE structures. However, since HDE is beyond this paper's scope on the basic structure of KG, we will not repeat them here.

5 Evaluation and Application

Finally, we first evaluated the usability of the method through user experiments. Then, we implemented a KG visualization analysis system based on the theoretical research and verification results to verify method's feasibility.

5.1 Evaluation of KG Visualization Analysis Methods

To verify the effectiveness of the KG visualization analysis method proposed in this paper, we recruited eight people with different knowledge of KG to conduct the information extraction task experiment [21]. The reviewer extracts information from the two equivalent visual representations based on the task then selects the most convenient and effective view. The accuracy difference (AD), completion time difference (TD), and satisfaction ratio (SR) of user information extraction were used for quantitative analysis. The experimental materials are all derived from the KG of Beijing culture. For example, "please refer to Fig. 4a and Fig. 4b to find the elements under a specific building". The specific tasks and results are shown in Table 5.

Table 5. Evaluate tasks and test results

KG structure	Control group	Tasks performed	AD	CTD/s	SR
2DE - corresponding relationship	Line chart - Node link diagram	Did the number of young actors go up or down from 1940 to 1980?	50%	−22.04	8:0
3DE - hierarchical relationship	Tree graph - Node link diagram	Find out the architectural elements of the Olympic Sports Center in the 1980s	25%	−7.19	7:1
3DE - shared relationship	Polar diagram - Node link diagram	What are the Lantern Festival delicacies in the picture?	37.5%	−14.27	7:1
3DE-circular relationship	Chord diagram - Node link diagram	Which time-honored brand have Lu Xun and Hu Shi been to?	0%	−0.175	4:4

The experimental results show that most AD is greater than 0. All CTD is less than 0. Most users are satisfied with the optimized view. We can conclude that almost all the visual expressions generated by the visualization analysis method proposed in this paper are more intuitive and diversified than the node-link graph. These results further verify the effectiveness of this method.

5.2 Application of KG Visualization Analysis Method

Relying on Web technology, we integrated the algorithm in Sect. 4 and implemented a web application for visualization analysis of KG. Users can use this application to complete all processes, from structural configuration to KG analysis. This application mainly includes data structure configuration and visual analysis board module.

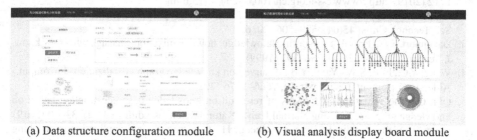

(a) Data structure configuration module   (b) Visual analysis display board module

Fig. 10. Application of KG visualization analysis

Data Structure Configuration Module. This module (see Fig. 10a) sets up the basic structure of KG. First, users can select the basic structure of KG and view the structure introduction, the corresponding visual information chart, and the analysis latitude. Secondly, users can drag and drop the previously uploaded entity and relation-ship, then put them in the corresponding position to customize the visualization analysis content of KG. After completing the configuration, users can enter the visual analysis display board module to start the detailed analysis.

Visual Analysis Display Board Module. We can manage the visualization form of KG generated by the application in this module (see Fig. 10b). Among them, we use 3D force-oriented layout to display the relationship structure between the entities of KG to relieve the visual burden brought by the traditional node-link graph. The upper part is the visualization diagram currently selected for analysis. The lower part is the thumbnail display of all the visualization charts corresponding to KG. Users can switch the display to analyze comprehensively.

6 Summary and Prospect

Considering that the current KG has high barriers to understanding, limited interpretation dimensions, and single expression forms, this paper extracts the basic structure and visualization analysis dimensions of KG. It then summarizes the visualization analysis method of KG and finally verifies, evaluates, and applies this method, which can assist the public to understand KG more comprehensively. However, there are still shortcomings in this paper. For example, we did not discuss and verify the HDE structure in depth. In the future, we should make a more comprehensive summary of large complex networks and extract a set of methods for deconstructing large complex networks. We can also consider conducting experimental verification on a broader dataset and publicizing the relevant results in order to expand the application in the relevant field.

References

1. Huang, H.Q., Yu, J., Liao, X.: Review on knowledge graphs. Comput. Syst. Appl. **28**(06), 1–12 (2019). http://www.c-s-a.org.cn/1003-3254/6915.html
2. Xu, Z.L., Sheng, Y.P., He, L.R.: Review on knowledge graph techniques. J. Univ. Electron. Sci. Technol. China **45**(04), 589–606 (2016)
3. Davenport, T.H., Prusak, L.: Working knowledge: how organizations manage what they know. J. Technol. Transfer **26**(4), 396–397 (1999)
4. Wang, Y.C., Luo, S.W., Yang, Y.B.: A survey on knowledge graph visualization. J. Comput.-Aided Des. Comput. Graph. **31**(10), 1666–1676 (2019)
5. Qiu, G.: The visualization analysis of research on knowledge base construction based on knowledge graph. J. Xinjiang Normal Univ. (Natural Sciences Edition) **38**(2), 33–40 (2019)
6. Weng, J., Gao, Y., Qiu, J., Ding, G., Zheng, H.: Construction and application of teaching system based on crowdsourcing knowledge graph. In: Zhu, X., Qin, B., Zhu, X., Liu, M., Qian, L. (eds.) Knowledge Graph and Semantic Computing: Knowledge Computing and Language Understanding: 4th China Conference, CCKS 2019, Hangzhou, China, August 24–27, 2019, Revised Selected Papers, pp. 25–37. Springer Singapore, Singapore (2019). https://doi.org/10.1007/978-981-15-1956-7_3
7. Liang, X.T., Feng, G.H.: Review of knowledge mapping construction method. Libr. J. **05**, 12–18 (2013)
8. Wang, Z.C.: Design and Implementation of Visual Analysis System for Knowledge Graph. Dalian University of Technology (2019)
9. Yang, Z.: Design and Implementation of Service Platform for Vertical Knowledge Graph Visualization. Beijing University of Posts and Telecommunications (2019)
10. Yang, S.L., Han, R.Z.: Studies in the status quo of knowledge mapping application research abroad. Inf. Document. Serv. **06**, 15–20 (2013)
11. Yang, Y.J., Xu, B., Hu, J.W., Tong, M.H., Zhang, P., Zheng, L.: Accurate and efficient method for constructing domain knowledge graph. Ruan Jian Xue Bao/J. Softw. **29**(10), 2931−2947 (2018). http://www.jos.org.cn/1000-9825/5552.htm. ((in Chinese))
12. Eck, N., Waltman, L.: Software survey: VOSviewer, a computer program for bibliometric mapping. Scientometrics **84**(2), 523–538 (2010)
13. Xiao, M., Qiu, X.H., Huang, J.: Comparison of software tools for mapping knowledge domain. Libr. J. **32**(3), 61–69 (2013)
14. Li, M.X., Wang, S.: Research context and theme analysis of mapping knowledge domains in china in recent ten years. Document. Inf. Knowl. **4**, 93–101 (2016)
15. Kohlhammer, J., Keim, D., Pohl, M., et al.: Solving problems with visual analytics. Procedia Comput. **7**, 117–120 (2012)
16. Yang, J., Zhang, Q.Y., Wang, L.M.: XML document information hiding based on sub-element permutation and combination. Comput. Eng. **35**(20), 153–156 (2009)
17. Chynał, P., Sobecki, J.: Eyetracking Evaluation of Different Chart Types Used for Web-Based System Data Visualization. Network Intelligence Conference. IEEE (2011)
18. Shneiderman, B.: The eyes have it: a task by data type taxonomy for information visualizations. In: The Craft of Information Visualization, pp. 364–371. Elsevier (2003). https://doi.org/10.1016/B978-155860915-0/50046-9
19. Zheng, Y.F., Zhao, Y.N., Bai, X.: Survey of big data visualization in education. J. Front. Comput. Sci. Technol. **15**(03), 403–422 (2021)
20. Zhu, P.F.: Research of Thematic Application Analysis Based on Data Visualization. Beijing University of Posts and Telecommunications (2019)
21. Jiang, T.T., Fan, S.X., Wang, H.: Impact of faceted search in academic library OPAC on the user experience: based on the comparison of the experimental analysis. Library Inf. Serv. **000**(004), 114–121 (2015)

PatentMiner: Patent Vacancy Mining via Context-Enhanced and Knowledge-Guided Graph Attention

Gaochen Wu[1], Bin Xu[1(✉)], Yuxin Qin[1], Fei Kong[2], Bangchang Liu[2], Hongwen Zhao[2], and Dejie Chang[2]

[1] Department of Computer Science and Technology, Tsinghua University, Beijing, China
{wgc2019,xubin}@tsinghua.edu.cn, tyx16@mails.tsinghua.edu.cn
[2] Beijing MoreHealth Technology Group Co. Ltd, Beijing, China
{kongfei,liubangchang,zhaohongwen,changdejie}@miao.cn

Abstract. Although there are a small number of work to conduct patent research by building knowledge graph, but without constructing patent knowledge graph using patent documents and combining latest natural language processing methods to mine hidden rich semantic relationships in existing patents and predict new possible patents. In this paper, we propose a new patent vacancy prediction approach named **PatentMiner** to mine rich semantic knowledge and predict new potential patents based on knowledge graph (**KG**) and graph attention mechanism. Firstly, patent knowledge graph over time (e.g. year) is constructed by carrying out named entity recognition and relation extraction from patent documents. Secondly, Common Neighbor Method (**CNM**), Graph Attention Networks (**GAT**) and Context-enhanced Graph Attention Networks (**CGAT**) are proposed to perform link prediction in the constructed knowledge graph to dig out the potential triples. Finally, patents are defined on the knowledge graph by means of co-occurrence relationship, that is, each patent is represented as a fully connected subgraph containing all its entities and co-occurrence relationships of the patent in the knowledge graph; Furthermore, we propose a new patent prediction task which predicts a fully connected subgraph with newly added prediction links as a new patent. The experimental results demonstrate that our proposed patent prediction approach can correctly predict new patents and Context-enhanced Graph Attention Networks is much better than the baseline.

Keywords: Knowledge graph · Graph attention networks · Link prediction · Co-occurrence relationship

1 Introduction

Patent is a kind of intellectual property, which endows inventors with the exclusive right to the invention within a certain period, so that it can be widely used to promote progress of science and technology and development of industry. Patents are valuable knowledge and technical information resources with characteristics of large quantity and

B. Qin et al. (Eds.): CCKS 2021, CCIS 1466, pp. 227–239, 2021.
https://doi.org/10.1007/978-981-16-6471-7_17

wide content. Therefore, research on patents has very important theoretical value and high-practical significance to scientific research and enterprise development (Jokanovic et al. 2017).

Patent documents contain a wealth of entities and relationships, which enable us to construct the patent knowledge graph with knowledge extraction (Schlichtkrull et al. 2018), so as to dig out hidden semantic relationships in the existing patents by virtue of knowledge graph and graph neural network (Veličković et al. 2019; Zhou et al. 2018). Sarica et al. (2019) attempted to use natural language processing techniques to build an engineered knowledge graph with patent database as data resources and thereby provide convenience for patent retrieval.

However, they did not explore and utilize new latest technologies of knowledge graph, for example, TransE (Bordes et al. 2013) and graph neural networks (Velickovic et al. 2018), to deeply mine potential rich semantic relationships hidden in patents (Wu et al. 2021). Xu et al. (2015) used Freebase (Bollacker et al. 2008) and Mesh word table to create knowledge graph in the field of lung cancer, and then tagged patent literatures with the knowledge graph, and made emerging technology prediction based on networks between labels. However, they did not directly construct a patent knowledge graph based on patent documents to predict emerging technologies, nor did take advantage of co-occurrence relationship (Surwase et al. 2011) to define patents in the knowledge graph. To sum up, the research direction of new technology prediction based on patent knowledge graph is almost blank.

It is of great significance and value to determine directions of technological research and development strategies of enterprises to predict potential technical blank points that have not been applied for patents by mining existing massive patent documents. In this paper, we propose a patent vacancy prediction method called **PatentMiner** based on knowledge graph and graph neural networks to explore rich semantic information hidden in patents and predict potential possible new patents. Firstly, based on patent documents, the patent knowledge graph that changes with year is constructed through named entity recognition and relation extraction. Then, Common Neighbor Method (Taskar et al. 2003), Graph Attention Networks (Velickovic et al. 2018; Vaswani et al. 2017) and Context-enhanced (Peters et al. 2018; Devlin et al. 2019; Raford et al. 2018) Graph Attention Networks are proposed for performing link prediction on the constructed knowledge graph, so as to dig out potential triples which currently exist but have not been found. Finally, we define a patent in patent knowledge graph by using co-occurrence relationship, that is, each patent is represented as a fully-connected sub-graph containing all its entities and co-occurrence relationships in the patent knowledge graph. In addition, we propose a patent vacancy prediction task which predicts a fully-connected sub-graph with newly added prediction links as a new potential possible patent.

To demonstrate the effectiveness of the proposed patent prediction approach in this work, we take advantage of the USPTO patent database and select patent documents from 2010 to 2019 in the field of electronic communication to carry out a series of experiments. Experimental results show our proposed PatentMiner can accurately predict new patents, especially, prediction accuracy of Context-enhanced Graph Attention Networks (CGAT) is superior to the baseline (e.g. Common Neighbor Method), detailed in Table 3 and

Sect. 4.4, which fully demonstrate the capability of our proposed approach predicting new potential patents.

In summary, this paper has made the following contributions: **(1)** We propose a patent prediction approach called PatentMiner by combining knowledge graph and graph attention networks, which can accurately predict new potential possible patents. **(2)** We define a patent in patent knowledge graph by using co-occurrence relationship, that is, each patent is represented as a fully-connected subgraph of the knowledge graph containing all its entities and co-occurrence relationships, which quantifies accurately patent in knowledge graph. **(3)** A patent vacancy prediction task is proposed on patent knowledge graph, which predicts a fully connected subgraph containing newly added prediction links as a new possible patent. Meanwhile, we provide a baseline called common neighbor method and develop a state-of-the-art model called context-enhanced graph attention networks. **(4)** Experimental results demonstrate that our proposed approach (PatentMiner) can accurately predict new potential patents. The more important observation is that our proposed context-enhanced graph attention networks is much better than the baseline.

2   Related Work

Patent clustering and automatic classification are common topics in patent research. S. Jun (2014) constructed a combinatorial classification system by using data reduction and K-means clustering to solve the problem of sparseness in document clustering. Wu et al. (2016) used self-organizing mapping and support vector machine classification models to design a patent classification approach based on patent quality.

Statistical and probabilistic models are widely used in patent analysis. R. Klinger et al. (2008) proposed an approach to extract names of organic compounds from scientific texts and patents by using conditional random field models and Bootstrapping method. M. Klallinger et al. (2015) developed a set of chemical substance recognition systems by combining conditional random field model and support vector machine, and using natural language processing methods such as rule matching and dictionary query. A. Suominen et al. (2017) analyzed 160,000 patents using implicit Dirichlet distribution (LDA) method, and classified them into different groups according to patent theme. They established knowledge portrait of company. Lee C et al. (2016) divided technology life cycle into several stages and used hidden Markov chain model to predict the probability that the technology included in a patent is in a specific stage in the life cycle.

In recent years, neural network methods have been widely used in patent research. Lee et al. (2017) used the number of patent application and combined deep belief neural networks to predict future performance of companies. A. Trappey et al. (2012) combined principal component analysis method and back propagation neural networks to improve analysis effect of patent quality and shorten time needed to determine and evaluate the quality of new patents. Paz-Marin et al. (2012) used evolutionary S-type unit neural networks and evolutionary product unit neural networks to predict research and development performance of European countries. B. Jokanovic et al. (2017) assessed economic value of patents based on different scientific and technical factors using over-limit learning machine.

Patent documents contain a large number of entities and various relationships between entities, which make patent documents good materials for constructing knowledge graph. Therefore, it will be a very attractive research direction to make use of knowledge graph to explore hidden rich semantic relationships in patent data. Sarica et al. (2019) attempted to use natural language processing methods to build an engineering knowledge graph using patent database as the data source, so as to provide convenience for patent retrieval. Xu et al. (2012) used Freebase (Bollacker et al. 2008) and Mesh word table to create knowledge graph in the field of lung cancer, and then tagged patent literatures with this graph, and finally made emerging technology prediction based on the networks between labels.

Although there are a small amount of research by building knowledge graph for studying patents, but we could not find work by combining patent knowledge graph and latest natural language processing methods, such as pretraining models (Peters 2018; Devlin et al. 2019; Radfordet al. 2018), attention mechanism (Vaswani et al. 2017) and graph neural network (Velickovic et al. 2018), to mine rich semantic relationships and knowledge hidden in patents and then forecast new patents. In this paper, we propose a patent vacancy prediction approach called **PatentMiner** based on knowledge graph and graph attention networks to perform link prediction and forecast patents.

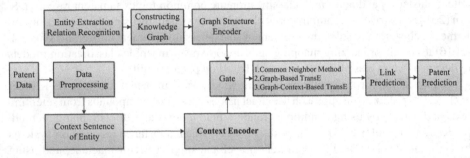

Fig. 1. PatentMiner - Our proposed patent vacancy prediction approach via context-enhanced and knowledge-guided graph attention mechanism by using gate mechanism to combine graph encoder and context encoder

3 PatentMiner

The overall structure of our proposed patent prediction approach named PatentMiner is shown in Fig. 1. Our approach consists of the following four steps: (1) Downloading and cleaning patent data, then constructing the patent knowledge graph through named entity recognition (NER) and relation extraction. (2) Embedding representations of entities and relationships using graph structure encoder (e.g., GAT) and contextual text encoder (e.g., BERT). Studying link prediction approaches to mine potential triples in the graph. (4) Representing a patent as a fully-connected subgraph containing all its entities and co-occurrence relationships in the knowledge graph by using the co-occurrence relationship between entities, and forecasting a fully connected subgraph containing newly added prediction links as a new potential possible patent which may be applied in the future.

3.1 Patent Knowledge Graph Construction

Patent documents contain both structured and unstructured information, and research on patents includes analysis on these two parts. In this study, we firstly collect USPTO patent data in a certain period of time (e.g., from 2010 to 2019), and divide these patents according to the year of patent application into groups. Then entities and relationships in these patents are extracted and added to the knowledge graph by year. Finally, we build a patent knowledge graph KG(t) of different years for the following link and patent prediction tasks.

Entity Recognition. Named entity recognition refers to identification of entities with specific meaning in text, such as name of person, name of organization, name of location, etc. In this paper, domain entities are extracted from patent documents in the way of domain dictionary comparison.

Relation Extraction. Relation extraction is the extraction of semantic relations between two or more entities from text. In this work, based on co-occurrence relations, each patent is represented as a fully-connected sub-graph containing all its entities and co-occurrence relations in the patent knowledge graph, and we forecast a new possible patent as a fully-connected subgraph with newly added predictive co-occurrence links. Therefore, we firstly study the most important co-occurrence relationship between entities, that is, if two entities appear in a same patent document, we consider these two entities having a co-occurrence relationship. This method borrows the concept of co-citation proposed by Small Henry et al. (2013), an American intelligence scientist, that is to say, when two literatures are cited by the same paper, there is an internal relation-ship between them (Small Henry et al. 1977; Small Henry and Boyack Kevin et al. 2013; Small Henry and Boyack Kevin et al. 2014).

3.2 Graph and Context Embedding

After the construction of the patent knowledge graph in Sect. 3.1, in order to carry out link prediction on the graph to dig out potential triples, and then to further conduct patent prediction, it is necessary to encode the entities and relations in the graph. Contextual information and graph structure are crucial to accurately represent entities. In this study, graph structure encoder and context encoder are used to capture information of entities in the graph structure and the patent context text, respectively.

Graph Encoder. The graph structure encoder assigns a vector representation to each entity based on its position in the graph and the characteristic representation of its adjacent entities. In this study, we use graph attention mechanism to feature weighted sum of adjacent nodes as the vector representation of the target node. The details are shown as follows.

The constructed patent knowledge graph is formulated with a list of tuples $\left(e_i^{head}, r_i, e_i^{tail}\right)$ composing of a head entity e_i^{head}, a tail entity e_i^{tail}, and their relation r_i. We randomly initialize vectors $\overrightarrow{e_i}$ and $\overrightarrow{r_i}$ for e_i and r_i respectively.

Considering a single layer of graph attention on the graph, assuming that the input of this layer is a set of entity node $h = \left\{ \vec{h_1}, \vec{h_2}, ..., \vec{h_N} \right\}$, $\vec{h_i} \in R^F$, the output is a set of new features in Eq. (4b) $\bar{h} = \left\{ \vec{h_1'}, \vec{h_2'}, ..., \vec{h_N'} \right\}$, $\vec{h_i'} \in R^{F'}$, the input and output for here may have different dimensions F and F'. In order to convert input features h into output features $\bar{h}$, a general linear transformation is required, which is represented by a weight matrix $W_s \in R^{F' \times F}$. The importance score of node j to node i can be calculated using a unified attention mechanism as in Eq. (1) and Eq. (3):

$$s_{ij} = a\left(W_s \vec{h_i}, W_s \vec{h_j} \right) \tag{1}$$

In this paper, a in Eq. (1) is a single-layer network and its parameters are expressed in terms of $\vec{a} \in R^{2F'}$. Use function softmax to normalize the importance score s_{ij} in Eq. (1) to obtain the attention weight α_{ij} as in Eq. (2).

$$\alpha_{ij} = \text{softmax}(s_{ij}) = \frac{\exp(s_{ij})}{\sum_{k \in N_i} \exp(s_{ik})} \tag{2}$$

Where s_{ij} is the result of a single-layer network with $\vec{a}$ as the parameters and LeakyReLU as the activation function over the concatenation of $\vec{h_i} \oplus \vec{h_j}$ of $\vec{h_i}$ and $\vec{h_j}$ as in Eq. (3):

$$s_{ij} = \text{LeakyReLU}\left(\vec{a}^T \left[W_s \vec{h_i} \oplus W_s \vec{h_j} \right] \right) \tag{3}$$

Where W_s is a learnable weight matrix.

Finally, the representation of the node i in the patent knowledge graph can be calculated as in Eq. (4):

$$\begin{aligned} \vec{h_i'} &= Sigmoid \left\{ \sum_{j \in N_i} \alpha_{ij} \left(W_s \vec{h_j} \right) \right\} \quad (a) \\ \bar{h} &= \left\{ \vec{h_1'}, \vec{h_2'}, ..., \vec{h_N'} \right\}, \vec{h_i'} \in R^{F'} \quad (b) \end{aligned} \tag{4}$$

Context Encoder. Each entity e_i Corresponding to the node i in the constructed patent knowledge graph in Sect. 3.1 appears in a certain context text. It is assumed that the sentence containing the entity e_i is represented as $[w_1, w_2, ..., w_l]$, where l is the length of the sentence and w_i is the i-th word in the sentence. In this study, BERT (Devlin et al. 2019) is used to encode the sentence to obtain the contextual representation of the entity e, and the vectors of each word in the sentence are represented as $[H_1, H_2, ..., H_l] \in R^d$, where d is the dimension of the vector and H_i represents the hidden state of w_i. Then we compute a bilinear attention weight for each word w_i as in Eq. (5).

$$\mu_i = e^T W_f H_i, \quad \mu' = Soft \max(\mu) \tag{5}$$

Where W_f is a bilinear matrix. We finally get the context representation as in Eq. (6):

$$\bar{e} = \mu'^T H_i \tag{6}$$

Gate Mechanism. We use gate mechanism as in (Wang et al. 2019) to combine the graph-encoded representation $\bar{h}$ calculated in Eq. (4B) with the context-encoded representation $\bar{e}$ obtained in Eq. (6) to form the final representation E of the entity e as in Eq. (7):

$$
\begin{aligned}
g_e &= Sigmoid(\overleftrightarrow{g}_e), \quad\quad (a)\\
E &= g_e \odot \bar{h} + (1 - g_e) \odot \bar{e} \quad (b)
\end{aligned}
\tag{7}
$$

Where g_e is an entity-dependent gate function of which each element is in [0,1], $\overleftrightarrow{g}_e$ is a learnable parameter for each entity e, $\odot$ is an element-wise multiplication.

3.3 Link Prediction

In order to predict new potential possible patents on the constructed patent knowledge graph, we first make link prediction on the graph to dig out potential triples.

Link Prediction Task. In this work, the link prediction task on the knowledge graph is defined in the following way: given the patent document data up to time T (e.g., 2015), the patent knowledge graph at time T is constructed and contains entities and relations extracted from existing patents. The task of link prediction is to infer missing links (links that exist but are not observed currently) from existing nodes and edges, which represent possible emerging technologies in the future. It is feasible to make link prediction on the patent knowledge graph, because the entities and relations contained in the graph will increase gradually with the passage of time, and the existing entities and relations will not disappear.

Collecting patent data in a certain period of time (e.g., from 2010 to 2019), dividing these patents according to the year of patent application, extracting knowledge from these patent documents and adding them to the knowledge graph according to the year, finally the patent knowledge graph changing with year can be obtained.

Common Neighbor Method. The common neighbor method is based on assumption that if two people in a social network have many public acquaintances, they are more likely to meet each other than two people without any public contact. Therefore, the more neighbors two nodes in knowledge graph have in common, the higher the probability that there is a potential edge between them.

Considering two nodes x and y in the graph, and assuming that the sets of their neighbors are Γx and Γy respectively. A scoring function is defined as $s(x, y)$, representing the probability of a predicted link between x and y. In the common neighbor method, $s(x, y)$ is the number of common neighbor nodes between and y, which is formally expressed as in Eq. (8):

$$s(x, y) = |\Gamma x \cap \Gamma y| \tag{8}$$

A reasonable threshold can be set by analyzing the structure of the graph as ζ, such as counting exit degree and entry degree of nodes. When $s(x, y) > ζ$, it is assumed that there is a predictive link between x and y.

TransE. The fundamental principle of TransE (Bordes Antoine et al. 2013) is to represent both entities and relations as vectors, and to transform the problem of judging the existence of a tuple to verify whether the arithmetic expressions between entities and relations are valid. For example, for a triplet (h, r, t), the vectors h and t representing head entity and tail entity, and the vector r representing the relation should satisfy $h + r \approx t$. The loss function of this model can be defined as in Eq. (9):

$$L = \sum_{(h,r,t)\in K} \sum_{(h',r,t')\in K'} \max\left(0, \gamma + d(h+r, t) - d(h' + r, t')\right) \qquad (9)$$

Where γ is a positive constant and $d(u, v) = |u - v|_2^2$ is a distance function. After the model is optimized according to this loss function, each pair of triples (h, r, t) can be scored using the model. The higher the score obtains, the more likely this tuple (h, r, t) is to be true.

3.4 Patent Prediction

On the basis of link prediction, we can also forecast the future possible new patents.

Patent Definition. According to the co-occurrence relationship between entities, for a patent P, assuming that it contains a set of entities $E = \{e_1, e_2, ..., e_n\}$ with a total of n entities, therefore, the patent P in the patent knowledge graph should be expressed as a fully connected subgraph containing n nodes and $R = \left\{\frac{n(n-1)}{2}\right\}$ co-occurrence relations, so the patent P can be represented as $P = G(E, R)$, where E is the entity set of the patent, R is the set of all co-occurrence relationships on the entity set of the patent P.

Patent Prediction Task. For a patent knowledge graph of time t, link prediction is first performed on the graph, and then all predicted links are added to the graph to obtain a new patent knowledge graph. The new graph is likely to include some new patents that each is represented as a maximal fully connected subgraph with at least one predicted link. According to the patent definition, we can identify these predictive new patents and compare them with patents in subsequent years $t + \Delta t$ to confirm the accuracy of our proposed models forecasting new patents.

4 Experiments and Results

We verify effectiveness of our proposed approach in this part. Firstly, we select and download the USPTO patents from 2010 to 2019 in the field of electronic communication, and preprocess the patent data. Secondly, based on the patent data, we construct the patent knowledge graph changing with year. Finally, link prediction and patent prediction are carried out on the constructed patent knowledge graph.

4.1 Patent Data Preprocessing

We firstly download patent data including patent summary and CPC classification number from PatentsView, a database of the U.S. Patent and Trademark Office, and align the downloaded table data according to the unique patent number. Then, we select the patent documents containing H04L in the classification number from 2010 to 2019 according to the year and classification number. Finally, the title and abstract parts of each patent document are extracted and combined to form a separate text file. The number of patents by year are shown in Table 1.

Table 1. Statistics of constructed dynamic accumulative patent knowledge graph from 2010 to 2019 (year)

Cut-off year	Patents	Entities	Tuples
2010	16990	2401	120132
2011	17787	3578	151209
2012	20642	4698	169823
2013	23503	5733	180217
2014	28716	6817	197821
2015	29473	7284	206412
2016	31464	7905	220719
2017	33697	8601	258121
2018	32981	9054	273930
2019	35742	9901	291203

4.2 Constructing Patent KG (Year)

We download a glossary of terms for electronic communications from the U.S. Federal Standards website, which cover 19,803 domain entities in the field of electronic communications. Starting with the patent data in 2010, the patent documents are compared with glossary year by year to obtain knowledge graphs with an increasing number of entities and relationships. The detailed statistics of the knowledge graphs by different years are shown in Table 1.

4.3 Link Prediction Results

In this study, we use link prediction accuracy and patent prediction accuracy to measure the performance of our proposed methods. For the predicted new links, we examine whether they accurately appear in the graphs constructed using patents which are applied in subsequent years. For example, considering the constructed graph of 2010 (e.g., KG (2010)), which means that the graph is constructed using all patents up to 2010, supposing

Table 2. Results of link prediction

Cut-off year	Link prediction (Number of new links)			Accuracy of link prediction (%)		
	CNM	GAT	CGAT	CNM	GAT	CGAT
2010	9821	13021	16001	12.12	15.92	21.73
2011	10215	16723	18902	13.34	16.47	22.63
2012	10928	20812	21982	13.91	18.03	24.85
2013	11384	26914	27031	14.14	19.85	25.91
2014	11892	32312	31019	15.08	20.54	26.14
2015	12309	36912	38632	15.87	21.32	27.13
2016	13018	42453	45812	16.82	22.71	28.95
2017	14219	48912	51238	17.02	24.56	29.73
2018	16921	56834	64123	17.82	25.64	32.41
2019	18388	62981	72943	18.54	26.01	33.81

the set of new links predicted by the proposed approach over the graph of 2010 are represented as $R(2010)$ and the whole link set of the graph of the year (e.g., KG(Year)) as $R(Year)$, then the prediction link accuracy is calculated as in Eq. (10):

$$a_{link}(2010|Year) = \frac{|R(Year) \cap R(2010)|}{|R(2010)|} \tag{10}$$

Where $|R(Year) \cap R(2010)|$ represents the number of accurate prediction links.

In order to investigate effectiveness of our proposed approach, we use common neighbor method (**CNM**) as the baseline and verify graph attention networks (**GAT**) and context-enhanced graph attention networks (**CGAT**).

CNM. We count the maximum number of common neighbors of each pair of nodes in the graph as M. If a pair of nodes does not have a link connection, but their common neighbors $m \geq \left[\frac{M}{2}\right]$, a new link is considered to exist between this pair of nodes.

GAT. Firstly, we utilize a graph structure encoder with the attention mechanism to encode the entities and relations in the graph, and then the TransE method is used to train the prediction model. Finally, we take advantage of the model to score and forecast the potential links.

CGAT. On the basis of **GAT**, context encoder is added to enrich the representations of entities, such as BERT in this paper. We introduce a gate function to combine the node representations learned by graph encoder and context encoder to form a comprehensive representation for each entity.

According to the experimental results in Table 2, as the scale of the graph increases, the number of new links predicted by the three models quickly go up. Furthermore, the accuracy of link prediction improves year by year. Compared to the baseline, we observe

Table 3. Results of patent prediction

Cut-off year	Patent prediction (Number of new patents)			Accuracy of patent prediction (%)		
	CNM	GAT	CGAT	CNM	GAT	CGAT
2010	2081	3083	3729	7.32	10.23	11.23
2011	2871	3612	4011	8.14	10.96	11.92
2012	3067	3902	5874	8.95	11.32	12.37
2013	3571	4205	6920	10.21	11.47	12.78
2014	4018	4923	7810	11.94	12.03	14.01
2015	5219	5612	8451	12.29	12.57	14.81
2016	5598	6472	9012	12.63	12.83	15.23
2017	6201	7201	10983	12.91	13.42	16.02
2018	7034	8312	12394	13.81	14.71	16.72
2019	7561	9056	14834	14.05	15.23	17.31

that the number of new links found using GAT and CGAT increase significantly in every year from 2010 to 2019. More importantly, the link prediction accuracy of CGAT is significantly improved over the baseline on the same dataset, with an average increase of 4 points.

4.4 Patent Prediction Results

According to the definition of co-occurrence relationship, all entities and their co-occurrence relationships contained in each patent are represented as a fully connected subgraph in the knowledge graph. In this paper, the fully connected subgraph containing new predicted links is regarded as a new patent forecasted by the model.

Supposing the set of domain entities included in a new predicted patent $p = G(E, R)$ as E, if there is at least one patent $p' = G(E', R')$ applied in the future years, where its set of domain entities satisfy $E' \supseteq E$, then the predicted new patent p is considered to be valid. If the set of all new forecasted patents is P, and the valid set of new patents is P_0, then the accuracy of patent prediction is calculated as in Eq. (11).

$$a_{patent} = \frac{|P_0|}{|P|} \tag{11}$$

Experimental results are shown in Table 3. With the scale of the graph increasing, we observe that the number of new patents predicted by the three models increases, and the patent prediction accuracy also improves year by year.

Furthermore, for the patent prediction task, CGAT using both contextual representation and graph representation is better than GAT only based on graph encoder. Moreover, the maximum prediction accuracy of CGAT reaches 17.31% in 2019 with 3.26 points higher than the baseline and 2.06 points higher than GAT. The experimental results demonstrate that our proposed approach can predict new potential possible patent.

5 Conclusion

In this study, we propose a new patent vacancy prediction approach called PatentMiner via patent knowledge graph and graph attention. We define a patent on knowledge graph by using co-occurrence relationships, and a patent prediction task is proposed to predict the fully connected subgraph containing new predictive links as a new patent. Experimental results demonstrate that our proposed approach can correctly predict new patents. Meanwhile, there is still much room for improvement on the patent prediction task.

Acknowledgements. This work is supported by the National Key Research and Development Program of China (2017YFB1401903). It also got partial support from Beijing MoreHealth Technology Group Co. Ltd., and Beijing Key Lab of Networked Multimedia.

References

Jokanovic, B., Lalic, B., Milovancevic, M., et al.: Economic development evaluation based on scienceandpatents. Phys. Stat. Mech. Appl. **481**, 141–145 (2017)

Breitzman, A.F., Mogee, M.E.: The many applications of patent analysis. J. Inf. Sci. **28**(3), 187–205 (2002)

Schlichtkrull, M., Kipf, T.N., Bloem, P., van den Berg, R., Titov, I., Welling, M.: Modeling relational data with graph convolutional networks. In: Gangemi, A., et al. (eds.) ESWC 2018. LNCS, vol. 10843, pp. 593–607. Springer, Cham (2018). https://doi.org/10.1007/978-3-319-93417-4_38

Veličković, P.: The resurgence of structure in deep neural networks (Doctoral thesis) (2019). https://doi.org/10.17863/CAM.39380

Zhou, J., Cui, G., Zhang, Z., Yang, C., Liu, Z., Sun, M.: Graph Neural Networks: A Review of Methods and Applications. arXiv:abs/1812.08434 (2018)

Sarica, S., Jianxi, L., Kristin, L.W.: Technology Knowledge Graph Based on Patent Data. arXiv: abs/1906.00411 (2019), n. pag

Bordes, A., Usunier, N., Garcia-Duran, A., Weston, J., Yakhnenko, O.: Translating Embeddings for Modeling Multi-relational Data (2013)

Velickovic, P., Cucurull, G., Casanova, A., Romero, A., Liò, P., Bengio, Y.: Graph Attention Networks. arXiv:abs/1710.10903 (2018)

Shao, L., Zhao, Z., Xu, D.: Research on technology prediction method based on patent literature and knowledge map. Sci. Technol. Manage. Res. **35**(14), 134–140 (2015)

Bollacker, K., Evans, C., Paritosh, P., Sturge, T., Taylor, J.: Freebase: a collaboratively created graph database for structuring human knowledge. In: Proceedings of the 2008 ACM SIGMOD International Conference on Management of Data (2008)

Surwase, G., Sagar, A., Kademani, B., Bhanumurthy, K.: Co-citation Analysis: An Overview (2011)

Taskar, B., Wong, M.F., Abbeel, P., et al.: Link Prediction in Relational Data. Neural Information Processing Systems, pp. 659–666 (2003)

Vaswani, A., Shazeer, N., Parmar, N., et al.: Attention is All you Need. Neural information processing systems, pp. 5998–6008 (2017)

Peters, M.E., Neumann, M., Iyyer, M.: Deep contextualized word representations. arXiv PreprintarXiv (1802) (2018)

Devlin, J., et al.: BERT: Pre-training of Deep Bidirectional Transformers for Language Understanding. NAACL-HLT (2019)

Alec, R., Karthik, N., Tim, S., Ilya, S.: Improving language understanding with unsupervised learning. Technical report, OpenAI (2018)

Trappey, A.J.C., Trappey, C.V., Wu, C.Y., Lin, C.L.: A patent quality analysis for innovative technology and product development. Adv. Eng. Informatics **26**, 26–34 (2012)

Jun, S., Park, S.-S., Jang, D.-S.: Document clustering method using dimension reduction and support vector clustering to overcome sparseness. Expert Syst. Appl. **41**(7), 32043212 (2014)

Wu, J.L., Chang, P.C., Tsao, C.C.: A patent quality analysis and classification system using self-organizing maps with support vector machine. Appl. Soft Comput. **41**, 305–316 (2016)

Klinger, R., Kolářik, C., Fluck, J., et al.: Detection of iupac and iupac-like chemical names. Bioinformatics **24**(13), i268–i276 (2008)

Krallinger, M., Rabal, O., Leitner, F.: The chemdner corpus of chemicals and drugs and its annotation principles. J. Cheminf. **7**(Suppl1), S2 (2015)

Suominen, A., Toivanen, H., Seppänen, M.: Firms' knowledge profiles: mapping patent data with unsupervised learning. Forecast. Soc. Change **115**, 131–142 (2017)

Lee, C., Kim, J., Kwon, O.: Stochastic technology life cycle analysis using multiple patent indicators. Technol. Forecast. Soc. Chang. **106**, 53–64 (2016)

Lee, J., Jang, D., Park, S.: Deep learning-based corporate performance prediction model considering technical capability. Sustainability **9**(6), 899 (2017)

Trappey, A., Trappey, C., Wu, C.-Y., et al.: A patent quality analysis for innovative technology and product development. Adv. Eng. Inf. **26**(1), 26–34 (2012)

de la Paz-Marín, M., Campoy-Muñoz, P., Hervás-Martínez, C.: Non-linear multiclassifier model based on artificial intelligence to predict research and development performance in european countries. Technol. Forecast. Soc. Change **79**(9), 1731–1745 (2012)

Wang, Q., et al.: PaperRobot: Incremental Draft Generation of Scientific Ideas. arXiv:abs/1905. 07870 (2019), n. pag

Boyack, K., Small, H., Klavans, R.: Improving the accuracy of co-citation clustering using full text. J. Am. Soc. Inform. Sci. Technol. **64**, 1759–1767 (2013). https://doi.org/10.1002/asi.22896

Small, H., Boyack, K., Klavans, R.: Identifying emerging topics by combining direct citation and co-citation. In: Proceedings of ISSI 2013 - 14th International Society of Scientometrics and Informetrics Conference, vol. 1, pp. 928–940 (2013)

Small, H., Boyack, K., Klavans, R.: Identifying emerging topics in science and technology. Res. Policy **43**, 1450–1467 (2014). https://doi.org/10.1016/j.respol.2014.02.005

Small, H.: A co-citation model of a scientific specialty: a longitudinal study of collagen research. Soc. Stud. Sci. - SOC STUD SCI. **7**, 139–166 (1977). https://doi.org/10.1177/030631277700 700202

Wu, G., Xu, B., Chang, D., Liu, B.: A Multilingual Modeling Method for Span-Extraction Reading Comprehension. arXiv:abs/2105.14880 (2021)

Multi-task Feature Learning for Social Recommendation

Yuanyuan Zhang[1], Maosheng Sun[2], Xiaowei Zhang[1], and Yonglong Zhang[1(✉)]

[1] School of Information Engineering, Yangzhou University, Jiangsu, China
{xwzhang,ylzhang}@yzu.edu.cn
[2] Office of Informationization Construction and Administration,
Yangzhou University, Jiangsu, China
sms@yzu.edu.cn

Abstract. The purpose of the recommender system is to recommend personalized products or information for users. It is widely used in many scenarios to deal with information overload problems to improve user experience. As an existing popular recommendation method, collaborative filtering usually suffers from data sparsity and cold start problems. Therefore, researchers usually make use of side information, such as contexts or item attributes, to solve the problem and improve the performance of the recommender systems. In this paper, we consider social relationship and knowledge graph as side information, and propose a multi-task feature learning model, Social-MKR, which consists of recommendation module and knowledge graph embedding (KGE) module. In recommendation module, we build the social network among users based on the user-item interactions, and conduct the GCN model to obtain the specific user's neighborhood representation, which can be used as the input of the recommendation module. Like MKR, the KGE module is used to assist recommendation module by a cross&compression unit, which can learn high-order hidden features between items and entities. Extensive experiments on real-world datasets (e.g., movie,book and news) demonstrate that Social-MKR outperforms several state-of-the-art methods.

Keywords: Recommender systems · Social relationship · Knowledge graph · Multi-task

1 Introduction

With the advancement of network technology, people can easily access a large amount of online information, such as commodities, movies. At the same time,

This work was supported in part by the National Natural Science Foundation of China under Grant 61872313, in part by the Key Research Projects in Education Informatization in Jiangsu Province under Grant 20180012, in part by the Yangzhou Science and Technology under Grant YZ2020174 and Grant YZ2019133, and in part by the Open Project in the State Key Laboratory of Ocean Engineering, Shanghai Jiao Tong University under Grant 1907.

ⓒ Springer Nature Singapore Pte Ltd. 2021
B. Qin et al. (Eds.): CCKS 2021, CCIS 1466, pp. 240–252, 2021.
https://doi.org/10.1007/978-981-16-6471-7_18

the problem of information overload is getting more and more serious, which causes users to spend a lot of time getting the information they really want. To alleviate the above problem, recommender system has been widely deployed to achieve personalized information filtering. Among recommendation techniques, Collaborative Filtering (CF) [5] utilizes user-item interactions (e.g., click, rate, purchase) to make recommendations based on the common preferences of different users. However, CF-based methods usually suffer from data sparsity and cold start problems. To solve these limitations, the researchers pointed out that some side information should be incorporated into CF, for example, social networks [15], attributes [11], contexts [9], etc.

Knowledge Graph (KG) as important side information is also always used for the recommender system [13,14]. A KG is a type of directed heterogeneous graph in which nodes correspond to entities and edges correspond to relations. Existing KG-based recommender systems apply KGs in three ways: the embedding-based method, the path-based method, and the unified method. Among the embedding-based methods, the core of the MKR model [13] lies in the design of the cross&compression unit, which mainly utilizes the identity of the entities in the knowledge graph and the items in the recommender system to learn high-order hidden features between items and entities. However, it does not consider the social relationship between users, which is very useful in improving the performance of the recommender system.

Based on MKR, we introduce the social network on the user-end, and proposed a multi-task feature learning model, Social-MKR, which consists of recommendation module and knowledge graph embedding (KGE) module. In the recommendation module, we build the social network among users based on the user-item interactions, and conduct the GCN model to obtain the specific user's neighborhood representation, which can be used as the input of the recommendation module. In the KGE module, entities and relations are embed into continuous vector spaces while preserving their structure. Empirically, we evaluate our method in three recommendation scenarios (i.e., movie, book, and music). The results show that, compared with the state-of-the-art methods, the performance of Social-MKR in click-through rate prediction (CTR) and Top-K recommendation has been greatly improved. In addition, Social-MKR can effectively solve the problem of data sparsity.

In summary, our contributions in this paper are as follows:

- We build the social network among users based on the user-item interactions, and conduct the GCN model to obtain the specific user's neighborhood representation.
- We propose the Social-MKR model, an end-to-end framework utilizing social networks and KG to assist the recommender system. Social-MKR can capture user's potential preferences according to the constructed social network.
- We conduct experiments on three real-world recommendation scenarios, and the results demonstrate the efficacy of Social-MKR over several state-of-the-art methods.

2   Related Work

2.1   Collaborative Filtering

Collaborative filtering (CF) usually utilizes the historical feedback data of users and items to mine the correlation between users and the items themselves to make recommendations. Among them, the model-based CF method attempts to alleviate the sparsity problem by establishing a reasoning model. A common implementation is the latent factor model [7], which extracts the latent representation of the user and item from the high-dimensional user-item interaction matrix, and then uses the inner product or other methods to calculate the similarity between the user and item. Researchers often improve recommendation performance from two aspects. One way to improve is enhancing the embedding function, specifically incorporating more side information into CF, such as item content [3], social networks [15] and knowledge graphs [14]. Another improvement is to enhance the interaction function, for example, NeuMF [4], which uses a nonlinear neural network as the interaction function.

2.2   Knowledge Graph-Based Recommendation

To utilize the KG information, we usually apply knowledge graph embedding (KGE) algorithms to encode the KG into low-rank embedding. KGE algorithms can be divided into two classes: translation distance models, such as TransE [1], TransH [16], etc., and semantic matching models, such as DistMult [17]. However, previous works [11,19] usually directly use the raw latent vector of structural knowledge obtained through KGE technique to make recommendations. Embedding-based methods essentially leverage the information in the graph structure. Recently, some papers [2,13] have tried to improve the recommendation performance by adopting a multi-task learning strategy, which can jointly learn recommendation task under the guidance of graph-related task. The essence of the multi-task learning is that item embeddings in the recommendation module share features with the associated entity embeddings in the KG.

3   Our Approach

In this section, we present the proposed Social-MKR model, the framework of which is illustrated in Fig. 1. We first formulate the knowledge-graph-aware recommendation problem, then introduce the components of Social-MKR model in detail, including neighborhood embedding module and recommendation module. Finally, we introduce the complete learning algorithm for Social-MKR.

3.1   Problem Formulation

In a typical recommendation scenario, let $U = \{u_1, u_2, \cdots, u_M\}$ and $V = \{v_1, v_2, \cdots, v_N\}$ represent the sets of users and items, respectively. For simplicity, we treat $u_j \in U$ and j as equivalent. According to whether the user has browsed a certain item, let $Y = \{y_{uv} | u \in U, v \in V\}$ denote the user-item interaction matrix, in which the element y_{uv} is defined as follows:

$$y_{uv} = \begin{cases} 1, & \text{if interaction } (u, v) \text{ is observed} \\ 0, & \text{otherwise} \end{cases} \tag{1}$$

where, a value of 1 for y_{uv} indicates that user u visited item v, such as behaviors of clicking, watching, browsing, etc.; otherwise $y_{uv} = 0$.

Besides the interaction matrix Y, a knowledge graph $\mathcal{G}$ is also provided, which is comprised of massive entity-relation-entity triples (h, r, t). Here $h \in \mathcal{E}$, $r \in \mathcal{R}$, and $t \in \mathcal{E}$ denote the head, relation, and tail of a triple, $\mathcal{E}$ and $\mathcal{R}$ denote the set of entities and relations in $\mathcal{G}$, respectively.

Given the user-item interaction matrix Y and the knowledge graph $\mathcal{G}$, our goal is to learn the probability $\hat{y}_{uv}$ of the user u visiting the candidate item v with which he has had no interaction before:

$$\hat{y}_{uv} = \mathcal{F}(u, v | \Theta, Y, \mathcal{G}) \tag{2}$$

where Θ denotes the model parameters of function $\mathcal{F}$.

3.2   MKR Brief

Here, we briefly introduce the MKR model, which is an end-to-end recommendation framework based on knowledge graphs. Its core lies in the design of cross&compression units, which mainly utilize the identity of the entities in the knowledge graph and the items in the recommender system. In MKR, the cross operation includes each possible feature interaction between item v and its associated entity e, the cross feature matrix of layer l is defined as follows:

$$\mathbf{C}_l = \mathbf{v}_l \mathbf{e}_l^\top = \begin{bmatrix} v_l^{(1)} e_l^{(1)} & \cdots & v_l^{(1)} e_l^{(d)} \\ \cdots & & \cdots \\ v_l^{(d)} e_l^{(1)} & \cdots & v_l^{(d)} e_l^{(d)} \end{bmatrix} \tag{3}$$

where $\mathbf{v}_l \in \mathbb{R}^d$ and $\mathbf{e}_l \in \mathbb{R}^d$ respectively represent latent features of the items and entities from layer l. The compression operation projects the cross feature matrix from $\mathbb{R}^{d \times d}$ space back to the feature spaces $\mathbb{R}^d$, and outputs the potential feature representations of items and entities in the next layer:

$$\begin{aligned} \mathbf{v}_{l+1} &= \mathbf{C}_l \mathbf{w}_l^{VV} + \mathbf{C}_l^\top \mathbf{w}_l^{EV} + \mathbf{b}_l^V = \mathbf{v}_l \mathbf{e}_l^\top \mathbf{w}_l^{VV} + \mathbf{e}_l \mathbf{v}_l^\top \mathbf{w}_l^{EV} + \mathbf{b}_l^V \\ \mathbf{e}_{l+1} &= \mathbf{C}_l \mathbf{w}_l^{VE} + \mathbf{C}_l^\top \mathbf{w}_l^{EE} + \mathbf{b}_l^E = \mathbf{v}_l \mathbf{e}_l^\top \mathbf{w}_l^{VE} + \mathbf{e}_l \mathbf{v}_l^\top \mathbf{w}_l^{EE} + \mathbf{b}_l^E \end{aligned} \tag{4}$$

where $\mathbf{w}_{l}^{\cdot} \in \mathbb{R}^d$ and $\mathbf{b}_l^{\cdot} \in \mathbb{R}^d$ are trainable weight and bias vectors.

In recommendation module, for item v, MKR uses L cross&compress units to extract its latent feature. Then, MKR calculates the predicted probability of user u visiting item v with user u's latent feature.

3.3 Framework

Social-MKR consists of two main components: recommendation module (including neighborhood embedding module) and KGE module. The recommendation module takes one user, its neighborhood and one item as input. We use multi-layer perceptrons (MLP) and cross&compress unit to extract short and dense features for users and items respectively. The extracted features are then fed into another NeuMF together to output the predicted probability. User neighborhood embedding module conducts the GCN model to obtain the specific user's neighborhood representation. KGE module is the same as MKR.

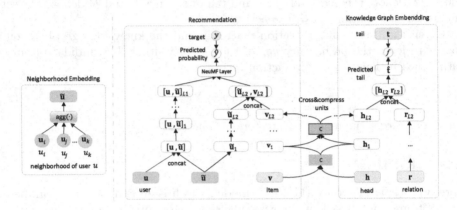

Fig. 1. Framework of Social-MKR.

3.4 Neighborhood Embedding Module

Construct a Social Network. The Social-MKR model constructs the social relationship ξ_{ij} between user i and user j according to the user-item interaction matrix Y:

$$\xi_{ij} = \begin{cases} 1, & \sum_{v \in V} y_{iv} \cdot y_{jv} \geq k \\ 0, & \text{otherwise} \end{cases} \tag{5}$$

where k is the interaction threshold, $\xi_{ij} = 1$ means that users u_i and u_j have visited k or more of the same items at the same time; otherwise, $\xi_{ij} = 0$.

Preference Propagation. It is generally believed that humans usually acquire and disseminate knowledge through acquaintances such as friends, colleagues, or partners, which means that user's potential social networks may play a fundamental role in helping them filter information [8].

Social networks usually contains fruitful facts and connectivity information between users. For example, as illustrated in Fig. 2, user b is linked with user a, user d and user c, while user c is further linked with the book and watch he has purchased. These complicated connections in Social networks provide us a

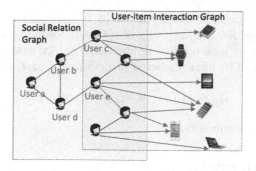

Fig. 2. Illustration of the social network in the recommender system.

deep and latent perspective to explore user preferences. For example, if a user has purchased a certain item, then the probability of his friend buying the item becomes greater.

To characterize user's hierarchically extended preferences in social networks, we represent user u's neighbor information by aggregation operation (see Fig. 1) as follows:

$$N_u = \{u_j | \xi_{uj} = 1, u_j \in U\} \tag{6}$$

$$\widetilde{\mathbf{u}} = \mathrm{agg}(N_u) = \frac{1}{|N_u|} \sum_{j \in N_u} \mathbf{u}_j \tag{7}$$

where N_u represents the set of neighbors of user u, $\mathbf{u}_j$ is the embedding vector of user j, and $\mathrm{agg}(\cdot)$ is an aggregation function.

3.5 Recommendation Module

Three raw feature vectors $\mathbf{u}$, $\widetilde{\mathbf{u}}$ and $\mathbf{v}$ are taken as the input of the recommendation module which describe user u, u's neighborhood and item v, respectively. $\mathbf{u}$ and $\mathbf{v}$ can be customized as one-hot ID [4], attributes [11], bag-of-words [10], or their combinations, based on the application scenario. Given user u and his neighbors set N_u, we use an $L1$-layer MLP to extract their common latent feature and an $L2$-layer MLP to extract u's neighborhood latent condensed feature:

$$[\mathbf{u}, \widetilde{\mathbf{u}}]_{L1} = \mathcal{M}(\mathcal{M}(\cdots \mathcal{M}([\mathbf{u}, \widetilde{\mathbf{u}}]))) = \mathcal{M}^{L1}([\mathbf{u}, \widetilde{\mathbf{u}}]) \tag{8}$$

$$\widetilde{\mathbf{u}}_{L2} = \mathcal{M}(\mathcal{M}(\cdots \mathcal{M}(\widetilde{\mathbf{u}}))) = \mathcal{M}^{L2}(\widetilde{\mathbf{u}}) \tag{9}$$

where $\mathcal{M}(\mathbf{x}) = \sigma(\mathbf{W}\mathbf{x} + \mathbf{b})$ is a fully-connected neural network layer with weight $\mathbf{W}$, bias $\mathbf{b}$, and nonlinear activation function $\sigma(\cdot)$. For item v, we use $L2$ cross&compress units to extract its feature:

$$\mathbf{v}_{L2} = \mathbb{E}_{e \sim S(v)} \left[\mathbf{C}^{L2}(\mathbf{v}, \mathbf{e})[\mathbf{v}] \right] \tag{10}$$

where $S(v)$ is the set of associated entities of item v.

After obtaining common latent feature $[\mathbf{u}, \widetilde{\mathbf{u}}]_{L1}$ of user u and his neighborhood, common latent feature $[\widetilde{\mathbf{u}}_{L2}, \mathbf{v}_{L2}]$ of u's neighborhood and item v, we combine the two pathways by a predicting function f_{RS}, for example, inner product or an NeuMF layer. The final predicted probability of user u visiting item v is:

$$\hat{y}_{uv} = \sigma\Big(f_{RS}\big([\mathbf{u}, \widetilde{\mathbf{u}}]_{L1}, [\widetilde{\mathbf{u}}_{L2}, \mathbf{v}_{L2}]\big)\Big) \tag{11}$$

where $[\,,\,]$ means concatenation.

3.6 Learning Algorithm

The complete loss function of Social-MKR is as follows:

$$
\begin{aligned}
\mathcal{L} &= \mathcal{L}_{RS} + \mathcal{L}_{KG} + \mathcal{L}_{REG} \\
&= \sum_{u \in U, v \in V} \mathcal{J}(\hat{y}_{uv}, y_{uv}) \\
&\quad - \lambda_1 \Big(\sum_{(h,r,t) \in \mathcal{G}} score(h,r,t) - \sum_{(h',r,t') \notin \mathcal{G}} score(h',r,t') \Big) \\
&\quad + \lambda_2 \|\mathbf{W}\|_2^2
\end{aligned}
\tag{12}
$$

In Eq. 12, the first term is the loss of the recommender system, where $\hat{y}_{uv}$ is predicted probability of user u visiting item v, and $\mathcal{J}$ is the cross-entropy function. The second term is the loss of KGE module, in which (h, r, t) and (h', r, t') represent true triples and false triples, respectively, and $score(\cdot)$ is the score function of the knowledge graph feature learning module. Last term is the regularization term, which can prevent overfitting.

The learning algorithm of Social-MKR is presented in Algorithm 1. Before model training, we first build user social relationships and similar user sets (lines 2–3), Next, we perform neighbor aggregation operations (line 4). The model training process includes two parts, namely the recommendation task (lines 7–11) and the KGE task (Lines 13–15).

4 Experiments

In this section, we evaluate the performance of Social-MKR in three real-world recommendation scenarios: movie, book and music.

4.1 Datasets

We utilize three datasets including MovieLens-1M[1], Book-Crossing[2] and Last.FM[3] dataset in our experiments. The basic statistics of the three datasets are presented in Table 1.

---

[1] https://grouplens.org/datasets/movielens/1m/.
[2] http://www.informatik.uni-freiburg.de/~cziegler/BX/.
[3] https://grouplens.org/datasets/hetrec-2011/.

Algorithm 1: Multi-task learning algorithm for Social-MKR
_____
Input: Interaction matrix Y, knowledge graph $\mathcal{G}$, $u \in U$ $v \in V$
Output: Prediction function $\mathcal{F}(u, v | \Theta, Y, \mathcal{G})$
1 Initialize all parameters
2 $\xi_{ij} \leftarrow Rel(Y, i, j)$ for $\forall u_i, u_j \in U$
3 $N_u = \{u_j | \xi_{uj} = 1, u_j \in U\}$ for $\forall u \in U$
4 $\tilde{u} = agg(N_u)$
5 **for** number of training iteration **do**
6   //recommendation task
7   **for** t steps **do**
8          $\tilde{Y} \leftarrow \{(u, v) | y_{uv} = 1\} \bigcup \{(u', v') | y_{u'v'} = 0\}$, $|(u, v)| = s_1$, $|(u, v)| = s_2$;
9   Sample $e \sim S(v)$ for each item v in the $\tilde{Y}$;
10   Update parameters of $\mathcal{F}$ by gradient descent on Eq.(3-4),(8-12);
11   **end for**
12   //knowledge graph embedding task
13     $\tilde{\mathcal{G}} \leftarrow \{(h, r, t) | (h, r, t) \in \mathcal{G}\} \bigcup \{(h', r, t') | (h', r, t') \notin \mathcal{G}\}$, $|(h, r, t)| = s_1$, $|(h', r, t')| = s_2$;
14   Sample $v \sim S(h)$ for each head h in the $\tilde{\mathcal{G}}$;
15   Update parameters of $\mathcal{F}$ by gradient descent on Eq.(1)-(3), (7-9) in the literature [13];
16 **end for**
_____

Table 1. Basic statistics and hyper-parameter settings for the three datasets.

Dataset	Users	Items	Interactions	KG triples	Hyper-parameters
MovieLens-1M	6,036	2,347	753,775	20,195	$L1 = 3, L2 = 1, d = 8, t = 3, \lambda_1 = 0.5$
Book-Crossing	17,860	14,910	139,746	19,793	$L1 = 1, L2 = 1, d = 8, t = 2, \lambda_1 = 0.1$
Last.FM	1,872	3,846	42,346	15,518	$L1 = 2, L2 = 2, d = 4, t = 2, \lambda_1 = 0.1$

4.2 Experiments Setup

We set f_{RS} as inner product, and $\lambda_2 = 10^{-6}$ for all three datasets, and other hyper-parameter are given in Table 1. The settings of hyper-parameters are determined by optimizing AUC on a validation set.

We use the below two evaluation systems to evaluate the performance of our method: (1) In click-through rate (CTR) prediction, we output the predicted click probability by applying the trained model to each interaction in the test set. We use two metrics, AUC and Accuracy, to evaluate the performance of CTR prediction. (2) In top-K recommendation, we select K items with highest predicted click probability for each user in the test set by using the trained model, and use Precision@K and Recall@K as two metrics.

4.3 Empirical Study

We conduct an empirical study to investigate the correlation between the size of the interaction threshold k and the performance of the experiment. Specifically, we aim to reveal how AUC changes with different interaction threshold. The results are shown in Fig. 3, which clearly shows that as the threshold increases, AUC first increases and then decreases. Analyzing the reasons, we found that if the interaction threshold is too low to accurately filter out similar users, the social network introduced will become "noise" instead. Only by setting appropriate interaction threshold according to the actual conditions of different application scenarios can we correctly capture the preferences of different users. This is consistent with our intuitive feeling.

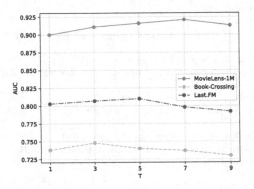

Fig. 3. The impact of threshold size on AUC.

4.4   Results

Comparison with Baselines. We compare the Social-MKR model with some representative recommendation algorithms (PER [18], RippleNet [12], LibFM [6], MKR [13], etc.). The results of all methods in CTR prediction and top-K recommendation are presented in Table 2 and Fig. 4, 5, respectively. We have the following observations:

- PER does not perform well in movie, book, and music recommendations, because user-defined meta-paths are difficult to achieve optimal in reality.
- RippleNet performs better in movies recommendation than other two datasets. This is because RippleNet is sensitive to the data set. When the MovieLens-1M data set is relatively dense, RippleNet can make recommendations more accurately by propagating user's preferences on the knowledge graph.
- MKR performs best in all baselines, which shows that MKR can accurately capture user's interests and preferences by establishing a cross-compression unit.
- In general, our Social-MKR performs best among all methods on the three datasets. Specifically, Social-MKR achieves average AUC gains of 5.8%, 6.9% and 7.9% in movie, book and music recommendation, respectively, which prove the efficacy of the model incorporating social relationships. Social-MKR also achieves outstanding performance in top-K recommendation as shown in Fig. 4 and Fig. 5.

Table 2. The results of AUC and accuracy in CTR prediction.

Model	MovieLens-1M		Book-Crossing		Last.FM	
	AUC	ACC	AUC	ACC	AUC	ACC
PER	0.710(−22.9%)	0.664(−21.9%)	0.623(−16.7%)	0.588(−17.2%)	0.633(−21.9%)	0.596(−21.3%)
Ripplenet	0.920(−0.1%)	0.842(−0.9%)	0.729(−2.5%)	0.662(−6.8%)	0.768(−5.2%)	0.691(−8.7%)
LibFM	0.892(−3.1%)	0.812(−4.5%)	0.685(−8.4%)	0.640(−9.9%)	0.777(−4.1%)	0.709(−6.3%)
Wide& Deep	0.898(−2.5%)	0.820(−3.5%)	0.712(−4.8%)	0.624(−12.1%)	0.756(−6.7%)	0.688(−9.1%)
MKR	0.917(−0.4%)	0.843(−0.8%)	0.734(−1.9%)	0.704(−0.8%)	0.797(−1.6%)	0.752(−0.7%)
Social-MKR	**0.921**	**0.850**	**0.748**	**0.710**	**0.810**	**0.757**

Results in Sparse Scenarios. To evaluate the performance of the Social-MKR
model in sparse data scenarios, we observe how the AUC of all methods in CTR
prediction changes with the ratio of training set of MovieLens-1M changes. The
results are shown in Table 3. Compared with using the complete training set
(r = 100%), when r = 10%, the AUC score decreases by 15.8%, 8.4%, 10.2%,
12.2%, 5.3% for PER, RippleNet, LibFM, Wide&Deep and MKR, respectively.
On the contrary, the Social-MKR model only decreased by 5.1%, which is less
than other baseline models. This shows that the introduction of social networks
can effectively alleviate the problem of data sparsity.

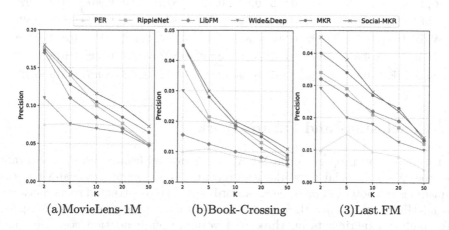

(a)MovieLens-1M   (b)Book-Crossing   (3)Last.FM

Fig. 4. The results of Precision@K in top-K recommendation.

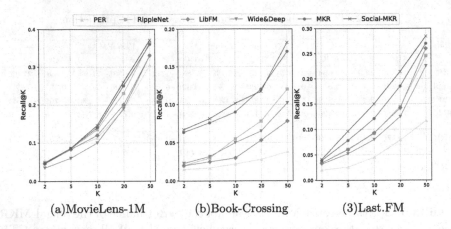

Fig. 5. The results of Recall@K in top-K recommendation.

Table 3. Results of AUC on MovieLens-1M in CTR prediction with different ratios of training set r.

Model	r									
	10%	20%	30%	40%	50%	60%	70%	80%	90%	100%
PER	0.598	0.607	0.621	0.638	0.647	0.662	0.675	0.688	0.697	0.710
Ripplenet	0.843	0.851	0.859	0.862	0.870	0.878	0.890	0.901	0.912	0.920
LibFM	0.801	0.810	0.816	0.829	0.837	0.850	0.864	0.875	0.886	0.892
Wide&Deep	0.788	0.802	0.809	0.815	0.821	0.840	0.858	0.876	0.884	0.898
MKR	0.868	0.874	0.881	0.882	0.889	0.897	0.903	0.908	0.913	0.917
Social-MKR	0.874	0.880	0.887	0.896	0.904	0.911	0.913	0.917	0.919	0.921

5   Conclusions and Future Work

In this paper, we propose Social-MKR, an end-to-end framework that naturally incorporates the social networks into recommender systems. Social-MKR can capture user's potential preferences according to the constructed social network. Social-MKR realizes click-through rate prediction by sharing user's feature space. We conduct experiments on three real-world recommendation scenarios, and the results demonstrate that Social-MKR outperforms several state-of-the-art methods. Besides, Social-MKR can effectively alleviate data sparsity problems.

For future work, we plan to (1) investigate other methods of incorporating social networks; (2) utilize the multi-hop information between users in social networks to better explore the potential interests of users.

References

1. Bordes, A., Usunier, N., Garcia-Duran, A., Weston, J., Yakhnenko, O.: Translating embeddings for modeling multi-relational data. In: Neural Information Processing Systems (NIPS), pp. 1–9 (2013)
2. Cao, Y., Wang, X., He, X., Hu, Z., Chua, T.S.: Unifying knowledge graph learning and recommendation: Towards a better understanding of user preferences. In: The World Wide Web Conference, pp. 151–161 (2019)
3. Chen, J., Zhang, H., He, X., Nie, L., Liu, W., Chua, T.S.: Attentive collaborative filtering: Multimedia recommendation with item-and component-level attention. In: Proceedings of the 40th International ACM SIGIR Conference on Research and Development in Information Retrieval, pp. 335–344 (2017)
4. He, X., Liao, L., Zhang, H., Nie, L., Hu, X., Chua, T.S.: Neural collaborative filtering. In: Proceedings of the 26th International Conference on World Wide Web, pp. 173–182 (2017)
5. Koren, Y., Bell, R., Volinsky, C.: Matrix factorization techniques for recommender systems. Computer **42**(8), 30–37 (2009)
6. Rendle, S.: Factorization machines with libfm. ACM Trans. Intell. Syst. Technol. (TIST) **3**(3), 1–22 (2012)
7. Salakhutdinov, R., Mnih, A.: Bayesian probabilistic matrix factorization using Markov chain Monte Carlo. In: Proceedings of the 25th International Conference on Machine Learning, pp. 880–887 (2008)
8. Sinha, R.R., Swearingen, K., et al.: Comparing recommendations made by online systems and friends. DELOS **106** (2001)
9. Sun, Y., Yuan, N.J., Xie, X., McDonald, K., Zhang, R.: Collaborative intent prediction with real-time contextual data. ACM Trans. Inform. Syst. (TOIS) **35**(4), 1–33 (2017)
10. Wang, H., Wang, N., Yeung, D.Y.: Collaborative deep learning for recommender systems. In: Proceedings of the 21th ACM SIGKDD International Conference on Knowledge Discovery and Data Mining, pp. 1235–1244 (2015)
11. Wang, H., Zhang, F., Hou, M., Xie, X., Guo, M., Liu, Q.: Shine: signed heterogeneous information network embedding for sentiment link prediction. In: Proceedings of the Eleventh ACM International Conference on Web Search and Data Mining, pp. 592–600 (2018)
12. Wang, H., et al.: Ripplenet: propagating user preferences on the knowledge graph for recommender systems. In: Proceedings of the 27th ACM International Conference on Information and Knowledge Management, pp. 417–426 (2018)
13. Wang, H., Zhang, F., Zhao, M., Li, W., Xie, X., Guo, M.: Multi-task feature learning for knowledge graph enhanced recommendation. In: The World Wide Web Conference, pp. 2000–2010 (2019)
14. Wang, X., He, X., Cao, Y., Liu, M., Chua, T.S.: Kgat: Knowledge graph attention network for recommendation. In: Proceedings of the 25th ACM SIGKDD International Conference on Knowledge Discovery & Data Mining, pp. 950–958 (2019)
15. Wang, X., He, X., Nie, L., Chua, T.S.: Item silk road: recommending items from information domains to social users. In: Proceedings of the 40th International ACM SIGIR Conference on Research and Development in Information Retrieval, pp. 185–194 (2017)
16. Wang, Z., Zhang, J., Feng, J., Chen, Z.: Knowledge graph embedding by translating on hyperplanes. In: Proceedings of the AAAI Conference on Artificial Intelligence, vol. 28 (2014)

17. Yang, B., Yih, W.t., He, X., Gao, J., Deng, L.: Embedding entities and relations for learning and inference in knowledge bases. arXiv preprint arXiv:1412.6575 (2014)
18. Yu, X., et al.: Personalized entity recommendation: a heterogeneous information network approach. In: Proceedings of the 7th ACM International Conference on Web Search and Data Mining, pp. 283–292 (2014)
19. Zhang, F., Yuan, N.J., Lian, D., Xie, X., Ma, W.Y.: Collaborative knowledge base embedding for recommender systems. In: Proceedings of the 22nd ACM SIGKDD International Conference on Knowledge Discovery and Data Mining, pp. 353–362 (2016)

Multi-stage Knowledge Propagation Network for Recommendation

Feng Xue(✉)[iD], Wenjie Zhou[iD], Zikun Hong[iD], and Kang Liu[iD]

Hefei University of Technology, Hefei, China
feng.xue@hfut.edu.cn

Abstract. As the knowledge graph can provide items with rich attributes, it has become an important way to alleviate cold start and sparsity in recommender systems. Recently, some knowledge graph based collaborative filtering methods use graph convolution networks to aggregate information from each item's neighbors to capture the semantic relatedness, and significantly outperform the state-of-the-art methods. However, in the process of knowledge graph convolution, only the item nodes can make use of knowledge, while the user nodes only contain the original ID information. This gap in information modeling makes it difficult for prediction function to capture the user preference for high-order attribute nodes in knowledge graph, which leads to the introduction of noise data. In order to give full play to the ability of knowledge graph convolution in mining high-order knowledge, we propose Multi-Stage Knowledge Propagation Networks (MSKPN), an end-to-end recommender framework which combines the graph convolution on both knowledge graph and user-item graph. It uses the collaborative signal latent in user-item interactions to build an information propagation channel between the user nodes and item nodes, so as to complement user representations. We conduct extensive experiments on two public datasets, demonstrating that our MSKPN model significantly outperforms other state-of-the-art models. Further analyses are provided to verify the rationality of our model.

Keywords: Recommender system · Knowledge graph · Graph convolution networks · Collaborative filtering

1 Introduction

With the advance of Internet technology, modern recommender system (RS) becomes popular and has been widely used in online applications, such as news, movies, and commodities. The main task of RS is to find items that a certain user may prefer. Among these algorithms of recommendation, Collaborative Filtering (CF) is the most dominant one which is deeply researched and widely used [1,2]. A core assumption of CF believes that several users with similar historical choices may have similar interests and preferences, thus making similar choices in the future. By modeling the historical interaction data, we can obtain the

© Springer Nature Singapore Pte Ltd. 2021
B. Qin et al. (Eds.): CCKS 2021, CCIS 1466, pp. 253–264, 2021.
https://doi.org/10.1007/978-981-16-6471-7_19

representation vectors of a user and an item, then the similarity score between the two vectors is calculated by a designed function, which is regarded as the possibility of interaction. However, CF algorithms often run into difficulties in case of sparsity as well as the cold start, since CF mainly estimates the user's preference by reconstructing the user-item interaction from the historical data.

In order to address the above issues of CF, many works [3,4] have devoted to utilize side information. In the field of recommendation, side information generally refers to data other than interactive records, such as user social connections, item attributes. Knowledge Graph (KG) is a type of directed heterogeneous graph in which nodes represent real-world entities and edges indicate their connections. Due to its superiority, many efforts [5,6] integrate KG into recommendation algorithm as side information, making impressive improvement of performance. KG can provide both entities and their relations the way to learn expressive semantic information to alleviate cold start and sparsity problem in recommendation. Furthermore, relations with various types in KG are helpful for increasing the diversity of recommendation by extending users' interests.

Early works use Knowledge Graph Embedding (KGE) [7] to learn low-dimension representation vectors of entities and relations, which are then fed into recommender framework [8]. Nevertheless, commonly used KGE methods focus on modeling rigorous semantic relatedness, which are more suitable for in-graph applications, such as knowledge graph completion and link prediction rather than recommendation. Besides these KGE based methods, there are works [9,10] combine the semantic learning of KG with the modeling and optimization of RS. Recently, graph neural networks (GNNs), especially graph convolution networks (GCNs), have made effective progress in many research fields. Some works [11,12] use GCNs to generate item embeddings on KG, improving the quality of recommendation with the effect of knowledge aggregation.

Although these knowledge graph convolution based methods have achieved great performance improvement, only the item nodes therein have the opportunity to make use of the semantic and association knowledge inside the KG, while the user embeddings denote nothing but user ID information. When stacking multiple convolution layers, the information gap between the two kinds of embeddings weakens the effect of utilizing knowledge from multi-hop neighbor nodes. To solve this problem, the user embeddings are supposed to be complemented in some manners. Wang [13] constructs a bipartite graph with the interactions between users and items, and then applies GCNs on it to capture the collaborative signal hidden in the interactions. We think the collaborative signal is suitable to play the role as a bridge between users and items to enhance the user embeddings. Therefore, in this paper, we combine the graph convolutions on both bipartite graph and knowledge graph, and propose an end-to-end recommendation model called MSKPN. Firstly, we use an attention-based knowledge graph convolution network to aggregate the high-order knowledge into item nodes, and then feed all user nodes on the bipartite graph into another single layer GCN, so that the knowledge can be further propagated to user nodes. The combination of the above two steps, to a certain extent, alleviates the problem of performance decline caused by the information gap, to make better use of the semantics and associated information in the knowledge graph.

We conduct extensive experiments on two public datasets, demonstrating that our MSKPN model significantly outperforms other state-of-the-art models. The contributions of this work are summarized as follows:

- we highlight the importance of high-order knowledge and empirically combine graph convolution on both item's knowledge graph and user-item bipartite graph in RS to make better use of knowledge.
- We develop a new method called Multi-Stage Knowledge Propagation Network (MSKPN) which is capable of aggregating high-order knowledge and propagating it into representations of users and items simultaneously.
- We conduct extensive experiments on two public benchmarks, demonstrating the effectiveness of our MSKPN model.

2   Methodology

In this section, we present the proposed MSKPN model, which extracts the semantic and structural information hidden in knowledge graph and integrates it into item embeddings, and then propagates the knowledge from item nodes to user nodes. Figure 1 illustrates the overall architecture, which consists of three main components: (1) an attention-based knowledge graph convolution step that outputs an item's embedding by injecting the knowledge of its multi-hop neighbors; (2) a secondary knowledge propagation step that utilizes a user's interactions to enhance the user ID embedding; and (3) a prediction step that calculate the inner product to capture the affinity score of a target user-item pair.

2.1   Attention-Based Knowledge Graph Convolution

The input to our model contains a user-item pair and a knowledge graph where each item can also be regarded as an entity. In order to capture the high-order semantic and structural information contained in knowledge graph, we use a pre-defined convolution operation based on attention mechanism to extract knowledge. Specifically, a sub-graph of an item and its neighbor attribute nodes are fed into the convolution layer, together with the input of the target user's embedding. Wherein, the user, the item and the attributes nodes are all embedded using their ID mapping.

For simplicity, we first explain this step in a single layer manner that only the target item and its direct neighbors are taken into consideration. For each neighbor node of an item, there is a weight score to characterize its importance in knowledge propagation. The weight is calculated by attention mechanism in order to capture a user's potential interests. As shown in Fig. 1, we use u to represent a target user, i and n_j are entities that represent a target item and one

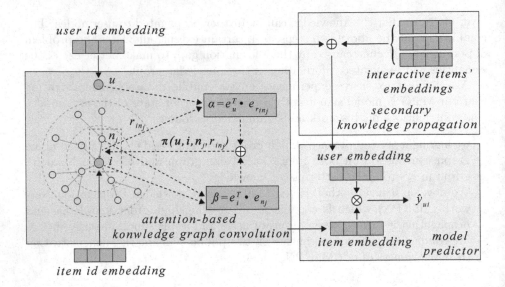

Fig. 1. Illustration of MSKPN model architecture.

of its neighbors in knowledge graph, and r_{in_j} represents the directed connection from n_j to i. The attention score is calculated as follows:

$$\pi(u, i, n_j, r_{in_j}) = \mathbf{e}_u^\top \cdot \mathbf{e}_{r_{in_j}} + \mathbf{e}_i^\top \cdot \mathbf{e}_{n_j} \tag{1}$$

where $\mathbf{e}_u \in \mathbf{R}^d$ and $\mathbf{e}_i \in \mathbf{R}^d$ are the ID embeddings for u and i; $\mathbf{e}_{n_j} \in \mathbf{R}^d$ and $\mathbf{e}_{r_{in_j}} \in \mathbf{R}^d$ are the embeddings of entity n_j and relation r_{in_j} respectively; d is the dimention of embeddings. The attention formula we use consists of two symmetrical parts, the former part models the importance of the relation r_{in_j} to the user u, and the latter part captures the inherent feature correlation between the head and tail entities. Finally, we use a softmax function to normalize all the weights of the information flow:

$$\tilde{\pi}(u, i, n_j, r_{in_j}) = \frac{\exp(\pi(u, i, n_j, r_{in_j}))}{\sum_{n_k \in N(i)} \exp(\pi(u, i, n_k, r_{in_k}))} \tag{2}$$

where $N(i)$ is the set of item i's neighbor attribute nodes in knowledge graph.

The new embedding of item i is composed of its ID embedding and the combination of its neighbors' embeddings so that the representation of i can capture the semantic and structure knowledge. The formula is as follows:

$$\mathbf{e}_i^* = \mathbf{e}_i + \sum_{n_j \in N(i)} \tilde{\pi}(u, i, n_j, r_{in_j})\mathbf{e}_{n_j} \tag{3}$$

where $\mathbf{e}_i^*$ is the item i's embedding which has aggregated the knowledge.

So far, we have clarified how to aggregate information from an item's direct neighbors in the knowledge graph. And an iterative method can be used to add multi-hop neighbors into convolution. We propagate the embedding of each entity to its direct neighbors so that each node contains the information of its own and its neighbors. Then we repeat this procedure for several times, thus each entity can receive knowledge from the information flow of its high-order neighbors.

2.2   Secondary Knowledge Propagation

After the knowledge graph convolution described in Sect. 2.1, all the item representations contain high-order knowledge. However, the user representations only contain the original ID information. The imbalance of the information contained in item embeddings and user embeddings makes it difficult for the subsequent prediction step to model the user's preference for the knowledge contained in the item. To complement the user embeddings, we construct a user-item bipartite graph and use another GCN to further capture the hidden collaborative signal so that the knowledge learned by the previous step can be propagated through the information flow from item nodes to user nodes. For a given user u, we get her new embedding as follows:

$$\mathbf{e}_u^* = \mathbf{e}_u + \sum_{i \in N(u)} \frac{1}{\sqrt{|N(u)|} \cdot \sqrt{|N(i)|}} \mathbf{e}_i^* \tag{4}$$

where $N(u)$ is the set of user u' interactive items, $N(i)$ is the set of item i' interactive users.

The final user embedding is also composed of two parts, one is her original user ID embedding, the other is the aggregation embedding composed of its all interactive items. Here, we employ symmetric sqrt normalization $\frac{1}{\sqrt{|N(u)|} \cdot \sqrt{|N(i)|}}$ on each item embedding to prevent user embedding from over smooth due to the excessive number of neighbors. Compared with normalization method that only at the left side or at the right side, symmetric sqrt normalization can take both the target node's and the neighbor node's popularity into consideration. It is more conducive to preserving the personalized collaborative signal [14].

2.3   Model Predictor and Optimization

For a given user-item pair, after the above two convolution steps, we get the final user embedding $\mathbf{e}_u^*$ and item embedding $\mathbf{e}_i^*$, then we calculate the probability of interation between the target user and item:

$$\hat{y}_{ui} = f(\mathbf{e}_u^*, \mathbf{e}_i^*) \tag{5}$$

here, f is the inner product function.

To optimize the model, we use point-wise loss which has been widely used in machine learning and recommender system. Specifically, we learn the predicted score for each user-item pair in the training set and try to make the score close to the observed value, where 1 for positive examples and 0 for the negative ones. We have the objective function as:

$$L = -\frac{1}{N}\left(\sum_{(u,i)\in Y^+} \log \sigma(\hat{y}_{ui}) + \sum_{(u,i)\in Y^-} \log\left(1 - \sigma(\hat{y}_{ui})\right) \right) + \lambda||\theta||^2 \qquad (6)$$

Where N is the size of training set, θ denotes all learnable parameters and λ is the coefficient of L_2 regularization. Y^+ indicates the set of observed interactions. Negative samples Y^- are randomly sampled from the dataset which is of the same size as the positive set. We then iteratively optmize the loss function by mini-batch Adam optimizer.

3 Experiments

In this section, we conduct plenty of experiments on two real-world datasets to evaluate our MSKPN model and answer questions as follows:

- **RQ1**: How does our model perform compared with other state-of-the-art collaborative filtering models, especially knowledge graph-based methods?
- **RQ2**: How does our special component (secondary knowledge propagation step) affects the influence of high-order knowledge on model performance?

3.1 Dataset Description

We conduct comprehensive experiment on two benchmark datasets: MovieLens-20M and Last-FM.

MovieLen-20M[1]: MovieLens is a widely used benchmark dataset to investigate the performance of recommendation algorithms. In this experiment, we select the largest version including 20 million ratings.

Last-FM[2]: This is the musician listening dataset collected from last.fm online music systems.

We use the KG constructed by KGCN. [11]. The datasets have been transformed into implicit feedback. For movielens-20M, we only use ratings greater than 4 as positive instances, while take all ratings in Last.FM as positive samples due to its sparsity. For each user, we randomly sample the same number of negative items as the user interacted. We hold out the 60%, 20% and 20% instances to construct the training, evaluation and test sets. Table 1 shows the statistics of the two datasets.

3.2 Experimental Settings

Evaluation Protocols. We use two metrics to evaluate our trained models: AUC and F1, which are commonly used to evaluate the performance in recommendation field.

---

[1] https://grouplens.org/datasets/movielens/.
[2] https://grouplens.org/datasets/hetrec-2011/.

Table 1. Basic statistics of the two datasets.

	MovieLens	Last.FM
#Users	138,159	1,872
#Items	16,954	3,846
#Interactions	13,501,622	42,346
#Entities	102,569	9,366
#Relations	32	60
#Triples	499,474	15,518

Baselines. We compare the proposed method with the following baselines:

- **SVD** [15]. It factorized the user-item interaction matrix into the product of two lower dimensionality rectangular matrices. The row or column of the two matrices correspond to a specific user or item.
- **LibFM** [16]. FMs (Factorization Machines) are supervised learning approaches that can model the second-order interactions of features. LibFM is a software implementation for FMs.
- **PER** [6]. It represents the connectivity between users and items by meta-path.
- **CKE** [8]. It is an integrated framework that jointly learns the representations of CF and knowledge base (text, pictures).
- **RippleNet** [9]. This is a memory-network-like approach that propagates users' preferences on KG for recommendation.
- **KGCN** [11]. This is a state-of-the-art GCN-based method that captures inter-item relatedness effectively by mining their associated attributes on the KG.

3.3 Performance Comparison (RQ1)

Table 2. Overall performance comparison.

Method	Movielens		Last.FM	
	AUC	F1	AUC	F1
SVD	0.963	0.919	0.769	0.696
LibFM	0.959	0.906	0.778	0.710
CKE	0.924	0.871	0.744	0.673
PER	0.832	0.788	0.633	0.596
RippleNet	0.968	0.912	0.780	0.702
KGCN	0.978	0.932	0.794	0.719
MSKPN	**0.983**	**0.943**	**0.854**	**0.778**
%Improv.	0.51%	1.18%	7.56%	8.21%

Overall Comparision. Table 2 reports the experimental results on the two datasets. We have the following observations:

- PER achieves poor performance on the two datasets. This indicates that handcrafted meta-paths cannot take full advantage of the KG. In addition, the quality of meta-path relies heavily on domain knowledge, further limiting the performance.
- CKE consistently outperforms PER across all cases, demonstrating the superiority of simultaneous learning of KGE and recommendation model, since CKE integrates CF with item's embedding mined from knowledge base.
- The two KG-free models, SVD and LibFM, perform well and are even better than the KG-based baseline CKE. Their high performance is due to extra consideration of interaction in the embedding step. So that they are able to capture collaborative signal to some extent. On the other hand, the relatively poor performance of CKE verifies TransR which is commonly used for KG completion is not entirely suitable for recommendation task.
- RippleNet and KGCN perform better than SVD and LibFM on the both datasets. This proves that the proper use of auxiliary information provided by the KG is capable of benefiting the performance improvement.
- MSKPN outperforms all the baselines across the two datasets with significant improvements, proofing that the design of our MSKPN model is meaningful. Compared with the best baseline KGCN, our MSKPN extra takes attribute nodes into consideration when calculating attention, which optimizes the weights of konwledge propagation in knowledge graph embedding step. For example, in a movie knowledge graph, two different actors play in the same movie. As attribute nodes, the two actors share the same relation with the movie nodes, but the influence of the two actors on user interest is different. Moreover, we particularly design Secondary Knowledge Propagation step to further model knowledge. The two different steps of information propagation integrate knowledge into the representation of items and users respectively so that balance the information content of two kinds of embeddings. It is helping for the prediction function because the inner product mainly calculates the similarity between two embeddings.

3.4   Effect of the Secondary Knowledge Propagation (RQ2)

An important advantage of GCNs is the ability of iteratively aggregating high-order neighbors' information into the central node. Taking high-order neighbors of item in the knowledge graph into account should be able to better model user preferences. However, the deeper H in KGCN seems to impair the performance of the recommendation model. We set the number of layer H as a hyper-parameter and carry out some leave-out experiments. The AUC of these experiments are shown in Table 3, where MSKPN-s model means removing the Secondary Knowledge Propagation step from MSKPN. From the results, we observe that the effects of the two models (KGCN and MSKPN-s) do not significantly improve and even decrease in most cases with the increase of H, which means that the two models can not make good use of the high-order knowledge.

Table 3. AUC results of three models with different number of layer.

H		1	2	3	4
MovieLens	KGCN	0.972	**0.976**	0.974	0.514
	MSKPN-s	**0.982**	0.981	0.981	0.975
	MSKPN	0.982	0.982	**0.983**	0.980
Last.FM	KGCN	**0.794**	0.723	0.545	0.534
	MSKPN-s	**0.832**	0.823	0.820	0.805
	MSKPN	0.843	0.851	**0.854**	0.853

One possible reason is that high-order neighbors of an item introduce both useful information and noise. The prediction function we use, inner product, is to measure the correlation or similarity between a user embedding and an item embedding to some extent. However, the poor user representation(the user id embedding) is not learned enough to model the user's correlation with high-order knowledge. Therefore, we particularly design the secondary knowledge propagation step, which uses historical interaction to construct a user-item bipartite graph and feeds all users to a single layer GCN based on the bipartite graph. In this way, we can make use of the collaborative signal hidden in interaction and complement user embeddings with the linear combination of interactive items' embeddings so that the prediction function can capture the correlation between the target item's high-order neighbors and the interactive items' high-order neighbors. In this case, we can make better use of the knowledge. As we can see from Table 3, the AUC of MSKPN is superior to MSKPN-s and KGCN, and gets better with the increase of H. This result verifies our assumption and shows that the high-order knowledge information extractd by GCNs is beneficial to recommender model. When H is up to 4, model performance begins to decline, which is reasonable since a too deep neighbor might introduce noise more than useful information to the representation learning.

4   Related Work

We review existing works that are most related to our work.

4.1   Collaborative Filtering

CF is the most widely used method in today's recommender systems, which utilizes the user's historical interaction to model the relationship between users and items. Among CF methods, Matrix Factorization (MF) has attracted wide attention from researchers and industry since its excellent performance. MF maps users and items into low-dimensional vectors and uses inner product to reconstruct the interaction as a prediction. To further improve the model performance, SVD++ [15] and FISM [17] incorporate user's interacted items as implicit feedback to encode user's representation, yielding convincing performance, which are followed by many techniques for recommender systems. Recently, attention network is well researched in recommender systems. NAIS [18] and DeepICF [2] differentiate the contribution of each interacted item to form the target user's representation by training the implicit data, and achieves the state-of-the-art performance. To overcome the insufficient expressive ability of inner product, NCF [1] uses Deep Neural Network (DNN) to capture complex interactions between users and items, verifying that DNN is helpful for modeling high-order interactions between users and items.

4.2   Graph-Based Methods

In this paper, we utilize KG to improve the performance of RS. Early works mainly use path-based methods which adopt pre-defined connectivity pattern to capture the relatedness in KG [19,20]. These patterns called meta-path are sequences of entity type, such as user-movie-user-movie. PER [6] introduce meta-path based latent features to represent the connectivity between users and items. HERec [21] generates entity embedding by meta-path based random walk and fuses the embedding with MF to predict rating score of users on items. However, such meta-path based methods have poor versatility since the quality of meta-path heavily relies on domain knowledge. Another kind of methods leverage KGE [22] to learn the representation of entities and relations. CKE [8] adopts a heterogeneous network embedding method to extract structure information of both nodes and relations, and integrates it into collaborative filtering framework. Due to the development of GCNs, recent works simultaneously learn the parameters of KG and recommendation model by introducing graph neural network. KGCN [11] generates the embedding of an item by adopting GCNs on KG, where different attribute nodes sharing a same relation type contribute the same to the item aggregation. In this case, KGCN can not capture the effect of the attribute nodes' inherent feature. To better model the item embeddings, our Attention-based Knowledge Graph Convolution step takes both of the relation types and the node attributes into consideration. And we specially design the Secondary Knowledge Propagation step in order to solve the problem of model performance degradation caused by high-order noise.

5   Conclusions

In this work, we utilize KG which contains massive relation and entities to improve the performance of recommender system. We highlight the importance of high-order neighbors' knowledge in generating representations of items, and propose a novel secondary knowledge propagation layer to complement user embeddings in order to deal with the noise problem. We conduct extensive experiments on two public datasets and our MSKPN model significantly outperform state-of-the-art baselines, demonstrating the effectiveness of our model. In the future, we plan to incorporate more side information such as social network together with KG to further explore the information propagation mechanism of GCNs.

References

1. He, X., Liao, L., Zhang, H., Nie, L., Hu, X., Chua, T.S.: Neural collaborative filtering. In: Proceedings of the 26th International Conference on World Wide Web, pp. 173–182. WWW 2017, International World Wide Web Conferences Steering Committee, Republic and Canton of Geneva, CHE (2017). https://doi.org/10.1145/3038912.3052569
2. Xue, F., He, X., Wang, X., Xu, J., Liu, K., Hong, R.: Deep item-based collaborative filtering for top-n recommendation. ACM Trans. Inf. Syst. **37**(3), 1–25 (2019). https://doi.org/10.1145/3314578
3. Cheng, H.T., et al.: Wide & deep learning for recommender systems. In: Proceedings of the 1st Workshop on Deep Learning for Recommender Systems. ACM (2016). https://doi.org/10.1145/2988450.2988454
4. Elkahky, A.M., Song, Y., He, X.: A multi-view deep learning approach for cross domain user modeling in recommendation systems. In: Proceedings of the 24th International Conference on World Wide Web. International World Wide Web Conferences Steering Committee (2015). https://doi.org/10.1145/2736277.2741667
5. Catherine, R., Cohen, W.: Personalized recommendations using knowledge graphs: a probabilistic logic programming approach. In: Proceedings of the 10th ACM Conference on Recommender Systems. ACM, September 2016. https://doi.org/10.1145/2959100.2959131
6. Yu, X., et al.: Personalized entity recommendation: a heterogeneous information network approach. In: Proceedings of the 7th ACM International Conference on Web Search and Data Mining. ACM, February 2014. https://doi.org/10.1145/2556195.2556259
7. Lin, Y., Liu, Z., Sun, M., Liu, Y., Zhu, X.: Learning entity and relation embeddings for knowledge graph completion. In: Bonet, B., Koenig, S. (eds.) AAAI. pp. 2181–2187. AAAI Press (2015)
8. Zhang, F., Yuan, N.J., Lian, D., Xie, X., Ma, W.Y.: Collaborative knowledge base embedding for recommender systems. In: Proceedings of the 22nd ACM SIGKDD International Conference on Knowledge Discovery and Data Mining. ACM, August 2016. https://doi.org/10.1145/2939672.2939673
9. Wang, H., et al.: RippleNet: propagating user preferences on the knowledge graph for recommender systems. In: Proceedings of the 27th ACM International Conference on Information and Knowledge Management. ACM, October 2018. https://doi.org/10.1145/3269206.3271739
10. Wang, X., Wang, D., Xu, C., He, X., Cao, Y., Chua, T.S.: Explainable reasoning over knowledge graphs for recommendation. Proceedings of the AAAI Conference on Artificial Intelligence, vol. 33, pp. 5329–5336, July 2019. https://doi.org/10.1609/aaai.v33i01.33015329
11. Wang, H., Zhao, M., Xie, X., Li, W., Guo, M.: Knowledge graph convolutional networks for recommender systems. In: The World Wide Web Conference on WWW 2019. ACM Press (2019). https://doi.org/10.1145/3308558.3313417
12. Wang, H., et al.: Knowledge-aware graph neural networks with label smoothness regularization for recommender systems. In: Proceedings of the 25th ACM SIGKDD International Conference on Knowledge Discovery & Data Mining. ACM, July 2019. https://doi.org/10.1145/3292500.3330836
13. Wang, X., He, X., Wang, M., Feng, F., Chua, T.S.: Neural graph collaborative filtering. In: Proceedings of the 42nd International ACM SIGIR Conference on Research and Development in Information Retrieval. ACM, July 2019. https://doi.org/10.1145/3331184.3331267

14. Kipf, T.N., Welling, M.: Semi-supervised classification with graph convolutional networks. In: ICLR (Poster). OpenReview.net (2017)
15. Koren, Y.: Factorization meets the neighborhood: a multifaceted collaborative filtering model. In: Proceeding of the 14th ACM SIGKDD International Conference on Knowledge Discovery and Data Mining - KDD 2008. ACM Press (2008). https://doi.org/10.1145/1401890.1401944
16. Rendle, S.: Factorization machines with libFM. ACM Trans. Intell. Syst. Technol. **3**(3), 1–22 (2012). https://doi.org/10.1145/2168752.2168771
17. Kabbur, S., Ning, X., Karypis, G.: FISM: factored item similarity models for top-n recommender systems. In: Proceedings of the 19th ACM SIGKDD International Conference on Knowledge Discovery and Data Mining. ACM, August 2013. https://doi.org/10.1145/2487575.2487589
18. He, X., He, Z., Song, J., Liu, Z., Jiang, Y.G., Chua, T.S.: NAIS: neural attentive item similarity model for recommendation. IEEE Trans. Knowl. Data Eng. **30**(12), 2354–2366 (2018). https://doi.org/10.1109/tkde.2018.2831682
19. Zhao, H., Yao, Q., Li, J., Song, Y., Lee, D.L.: Meta-graph based recommendation fusion over heterogeneous information networks. In: Proceedings of the 23rd ACM SIGKDD International Conference on Knowledge Discovery and Data Mining. ACM (2017). https://doi.org/10.1145/3097983.3098063
20. Hu, B., Shi, C., Zhao, W.X., Yu, P.S.: Leveraging meta-path based context for top-n recommendation with a neural co-attention model. In: Proceedings of the 24th ACM SIGKDD International Conference on Knowledge Discovery & Data Mining. ACM, July 2018. https://doi.org/10.1145/3219819.3219965
21. Shi, C., Hu, B., Zhao, W.X., Yu, P.S.: Heterogeneous information network embedding for recommendation. IEEE Trans. Knowl. Data Eng. **31**(2), 357–370 (2019). https://doi.org/10.1109/tkde.2018.2833443
22. Wang, Q., Mao, Z., Wang, B., Guo, L.: Knowledge graph embedding: a survey of approaches and applications. IEEE Trans. Knowl. Data Eng. **29**(12), 2724–2743 (2017). https://doi.org/10.1109/tkde.2017.2754499

Knowledge Graph Open Resources

TGKG: New Data Graph Based on Game Ontology

Jianshun Sang, Wenqiang Liu$^{(\boxtimes)}$, Bei Wu, Hao Guo, Dongxiao Huang, and Yiqiao Jiang

Tencent Inc, Shenzhen, China
masonqliu@tencent.com

Abstract. With the advent of the big data era, knowledge graph embodies great advantages. Especially in game domain, building a knowledge graph is of great value. However, some open-domain knowledge bases is not designed for games and the amount of game data is relatively limited. In this paper, we propose a game ontology. In addition, in the process of constructing game knowledge graph, we found it is import to identify an entity is a game entity or not. Thus, we propose a new DLC entity identification model to help us use the ontology to build the graph.

Keywords: Game ontology · DLC identification · TGKG

Resource type: Ontology and Dataset
Permanent URL: https://github.com/sangjianshun/TGKG.

1 Introduction

With the rapid development of game domain, plenty of games from different game types (like mobile, pc and console) have sprung up. To embrace the vision of the Semantic Web, it is necessary to construct a knowledge graph based on huge amounts of game data. Indeed, there are some open-domain knowledge bases like Freebase [1], YAGO [2], DBpedia [3] and Wikidata [4] involving game data as well. Although these bases are being widely used by many semantic applications, they are not knowledge graphs specifically designed for games, and the amount of game data is relatively limited. To the best of our knowledge, there is no prior work on building knowledge graph in game domain.

In this paper, we present a new game ontology and a new game dataset TGKG (Tencent Game Knowledge Graph) based on the ontology we proposed. TGKG contains generally but not exclusively many game entities from different types (like mobile, pc and console), people entities, company entities and their relations (like developOf, ownOf, sameEditionOf etc.).

For pc and console games, many entities are crawled from different websites. These entities include not only game entities but also DLC (Downloadable Content) entities. DLC entity does not belong to the game entity and needs to be screened and identified from the crawled entities. Otherwise, downstream task like entity alignment will be seriously affected because of mistaking DLC entity as a game entity. Unfortunately, it is not possible to directly identify whether a crawled entity is a DLC entity or a game entity on most websites. To solve this problem, we proposed a novel method to identify whether an entity is DLC entity or not. We summarize contributions of this paper as follows:

- To our best knowledge, we are the first to present a game ontology. The result is a classification scheme defined a hierarchy of classifications which includes different types of entities and their relations.
- We translate the game data extracted from different websites to RDF using our structure ontology. In addition, We construct a game knowledge graph named Tencent Game Knowledge Graph by these RDF.
- To solve the impact of DLC entity on construction of TGKG, we propose a new method to identify whether a crawled entity is a DLC entity or a game entity.

2   Development Methods

In this part, we will introduce our work in detail. We first follow the Resource Description Framework (RDF) model [5] and explain the TGKG definition.

TGKG Definition: TGKG is an RDF graph consisting of TGKG facts, where a TGKG fact is represented by an RDF triple to indicate the relation among game entity, people entity and company entity. For instance,

$$< tgkg : 113336883, tgkg : publishOf, tgkg : 1203918647104 > \qquad (1)$$

where tgkg is the IRI prefix[1], tgkg:113336883 is a game entity named PUBG Mobile, and tgkg:1203918647104 is a company entity named Tencent. This tgkg fact shows that the company Tencent publish the game PUBG Mobile.

2.1   Game Ontology

An overview of the major classes in the game ontology we proposed is given in Fig. 1. The proposed game ontology uses the OWL[2] to describe a conceptual model of the entities and their relations. OWL is a formal ontology language standardized by the World Wide Web Consortium (W3C).

The central concepts in the ontology are the classes tgkg:game, tgkg:people and tgkg:company. tgkg:people is people who designed a game or founded a company which is a studio of a game, namely, tgkg:people is people who has a relation with a game or a game company. tgkg:company is company which published or developed a game like Tencent (defined in TGKG as tgkg:113336883).

---

[1] http://www.semanticweb.org/databrain/ontologies/tgkgdata/.
[2] https://www.w3.org/TR/owl2-overview/.

Among all three entities, tgkg:game is the most important entity. According to our understanding of the game field, we divide game entities into four major categories (DLC entities, pc game entities, console game entities, mobile game entities). What needs to be pointed out is DLC entity does not belong to the game entity, therefore we have to separate the DLC entity and game entity. In fact, DLC entity cannot exist without a game entity. Thus, we add a relation isBasedof between DLC entity and game entity (pc game entity and console game entity). In addition, we add two subclasses named ios game entity and android game entity to the mobile game entity.

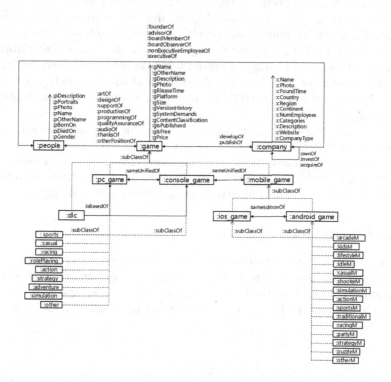

Fig. 1. An overview of the major classes in the game ontology

In the game field, genre is a very important feature. Genre reflects the relationship between different game entities to a certain extent. Nevertheless, different game entities have different genre categories. Thus, for pc game entities and console game entities, we add 9 subclasses (e.g. rolePlaying entity, adventure entity.) in these game entities based on the genre they belong. For ios game entities and android game entities, we add 15 subclasses (e.g. kidsM entity, lifestyleM entity, shooterM entity) based on the genre which mobile game entities belong. In fact, there are some genres which both pc, console games and mobile games jointly owned, therefore, we add a letter 'M' at the end of the subclasses to distinguish the genre entities from different game types. That is to say, the subclass with the letter 'M' at the end is the subclass of ios game entities and android game entities.

Based on the game entities defined in TGKG, we do two levels of entity alignment. Firstly, for ios game entities and android game entities, we add a relation sameEditionOf between two game if they are the same edition. Secondly, for pc game entities, console game entities and mobile game entities, we add a relation sameUnifiedOf between two game if they are a series of a game. In addition, we utilize the model we proposed to identify an entity is DLC entity or game entity. If we get a DLC entity of a game entity from our model, we add a relation isBasedOf between the DLC entity and pc or console game entity.

According to the data we crawled from different websites, we summarize 9 relationships (e.g. designOf, supportOf, programmingOf ect.) between people entity and game entity, 2 relationships (namely, developOf and PublishOf) between game entity and company entity, 6 relationships (e.g. founderOf, advisorOf, executiveOf etc.) between people entity and company entity and 3 relationships (namely, ownOf, investOf and acquireOf.) between company entity and company entity.

What's more, based on our observations on game data we crawled, we conclude 8 attributes in the people entity, each of these attributes has letter 'p' as the first letter. For company entity, we summarize 11 attributes and each of these attributes has letter 'c' as the first letter. In addition, we conclude 13 attributes in the game entity, each of these attributes has letter 'g' as the first letter.

2.2 DLC Entity Identification

The general process of the method of identifying whether a crawled entity is a DLC entity or a game entity, mainly includes two parts: The first part is a recall module aims to do a clustering based on the entities. The second part is a neural network module aims to further identify whether the candidate entity is a DLC entity or not.

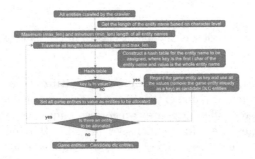

Fig. 2. Recall module

Recall Module. Figure 2 shows that the input of the recall module is the massive entities crawled by the crawler. Firstly, our recall module will count the length of these entity names based on the character level, thus, the maximum length and the minimum length of all entity names can be obtained. Secondly, we traverse all lengths between minimum length and maximum length to construct a hash table. The key of the hash table is a string composed of the first i characters of the entity name and the value is the whole entity name. Therefore, based on the construction strategy of the hash table, if the first i characters of the two entity names are the same, they will be aggregated together. This Recall module is based on this principle for rapid aggregation, and the time complexity can be reduced to $O(N * L)$, where N represents the number of entities crawler crawled and the magnitude is generally in the millions, what's more, L is generally a constant number.

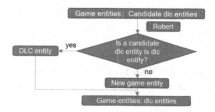

Fig. 3. Neural network module

Based on the observation of a large amount of entities, we found that the name of the DLC entity is usually an extension of the game entity name. Therefore, we assume that if the key of the hash table exists in the value of the hash table, it can be considered that the key is actually a game entity, and there is a high probability that these values of the hash table except for the game entity are DLC entities belonging to the game entity. In fact, it is also possible that there is a new game entity in the value of the hash table, so we regard the value as a candidate DLC entity set. If the key has already appeared in the value, the entities of the cluster will not enter the next iteration. The iteration will not end until all entities are assigned to the corresponding clusters based on the game entity. Finally, our recall module will generate all candidate DLC entities based on game entities.

Neural Network Module. To further improve the accuracy of the model, we design a neural network system shows in Fig. 3 to identify whether an entity is a DLC entity or not. We found that there is a description text in every entity crawled by crawler. From Fig. 4 we can see that the description text between DLC entity and game entity are different. That is to say, we can identify whether an entity is a game entity or not by the description of an entity. Therefore, we design a neural network module based on Robert model [6]. The input of the neural network module is the description text of an entity. After getting the distribution representation of the description's words, we average the representation to represent the description by the equation as follows

$$DocVector = \frac{1}{n}\sum_{i=1}^{n}\mathbf{x}_i \qquad (2)$$

and then we use a fully connected layer to output a two-dimensional vector. Based on the two-dimensional vector, we can do the DLC entity identification.

a. Game Entity   b. DLC Entity

Fig. 4. The description text between DLC entity and game entity. (a) is the description of a game entity. (b) is the description of a DLC entity.

3   Related Work

To bring the advantages of knowledge graph to the massive data, many open-domain knowledge bases like Freebase, YAGO, DBpedia and Wikidata were born. Although these bases are being widely used by many semantic applications, there is little game information within it. In addition, Zhishi.me [7] and XLore [8] are two early works on extracting open-domain knowldge from the Chinese Wikipedia, Hudong Baike and Baidu Baike. The difference between them is XLore automatically built an ontology using categories while Zhishi.me did not provide an ontology to describe the data crawled by crawler. By observations on game data we crawled, we manually constructed a game ontology and this schema can accurately describe the relation among game entities, people entities and company entities.

4   Conclusion and Future Work

This paper presents a game ontology, namely, a classification scheme defined a hierarchy of classifications which includes game entities, people entities, company entities and their relations. Based on the game ontology we proposed, we show the process of translating the game data crawled from different websites to RDF. In addition, we propose a new method to identify an entity whether a DLC entity or game entity. In future work, our plan is to integrate the TGKG we built with the open source knowledge graph.

References

1. Bollacker, K., Evans, C., Paritosh, P., Sturge, T., Taylor, J.: Freebase: a collaboratively created graph database for structuring human knowledge. In: SIGMOD 2008, pp. 1247–1250. ACM (2008)
2. Hoffart, J., Suchanek, F., Berberich, K., Weikum, K.: YAGO2: a spatially and temporally enhanced knowledge base from Wikipedia. Artif. Intell. **194**, 28–61 (2013)
3. Lehmann, J., et al.: DBpedia - a large-scale, multilingual knowledge base extracted from Wikipedia. Semant. Web J. **6**(2), 167–195 (2015)
4. Vrandečić, D., Krötzsch, M.: Wikidata: a free collaborative knowledgebase. Commun. ACM **57**(10), 78–85 (2014)
5. World Wide Web Consortium, et al.: RDF 1.1 concepts and abstract syntax (2014)
6. Liu, Y., Ott, M., Goyal, N., et al.: Roberta: a robustly optimized BERT pretraining approach. arXiv preprint arXiv:1907.11692 (2019)
7. Niu, X., Sun, X., Wang, H., Rong, S., Qi, G., Yu, Y.: Zhishi.me - weaving Chinese linking open data. In: Aroyo, L., et al. (eds.) ISWC 2011, Part II. LNCS, vol. 7032, pp. 205–220. Springer, Heidelberg (2011). https://doi.org/10.1007/978-3-642-25093-4_14
8. Wang, Z., Wang, Z., Li, J., Pan, J.Z.: Knowledge extraction from Chinese Wiki encyclopedias. J. Zhejiang Univ. Sci. C **13**(4), 268–280 (2012)

CSDQA: Diagram Question Answering in Computer Science

Shaowei Wang, Lingling Zhang[✉], Yi Yang, Xin Hu, Tao Qin, Bifan Wei, and Jun Liu

Xi'an Jiaotong University, Shaanxi, China
zhanglling@xjtu.edu.cn

Abstract. Visual Question Answering (VQA) has been a research focus of the computer vision community for recent years. Most of them are accomplished and verified on images of natural scenes. However, Diagram Question Answering (DQA) which is the task of answering natural language questions based on diagram is rarely noticed. Diagram is a more abstract carrier of knowledge and important resource composition in the multi-modal knowledge graph, research on it is of great significance for understanding the cognitive behavior of learners. In order to fill the scarcity of such data, this paper proposes the Computer Science Diagram Question Answering (CSDQA) dataset, which is the first geometric type of diagram dataset in this field. This dataset contains 1,294 diagrams with rich fine-grained annotations and 3,494 question-answer pairs, including multiple choice and true-or-false questions with two levels of difficulty. We have open sourced all the data in http://zscl.xjtudlc.com:888/CSDQA, hoping to provide convenience for researchers and make it the high-quality data foundation of DQA.

Keywords: Diagram · Question answering · Computer science

1 Introduction

Recent years, Visual Question Answering (VQA) received widespread attention as it is of great value in scenes such as aided-navigation and intelligent medical diagnosis. The task is to answer the natural language questions according to visual image context. Most studies concentrate on real-word scenes and are verified on large image datasets. For example, Zhang et al. [14] quantify domain shifts between popular VQA datasets such as COCO [4]. Hong et al. [1] propose a novel transformation driven visual reasoning task on TRANCE dataset. Similarly, VQA is commonly used in the education field to measure learners' mastery of knowledge. The difference is that the visual context becomes diagrams instead of natural images. Diagram is an abstract expression widely used in educational scenes, such as textbooks and slides. It's an essential component in the multi-model educational knowledge graph, often used to express information such as logic and concepts. Performing VQA on this more abstract visual form is helpful for computers to understand human cognitive behaviors and learning habits.

© Springer Nature Singapore Pte Ltd. 2021
B. Qin et al. (Eds.): CCKS 2021, CCIS 1466, pp. 274–280, 2021.
https://doi.org/10.1007/978-981-16-6471-7_21

However, there are still large gaps in the research related to the diagrams due to its more complex expressions and scarce data.

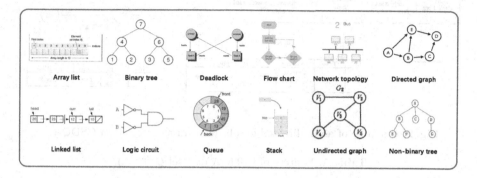

Fig. 1. Examples of each category diagrams in CSDQA.

Specifically, existing research mainly focus on the type of diagrams from natural subjects such as Biology and Geography. Morris et al. introduced SlideImages [7], a dataset which contains illustrations from educational presentations. Kembhavi et al. constructed AI2 Diagrams [2] (AI2D) dataset which contains diagrams from elementary school science textbooks. They also presented Textbook Question Answering [3] (TQA) dataset containing diagrams from life, earth and physics textbooks. They introduced the Diagram Parse Graphs to parse the structure of the diagrams, and cooperate with the natural language processing model to assist in solving questions. The second type of diagrams is mainly composed of graphic objects, such as circles, rectangles, and triangles. This type of diagram has abstract graphical representation and high-level logical relations, to understand it is a huge challenge, and related research is still blank.

In order to promote related research on the above-mentioned graphic type of diagrams, we build a novel Computer Science Diagram Question Answering (CSDQA) dataset. We summarize contributions of this paper as follows:

- We categorize the existing diagrams data types, and propose a new graphic type diagrams. On this type of data, we propose the novel DQA task, which is to complete the question answering task with the accompanying text as the natural language basis and the diagram as the computer vision basis.
- As shown in Fig. 1, we construct CSDQA dataset with over 1,000 diagrams of 12 categories. To our best knowledge, CSDQA is the first graphic diagram dataset and the first (Diagram Question Answering) DQA dataset in Computer Science domain which provides opportunities to expand the research methods and fields of VQA.
- We open source all the data in CSDQA. Researchers can easily obtain data and complete a variety of diagram understanding tasks including DQA.

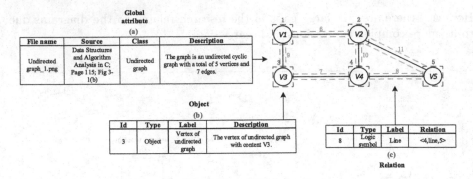

Fig. 2. Example of an undirected graph diagram annotation in CSDQA.

Table 1. Sources of CSDQA dataset.([1,2,3,4,5])

English textbook	*Data Structure and Algorithm Analysis in C*[8]
	Algorithms and Data Structures: The Basic Toolbox[6]
Chinese textbook	数据结构高分笔记 *(High Score Notes of Data Structure)*[9]
	数据结构C语言版 *(Data Structure C version)*[13]
	计算机操作系统 *(Computer Operating System)*[12]
	计算机组成原理 *(Principles of Computer Organization)*[11]
	数字逻辑电路 *(Digital Logic Circuit)*[5]
Blog	Zhihu[1], Chinese Software Developer Network[2], Douban[3]
Encyclopedia	Baidu pedia[4], Wiki pedia[5]

2 Dataset

2.1 Diagrams Collection and Annotation

Due to the limitations of the scenes where the diagram appears, we adopt a multi-source semi-supervised method to collect them. In terms of sources selection, we choose three high-quality sources: textbooks, blogs and encyclopedias, one regular web crawling source (See Table 1 for details). For web crawling source, the quality of the data obtained is often low, and a large amount of irrelevant data needs to be cleaned. For this type of data, we use the acquired samples to train a classifier (diagram-natural images classifier) to filter out the low-quality data. We use VGG16 [10] as the backbone and the images in the COCO [4] as negative samples. The accuracy of the trained model can reach over 99.46%. After the above process, all the collected samples meet the quality requirements. Each diagram in CSDQA then has two parts of the annotation: Global Attribute, Objects and Relations. We introduce them below in detail.

Global Attribute. This part describes the knowledge expressed in the diagram in a macro view. As shown in Fig. 1a, the annotation content is formulated as: <**Source, Class, Description**>, where **Source** records the place that the diagram is collected from; **Class** represents the knowledge to which the diagram belongs; **Description** is a summary of the knowledge conveyed by the diagram.

Object and Relation Attribute. Objects in the diagrams of CSQDA are mostly geometric shapes attached with text information as supplement descriptions. As shown in Fig. 1b, the annotation content is formulated as: <**Id, Type, Label, Description**>, where **Type** is used to distinguish objects from non-objects; **Label** is the subcategory of the object under the knowledge unit; **Description** contains other related information of the objects, such as the weight of the node the head node of the queue, or the text around the object. The expression form of the relation in diagram is to use logic symbols such as lines and arrows to connect different objects. Similarly, the annotation content of relation is formulated as: <**Id, Type, Label, Relation**>, where **Relation** indicates the objects and the logic symbol involved in a triple, such as < 4,line,5 >.

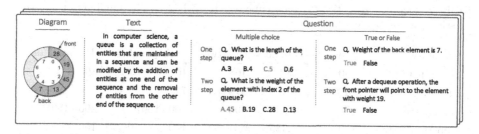

Fig. 3. Example of an queue diagram Q&A pairs annotation in CSDQA.

2.2 Question-Answer Pairs Generation

DQA is a machine comprehension task where the contexts are the diagram and the text, and the question is in natural language form. CSDQA dataset is used as a basis to verify such task. In addition to the diagram introduced above, we annotate relevant knowledge as the text context and Q&A pairs.

As shown in Fig. 2, we crawl text context using diagrams categories as keywords from Wikipedia. For Q&A pairs, they are divided into two types according to the number of answers: multiple choice questions and true-or-false questions.

---

[1] https://www.zhihu.com.
[2] https://www.csdn.net.
[3] https://www.douban.com.
[4] https://baike.baidu.com.
[5] https://www.wikipedia.org.

From the perspective of difficulty, questions can also be divided into simple questions (one-step reasoning) and complex questions (two-step reasoning). Specifically, a simple question only needs to be directly aligned on the diagram once according to the question to get the right answer. For example, for the simple true or false question in Fig. 2, the machine only needs to match 'back' with the elements indicated in the diagram to get the weight. But for complex questions, it needs to reason twice to get the correct answer. For example, for the complex true or false question in Fig. 2, the machine needs to understand the result of the queue after a dequeue operation for the first step, and in the second step, the machine needs to obtain the content of the head of the queue in this situation.

For more difficult two-step reasoning questions, we select three categories of diagram to annotate as a preliminary attempt, including queue, binary tree and stack. Among all the questions, the proportion of complex questions is 22.98%. All diagrams have at least one simple multiple question and one simple true-or-false question. To summarize, three categories diagrams in CSDQA correspond to four or six questions (including two simple questions), and the diagrams of the remaining categories correspond to two simple questions.

Table 2. Detailed statistics for each category in CSDQA (The numbers of two-step questions are in brackets).

Category	Diagrams	Objects	Relations	Multiple choice	True-or-False
Array list	100	583	468	100	100
Linked list	74	626	375	74	74
Binary tree	150	1,323	590	300 (150)	300 (150)
Non-Binary tree	150	1,489	651	150	150
Queue	150	1,261	444	303 (101)	303 (101)
Stack	150	540	403	300 (150)	300 (150)
Directed graph	71	695	377	71	71
Undirected graph	79	828	437	79	79
Deadlock	100	840	423	100	100
Flow chart	100	985	458	100	100
Logic circuit	70	913	432	70	70
Network topology	100	1,593	517	100	100
Total	1,294	11,776	5,675	1,747 (401)	1,747 (401)

2.3   Statistic

CSDQA dataset contains a total of 1,294 diagrams in 12 categories from five undergraduate courses: Data structure, Principles of Computer Networks, Computer Architecture, Digital Logic Circuit, and Computer Operating System.

Detailed statistics are shown in Table 1. We manually set up a ten-fold cross-validation division for simple and complex questions to facilitate full verification of tasks.

3 Conclusion

We introduce the characteristics of two types of diagrams, and construct the CSDQA dataset of geometrical diagrams. CSDQA has a wealth of annotations, which can be used as a test cornerstone for research on diagram classification, object detection, and DQA. All the data are open sourced, for conveniently used by researchers. At present, related research of multi-modality is getting more and more attention. The tasks in the education field such as DQA are consistent with the research in this scenario. We will supplement the content of the dataset in the future to better support diagram understanding research, further expand content of multi-modal research, eventually promote the progress of the computer vision community.

Acknowledgment. This work was supported by National Key Research and Development Program of China (2020AAA0108800), National Natural Science Foundation of China (62050194, 61937001, and 61877050), Innovative Research Group of the National Natural Science Foundation of China (61721002), Innovation Research Team of Ministry of Education (IRT 17R86), Project of China Knowledge Centre for Engineering Science and Technology, Consulting research project of Chinese academy of engineering "The Online and Offline Mixed Educational Service System for 'The Belt and Road' Training in MOOC China", China Postdoctoral Science Foundation (2020M683493).

References

1. Hong, X., Lan, Y., Pang, L., Guo, J., Cheng, X.: Transformation driven visual reasoning. In: Proceedings of the IEEE/CVF Conference on Computer Vision and Pattern Recognition, pp. 6903–6912 (2021)
2. Kembhavi, A., Salvato, M., Kolve, E., Seo, M., Hajishirzi, H., Farhadi, A.: A diagram is worth a dozen images. In: European Conference on Computer Vision, pp. 235–251 (2016)
3. Kembhavi, A., Seo, M., Schwenk, D., Choi, J., Farhadi, A., Hajishirzi, H.: Are you smarter than a sixth grader? Textbook question answering for multimodal machine comprehension. In: Proceedings of the IEEE Conference on Computer Vision and Pattern Recognition, pp. 4999–5007 (2017)
4. Lin, T.-Y., et al.: Microsoft COCO: common objects in context. In: Fleet, David, Pajdla, Tomas, Schiele, Bernt, Tuytelaars, Tinne (eds.) ECCV 2014. LNCS, vol. 8693, pp. 740–755. Springer, Cham (2014). https://doi.org/10.1007/978-3-319-10602-1_48
5. Liu, C.: Digital Logic Circuit. National Defense Industry Press (2002)
6. Mehlhorn, K., Sanders, P.: Algorithms and Data Structures: The Basic Toolbox. Springer Science & Business Media, Berlin (2008). https://doi.org/10.1007/978-3-540-77978-0

7. Morris, D., Müller-Budack, E., Ewerth, R.: Slideimages: a dataset for educational image classification. In: European Conference on Information Retrieval, pp. 289–296 (2020)
8. Shaffer, C.A.: Data Structures and Algorithm Analysis, edn. 3.2, update 0–3, Virginia Tech, Blacksburg (2012)
9. Shuai, H.: High Score Notes of Data Structure. China Machine Press, Beijing (2018)
10. Simonyan, K., Zisserman, A.: Very deep convolutional networks for large-scale image recognition. In: Proceedings of the IEEE Conference on Computer Vision and Pattern Recognition (2014)
11. Tang, S., Liu, X., Wang, C.: Principles of Computer Organization. Higher Education Press, Beijing (2000)
12. Tang, X., Liang, H., Zhe, F., Tang, Z.: Computer Operating System. Xidian University Press, Shanxi: (2007)
13. Yan, W., Wu, M.: Data Structure C Version. TsingHua University Press, Beijing (2002)
14. Zhang, M., Maidment, T., Diab, A., Kovashka, A., Hwa, R.: Domain-robust VQA with diverse datasets and methods but no target labels. In: Proceedings of the IEEE/CVF Conference on Computer Vision and Pattern Recognition, pp. 7046–7056 (2021)

MOOPer: A Large-Scale Dataset of Practice-Oriented Online Learning

Kunjia Liu[1], Xiang Zhao[1(✉)], Jiuyang Tang[1], Weixin Zeng[1], Jinzhi Liao[1], Feng Tian[2], Qinghua Zheng[2], Jingquan Huang[3], and Ao Dai[4]

[1] Science and Technology on Information Systems Engineering Laboratory, National University of Defense Technology, Changsha, China
xiangzhao@nudt.edu.cn
[2] Department of Computer Science, Xi'an Jiaotong University, Xi'an, China
[3] Intelligence Engine Technology Co. Ltd., Beijing, China
[4] Science and Technology on Parallel and Distributed Processing Laboratory, National University of Defense Technology, Changsha, China

Abstract. With the booming of online education, abundant data are collected to record the learning process, which facilitates the development of related areas. However, the publicly available datasets in this setting are mainly designed for a single specific task, hindering the joint research from different perspectives. Moreover, most of them collect the video-watching or course-enrollment log data, lacking of explicit user feedbacks of knowledge mastery. Therefore, we present MOOPer, a practice-centered dataset, focusing on the problem-solving process in online learning scenarios, with abundant side information organized as knowledge graph. Flexible data parts make it versatile in supporting various tasks, e.g., learning materials recommendation, dropout prediction and so on. Lastly, we take knowledge tracing task as an example to demonstrate the possible use of MOOPer. Since MOOPer supports multiple tasks, we further explore the advantage of combining tasks from different areas, namely, Deep Knowledge Tracing and Knowledge Graph Embedding. Results show that the fusion model improves the performance by over 9.5%, which proves the potential of MOOPer's versatility. The dataset is now available at https://www.educoder.net/ch/rest.

Keywords: MOOP · Online learning · Domain knowledge graph · Learning interaction

1 Introduction

The rapid development of Intelligent Tutoring System (ITS) and Massive Open Online Courses (MOOC) not only provides pedagogical advantages for educational revolution, but also promotes the development of various research areas. As practice plays an important role in knowledge mastery, Massive Open Online Practice (MOOP) is also proposed. As machine learning-based methods achieve state-of-the-art performance in tasks like computer vision and natural language

B. Qin et al. (Eds.): CCKS 2021, CCIS 1466, pp. 281–287, 2021.
https://doi.org/10.1007/978-981-16-6471-7_22

process, there are many works concerned with developing methods for exploring data in educational settings to better understand students. The abundant learning behavior collections promote the study of dropout prediction, knowledge tracing and learning behavior modeling. Their common goal is to better understand students' learning process by deeply looking into the interaction data, and to gain insights into the settings to improve educational outcomes.

However, most of the public MOOC datasets are designed for a specific task, hindering the joint research from different perspectives. ASSISTments09 dataset [3] is the benchmark dataset for knowledge tracing, which contains considerable size of elementary school math problem-solving records. KDDCUP15 is the dataset released in KDDCUP 2015, which is aimed at predicting user dropout by their behavior logs in online education settings. Although large-scale and high-coverage MOOCCube [7] is designed for multiple research interests, it can only model user behaviors by video-watching or course-enrollment activity, but fails to get more direct feedbacks to reflect their knowledge mastery. Therefore, we present MOOPer, a practice-oriented large-scale online learning dataset, collected from *Educoder.net*, platform for computer science subjects in China.

In summary, MOOPer has the following advantages over its counterparts:

- **Large-scale** MOOPer includes over 600 courses, 1,360 exercises, 4,550 challenges, and 2.5 million interaction records of 46,743 students from 2018 to 2019. Besides, abundant side information of learning materials and demographic message of teachers and students, are provided as well.
- **Practice-driven** In *Educoder.net*, users take carefully-designed exercises regarding to what they have learned in class. Therefore, their mastery of specific knowledge topic can be directly measured based on their problem-tackling performance, which provides useful information for following-up tutoring or recommendation.
- **Graph-organized** The dataset is organized as a knowledge graph, where the participants and learning objects are taken as entities. The intricate correlations between them can thus be well-described in a more intuitional way. Moreover, the graph-organized dataset facilitates the mutual promotion between knowledge graph research and educational data mining.

Therefore, our contribution can be summarized into three ingredients: a) we provide large-scale and comprehensive practice-centered online dataset b) we provide well-organized knowledge graph to provide abundant heterogenous information c) the dataset is designed to support multiple research interests, facilitating joint study of different tasks. To the best of our knowledge, this is among the first to present a dataset in practice-oriented online learning that involves a knowledge graph, which exhibits unique characteristics against existing datasets.

2 The Dataset

2.1 An Overview of MOOPer

The dataset is centered on interaction between users and challenges, where users' whole problem-solving processes are carefully extracted into three groups - user behavior, system feedback and user feedback. Details like the compile result of their code, how many times users have tried before succeeding, how users like the challenge they are dealing with and whether or not they have referred to the suggested answer, are provided. Besides, comprehensive side information is organized to build a whole picture of users' learning conditions. The various relationships between entities are organized accordingly, with their heterogenous nature taken into consideration. Figure 1 illustrates the basic structure of MOOPer.

Fig. 1. Overview of MOOPer.

Due to the rich interactions and side information provided, MOOPer dataset is quite versatile to support various tasks through free combination of different building blocks. By taking more comprehensive data into consideration, MOOPer dataset can provide more possibilities. There are some examples below.

User Behavior + System Feedback = Knowledge Tracing
User Behavior + User Feedback = Recommendation
User Behavior + Side Information = Dropout Prediction

2.2 Interaction Data Collection

The interactions between users and learning materials are carefully recorded, which can be categorized into three groups, namely, user behavior, system feedback and user feedback.

User behavior focuses on problem-solving process - how many times it takes to solve the problem, how many times they have tried to make it, and whether or not they have referred to the suggested answer, and so on. Such detailed records restore the practice scene to the best, as a result, providing as much information as possible for further research. Noteworthy, all records are anonymized to protect users' privacy.

User feedback provides users' rating of their interacted challenges. This kind of information implies users' preference in learning materials with different difficulties and question types. Besides, their discussions in forum are provided as well. The chit-chat content can be used to investigate their learning condition and learning satisfaction, while the questions and answers among peers reflect their 'blind spot' in knowledge mastery. Moreover, user active level in forum is also an important indication to speculate their psychological state and learning style [4].

System feedback provides the feedback of users' submission - are there syntax errors in their code, the compile result of their submitted code, the difference between the actual outputs and the expected ones (only if there are any) and so on. Such information provides deeper insights into users' learning ability and knowledge mastery.

2.3   Knowledge Graph Construction

It is proved that abundant side information is very useful in real tasks [6]. Thus, learning materials, institutions and teachers that provide them, are all listed as entities. Plentiful attributes of the entities are also provided to get a comprehensive view of the data. Noteworthy, the description and text content of learning materials are also attached as the last piece of the puzzle. The structure of knowledge graph is shown in the left part of Fig. 1.

There is a natural hierarchical structure between learning materials. A course may include several chapters following with exercises, while an exercise consists of several challenges with different topics indicating its knowledge components. Thereinto, challenges are also the fundamental building blocks of interaction data. Besides, exercise from a course might be forked to another course, and challenges in different exercise might share the same knowledge topic. These inter-references further extend the tree-like structure to a graph. In fact, the topics tagged along challenges are not standardized in the first place. We compute the Levenshtein Distance between them, and set a threshold to cluster similar knowledge topics.

3   Data Analysis and Application

This section will provide a deeper understanding of the presented dataset in a statistical view, then compare MOOPer with other similar datasets, and give an example application of MOOPer at last.

3.1 Statistics of MOOPer

MOOPer contains abundant heterogenous information and large-scale practice-oriented interaction data. There are over 2.5 million problem-solving records with detailed side information to provide a whole picture of users' learning conditions. Moreover, more than 82% users have interacted more than 20 challenges, while more than 75% challenges have more than 20 interactions.

The side information is organized as knowledge graph with 11 classes of entities and 10 classes of relations. The whole list of entities and relations can be referred to our published data repository.

Table 1. Comparison with other datasets.

Dataset	Course	Challenge	Topics	Student	Challenge interaction
ASSISTments09	–	26,688	124	4,217	525,534
KDDCUP15	39	–	–	112,448	–
MOOCCube	706	–	114,563	199,199	–
MOOPer	**600**	**4,550**	**3,277**	**46,743**	**2,532,524**

3.2 Comparison with Other Datasets

ASSISTments09 [3] is collected for knowledge mastery assessment, mainly contains elementary school math problem-solving records. It is one of the most widely used dataset in knowledge tracing task. It provides abundant interaction data, but lacks the users' submission for further analysis. Besides, side information is also oversimplified. KDDCUP15 is the dropout prediction dataset, mainly provides users' browsing logs, without side information or direct feedbacks from users. MOOCCube [7] provides abundant side information about the learning materials, which makes it the most similar one to ours among above comparison. However, its interaction data are mainly about activities like video-watching and course-enrollment. These activities can only lead to implicit feedbacks from users, while the problem-solving process MOOPer provides can directly model the mastery and preference of users. The comparison with other datasets in educational settings is shown in Table 1.

3.3 Demonstration of Application

In this section, we conduct knowledge tracing as an inspiration for many other applications that MOOPer dataset can support. Knowledge tracing task models the mastery of students as they interact with coursework by predicting user performances at the next time step based on their previous practicing records. [5]

We apply two of the most influential models in this task, namely, Bayesian Knowledge Tracing (BKT) [2] and Deep Knowledge Tracing (DKT) [5]. The former uses Hidden Markov Model (HMM) to update the probabilities that a learner answers exercises of a given knowledge topic correctly or incorrectly, while the latter adopts Recurrent Neural Network (RNN). For challenges containing several topics, we randomly choose one to represent it. Furthermore, since the versatility of MOOPer, we explore the combination of DKT and TransE [1], a benchmark model in Knowledge Graph Embedding (KGE) task. The fusion model achieves 9.5% improvement in AUC compared with DKT. The results are listed in Table 2.

Table 2. Results of deep knowledge tracing.

Method	AUC	Recall	F1-score
BKT	0.7097	0.7256	0.7245
DKT	0.7322	0.7652	0.7802
DKT + TransE	**0.8093**	**0.7733**	**0.7849**

The improvement of DKT+TransE comes from the additional information of knowledge graph structure, which shows the potential of incorporating more data types. Since the rich interaction data in MOOPer are not fully explored yet, as future work, it is of interest to further exploit the system output data in MOOPer to improve the performance.

4　Conclusion

We present MOOPer, a practice-centered dataset, focusing on the problem-solving process in online learning, with abundant side information organized as knowledge graph. Flexible parts makes it versatile in supporting various tasks. Then we provide an example application by conducting Bayesian Knowledge Tracing and Deep Knowledge Tracing. Lastly, we further prove the potential of MOOPer's versatility by combining Deep Knowledge Tracing and TransE, a benchmark model in Knowledge Graph Embedding task.

References

1. Bordes, A., Usunier, N., et al.: Translating embeddings for modeling multi-relational data. In: 27th Annual Conference on Neural Information Processing Systems 2013, Lake Tahoe, Nevada, United States, 5–8 December, pp. 2787–2795 (2013)
2. Corbett, A.T., Anderson, J.R.: Knowledge tracing: modeling the acquisition of procedural knowledge. User Model. User Adapt. Interact. **4**(4), 253–278 (1994)
3. Feng, M., Heffernan, N., Koedinger, K.: Addressing the assessment challenge with an online system that tutors as it assesses. User Model. User Adapt. Interact. **19**(3), 243–266 (2009)

4. Feng, W., Tang, J., Liu, T.X.: Understanding dropouts in MOOCs. In: Proceedings of the AAAI Conference on Artificial Intelligence, vol. 33, pp. 517–524. Honolulu, Hawaii, USA (2019)
5. Piech, C., Bassen, J., et al.: Deep knowledge tracing. In: Annual Conference on Neural Information Processing Systems 2015, Montreal, Quebec, Canada, 7–12 December 2015, pp. 505–513 (2015)
6. Vasile, F., Smirnova, E., Conneau, A.: Meta-Prod2Vec: product embeddings using side-information for recommendation. In: Proceedings of the 10th ACM Conference on Recommender Systems, Boston, MA, USA, 15–19 September 2016, pp. 225–232. ACM (2016)
7. Yu, J., Luo, G., et al.: MOOCCube: a large-scale data repository for NLP applications in moocs. In: Proceedings of the 58th Annual Meeting of the Association for Computational Linguistics, 5–10 July 2020, pp. 3135–3142 (2020, online)

MEED: A Multimodal Event Extraction Dataset

Shuo Wang[1,3], Qiushuo Zheng[2(✉)], Zherong Su[4], Chongning Na[5],
and Guilin Qi[1,6]

[1] School of Computer Science and Engineering, Southeast University, Nanjing, China
{wangs,gqi}@seu.edu.cn
[2] School of Cyber Science and Engineering, Southeast University, Nanjing, China
qiushuo_zheng@seu.edu.cn
[3] Southeast University-Monash University Joint Research Institute, Nanjing, China
[4] College of Software Engineering, Southeast University, Nanjing, China
[5] Zhejiang Lab, Hangzhou, China
na@zhejianglab.com
[6] Key Laboratory of Computer Network and Information Integration (Southeast University), Ministry of Education, Nanjing, China

Abstract. Multimodal tasks are gradually attracting the attention of the research community, and the lack of multimodal event extraction datasets restricts the development of multimodal event extraction. We introduce the new *Multimodal Event Extraction Dataset* (MEED) to fill the gap, we define event types and argument roles that can be used on multimodal data, then use controllable text generation to generate the textual modality based on visual event extraction dataset. In this paper, we aim to make full use of multimodal resources in the event extraction task by constructing a large-scale and high-quality multimodal event extraction dataset and promote researches in the field of multimodal event extraction.

Keywords: Multimodal dataset · Event extraction · Controllable text generation

Resource type: Dataset
Github Repository: https://github.com/a670531899/MEED

1 Introduction

Event Extraction (EE) is a task that extracting structured event information from unstructured data, specifically, extract specific events and corresponding arguments from data. As a traditional research topic of Natural Language Processing (NLP), it has many potential applications. For example, in the biomedical domain, doctors can use events extracted from medical records to inference the diagnosis of patience [8]. In the business and financial domain events extracted

© Springer Nature Singapore Pte Ltd. 2021
B. Qin et al. (Eds.): CCKS 2021, CCIS 1466, pp. 288–294, 2021.
https://doi.org/10.1007/978-981-16-6471-7_23

from different data can be used to inference what events may occur in the future and adjust strategy base on them [6]. Thus, event extraction plays an important role in diverse domains. Recently, contemporary society spreads news through multimedia, in this situation, the multimodal event extraction task has emerged in our sight.

Most event extraction methods extract event information from single-modal data, such as text [9], image [7,10] or video [1]. Thus, datasets of event extraction are mostly constructed for single modal such as ACE2005[1], RAMS [3] and SWiG [7]. [5] first proposed the multimodal event extraction task which aimed at extracting structured event information from multimodal news documents and constructed a M^2E^2 dataset for this task. Unfortunately, this M^2E^2 dataset only contains 6,167 sentences and 1,014 images which is too small to perform on such a complex multimodal task. As mentioned above, there is no large-scale and high-quality dataset for the multimodal event extraction task. To resolve this problem, we propose a novel framework to construct a multimodal event extraction dataset (MEED) in this paper. We summarize contributions of this paper as follows:

- We first define new multimodal event types and argument roles for the multimodal event extraction task. These events can be triggered by both visual data and textual data. Each event type has a superior event type and a set of argument roles.
- We propose a novel framework to construct the MEED dataset based on a large-scale grounding visual event extraction dataset SWiG. Each data in MEED has an image with sentences describing the event in the corresponding image with event trigger mentions and argument mentions annotated.
- We publish our constructed MEED dataset as an open resource in github.

2 MEED Construction

MEED is presented as plenty of pairs of image and text where texts are composed with a couple of sentences. Each image and its text are describing the same event with the same arguments. For event, one data has only one event type with an event trigger mention in each sentence of textual data. For argument, a sentence contains all argument entity mentions of its event and these entities are also grounded in the corresponding image.

Figure 1 illustrates the overview of our framework of constructing MEED which mainly includes four phases: **define event types, union arguments** and **generate text**.

2.1 Define Multimodal Event Types and Argument Roles

We define the multimodal event types and argument roles based on a large-scale visual event extraction dataset SWiG [7] because it has a rich variety of verbs

---

[1] https://catalog.ldc.upenn.edu/LDC2006T06.

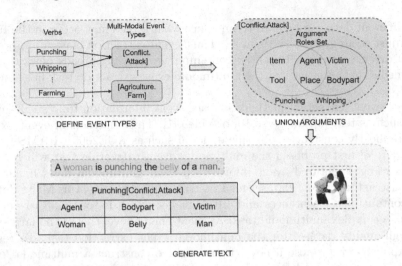

Fig. 1. Overview of our framework.

(visual event types) and argument roles. For each data in SWiG, there is an image with a unique verb label and a frame contains multiple argument roles with the corresponding argument labels. Considering that for textual modality, an event type only needs to be triggered by trigger words, while for visual modality it is needed to be able to determine the event type from an image. However, many verbs in SWiG can't be judged as events from images because they may be too generalized to represent a unique event. For example, an image with the verb *sit* may depict a meeting or having dinner as these two events are both displayed as a scene where several persons are sitting together. To ensure that each multimodal event type can be judged from both textual and visual modality, we deal with the verbs in SWiG and get the final multimodal event types. The details are as followed:

We first delete the verbs which are unable to represent a specific event and then cluster the remaining verbs into sets by observing if images of these verbs are representing similar events. Finally, we name each set with an event type and all verbs of the set are mapped into the new event type. For instance, the verbs *selling* and *buying* can't be distinguished visually because they are different expressions of the same event. So we can use an event type *Transaction. Transaction* to represent these two verbs. We follow these principles to ensure the rationality of our defined multimodal event types:

1) A sentence or an image can only trigger one single event, which makes sure that there will not be multiple events in a single data.
2) Cover as many verbs in SWiG as possible to ensure enough image data and rich variety.
3) The total event types should be as few as possible to reduce the pressure of training.

After defining multimodal event types, for the convenience of event classification, we cluster them into different sets according to their meanings and assign the sets with superior event types. For example, both the *Mow* and the *Weed* events can be classified as the superior event type *Agriculture*.

However, we can't just merge two visual verbs into a new event type because different verbs have different argument role sets. To solve this problem, we take the union set of all argument roles in each clustered visual verb set as the final argument roles set of the corresponding multimodal event type, which ensures that each event type has a complete argument role set.

2.2 Textual Modality Generation

To obtain the corresponding textual modality of images we follow [2] to generate textual descriptions for images in SWiG as a controllable text generation task [4] which allows us to restraint the generated textual descriptions must contain the same label mentions annotated in the original visual modality.

Specifically, we use the annotated argument entity labels in SWiG as the input objects of the controllable text generation model. In this way, the generated text must have the annotated argument entity mentions so that we just need to annotate these mention words with the corresponding argument roles in textual modality. For each pair of argument roles in the argument set of an event type if they have an ensured relation then we assign this relation to them as the relation input of the text generation model. Especially, to make sure that the generated textual descriptions contain the corresponding event trigger mentions, the verb label given in original visual modality must be assigned as a relation between argument roles which are the subject roles and object roles in the argument sets in most situations. For instance, *Operation.Repair* event contains *Agent* and *ObjectPart* arguments so we assign a relation between these two arguments which is

$$\langle Agent, Reparing, ObjectPart \rangle.$$

2.3 Events and Arguments Annotation

Multimodal data can be generated followed Sect. 2.1. As mentioned in Sect. 2.2, the argument entity mentions and their roles can be annotated based on the original visual annotations because they are the same in the twomodalities. Because the restriction we proposed in Sect. 2.2 ensures that the verb of each image appears in the generated text and each verb can represent a unique event type, these verb mentions in the text will be annotated as the event trigger words and the event type is the same as the verb's corresponding event type.

3   Statistics

In this section, we report the statistics of MEED and the format of examples in MEED is show in Fig. 2, sentences in gray boxes are the textual modality;

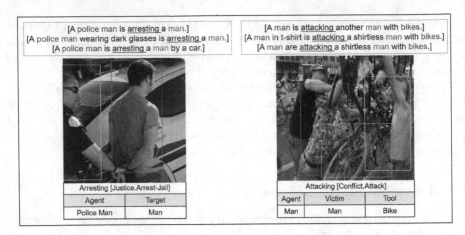

Fig. 2. Two examples from MEED.

each sentence is describing the same event in the image; words with underline are the trigger words and in different colors are the argument mentions which are also grounded in the image with boxes. The table below the image is the structured annotation of the event. As shown in Table 1, our MEED has 66 multimodal event types with 18 superior event types and 81 argument roles. There are at most 17 arguments, at least 1 argument (*Disaster.Storm* event only has one argument), and 4.3 arguments on average for one multimodal event type. The MEED contains 37,807 examples where one data contains an image with 3 sentences describing the image in different formats. Compared with the textual event extraction dataset ACE2005 and the multimodal event extraction dataset M^2E^2 [5], Our MEED dataset is not only rich in event types and argument roles but also has great advantages in the number of mentions of event triggers and arguments.

Table 1. Statistics of MEED and compare it with two other event extraction datasets.

Dataset	Event		Argument		Examples
	Types	Mentions	Roles	Mentions	
ACE2005	33	5,231	35	9,639	16,255
M^2E^2	8	1,297	15	1,966	6,168
MEED	66	113,421	81	362,807	37,807

4 Related Work

Event extraction is an important task in Natural Language Process (NLP) which extracts events and arguments in data to assist users in making further decisions.

To better research this task, researchers have proposed various event extraction datasets, such as in NLP ACE2005 is the most classic dataset in event extraction and [3] constructed a cross-sentence linking event extraction dataset. In Computer Vision (CV), [10] proposed imSitu to extract events and arguments in a large amount of images. But the positions of the arguments in imSitu are not annotated. [7] constructed SWiG which gave the position to each argument based on imSitu. However, these works only focused on a single modality and ignored the rich interaction between different modalities.

5 Conclusion and Future Work

This paper presents the process to construct a high-quality and large-scale multimodal event extraction dataset MEED with multimodal event types and argument roles. Our work generates textual modality based on a large-scale visual event extraction dataset SWiG and restrained the textual modality contains annotated verbs and argument mentions of the original dataset. Finally, we annotate the multimodal data with the multimodal event types and argument roles. In future work, our plan is to enrich the style of textual modality so that the dataset can be more diversified.

Acknowledgment. This work was supported by National Natural Science Foundation of China with Grant No. 61906037; the Fundamental Research Funds for the Central Universities with No. 224202k10011; the CCF-Baidu Open Fund with No. CCF BAIDU OF2020003.

References

1. Caba Heilbron, F., Escorcia, V., Ghanem, B., Carlos Niebles, J.: ActivityNet: a large-scale video benchmark for human activity understanding. In: CVPR, pp. 961–970 (2015)
2. Chen, S., Jin, Q., Wang, P., Wu, Q.: Say as you wish: fine-grained control of image caption generation with abstract scene graphs. In: CVPR, pp. 9962–9971 (2020)
3. Ebner, S., Xia, P., Culkin, R., Rawlins, K., Van Durme, B.: Multi-sentence argument linking. In: ACL, pp. 8057–8077 (2020)
4. Hu, Z., Yang, Z., Liang, X., Salakhutdinov, R., Xing, E.P.: Toward controlled generation of text. In: ICML, pp. 1587–1596 (2017)
5. Li, M., et al.: Cross-media structured common space for multimedia event extraction. In: ACL, pp. 2557–2568 (2020)
6. Nuij, W., Milea, V., Hogenboom, F., Frasincar, F., Kaymak, U.: An automated framework for incorporating news into stock trading strategies. IEEE Trans. Knowl. Data Eng. **26**(4), 823–835 (2013)
7. Pratt, S., Yatskar, M., Weihs, L., Farhadi, A., Kembhavi, A.: Grounded situation recognition. In: Vedaldi, A., Bischof, H., Brox, T., Frahm, J.-M. (eds.) ECCV 2020. LNCS, vol. 12349, pp. 314–332. Springer, Cham (2020). https://doi.org/10.1007/978-3-030-58548-8_19

8. Vanegas, J., Matos, S., González, F., Oliveira, J.: An overview of biomolecular event extraction from scientific documents. Comput. Math. Methods Med. **2015**, 571381 (2015)
9. Yang, S., Feng, D., Qiao, L., Kan, Z., Li, D.: Exploring pre-trained language models for event extraction and generation. In: ACL, pp. 5284–5294 (2019)
10. Yatskar, M., Zettlemoyer, L., Farhadi, A.: Situation recognition: visual semantic role labeling for image understanding. In: CVPR, pp. 5534–5542 (2016)

C-CLUE: A Benchmark of Classical Chinese Based on a Crowdsourcing System for Knowledge Graph Construction

Zijing Ji[1,3], Yuxin Shen[1,3], Yining Sun[2], Tian Yu[1], and Xin Wang[1,3(✉)]

[1] College of Intelligence and Computing, Tianjin University, Tianjin, China
{jizijing,shenyuxin,3019244138,wangx}@tju.edu.cn
[2] Qiushi Honors College, Tianjin University, Tianjin, China
sun_3019234250@tju.edu.cn
[3] Tianjin Key Laboratory of Cognitive Computing and Application, Tianjin, China

Abstract. Knowledge Graph Construction (KGC) aims to organize and visualize knowledge, which is based on tasks of Named Entity Recognition (NER) and Relation Extraction (RE). However, the difficulty of comprehension, caused by the differences in grammars and semantics between *classical and modern Chinese*, makes entity and relation annotations time-consuming and labour-intensive in *classical Chinese corpus*. In this paper, we design a novel crowdsourcing annotation system, which can gather collective intelligence as well as utilize *domain knowledge* to achieve efficient annotation and obtain fine-grained datasets with high quality. More specifically, we judge the *user professionalism*, calculated by online tests, considered in annotation results integration and rewards assignment, which plays a vital role in improving the accuracy of annotation. Moreover, we evaluate several pre-training language models, the state-of-the-art methods in Natural Language Processing (NLP), on the benchmark datasets obtained by the system over tasks of NER and RE. Benchmark datasets, implementation details, and evaluation processes are available at https://github.com/jizijing/C-CLUE. The access URL of the crowdsourcing annotation system is: http://152.136.45.252:60002/pages/login.html.

Keywords: Classical Chinese · Crowdsourcing annotation system · Knowledge graph construction · Natural language processing

1 Introduction

The Twenty-Four Histories are collections of classic Chinese historical documents from 2550 BC to 1644 AD, which records rich figures and events. Based on entity and relation extraction, we can construct a knowledge graph which can vividly demonstrate and effectively store the hard-to-understand classical Chinese corpus, convenient for gaining valuable information.

© Springer Nature Singapore Pte Ltd. 2021
B. Qin et al. (Eds.): CCKS 2021, CCIS 1466, pp. 295–301, 2021.
https://doi.org/10.1007/978-981-16-6471-7_24

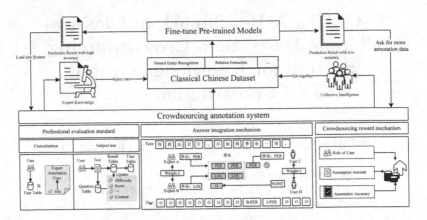

Fig. 1. The overall framework of our research.

Due to the difficulty of annotating entities and relations in classical Chinese corpus, caused by the great differences in semantics and grammars between classical and modern Chinese, a crowdsourcing annotation system should be introduced to gather collective intelligence as well as introduce domain-knowledge to achieve efficient and accurate annotations. According to the annotation results of the system, a series of relations and entities can be obtained to construct a classical Chinese benchmark set. Existing benchmark work mainly includes GLUE [4] and CLUE [5], a multi-task natural language understanding benchmark for English and its Chinese version. However in classical Chinese, to the best of our knowledge, only a dataset of NER task provided by the CCL2020 "Gulian Cup" Ancient Books NER Evaluation Competition [6] can be accessed, which just includes two types of entities, book titles and other proper names, and is not publicly available due to the confidentiality of competition data. Our benchmark consists of a fine-grained NER task and RE task built on corresponding datasets, which can be used to fine-tune the pre-training models, the state-of-the-art models of NLP, and evaluate the performance of the models over classical Chinese.

In this paper, we propose a crowdsourcing annotation system that introduces The Twenty-Four Histories and allows users to tag the entities and relations. The main difference between our system and the existing crowdsourcing system is that the annotation of classical Chinese corpus requires the introduction of expert domain knowledge, for which we introduce professionalism into the system. We judge the user professionalism through online tests, taking it into consideration during result integration stage and rewards allocation. In addition, unlike crowdsourcing systems that focus on task allocation, our system opens the general task to each user, that is, the content of The Twenty-Four Histories, and allows users to select the chapters of interest and annotate the same text differently, which can take full advantage of swarm intelligence. Based on the annotation results obtained from the system, we acquire a benchmark of classical Chinese which is composed of a dataset for NER task, including six types of entities, and a dataset for RE task, including seven categories and twenty-five subcategories of relations. Moreover, we fine-tune several pre-training models, the mainstream technology of NLP, evaluate their performance over the benchmark and release the experimental results as well as the implementation details. The overall framework of our research is shown in Fig. 1.

2 Crowdsourcing Annotation System Design

2.1 Professionalism Evaluation Method

Anhai Doan et al. [7] proposed that the task assignment mechanism based on the credibility degree of user may improve the overall utility of the crowdsourcing platform. In order to inject domain-knowledge into our annotation system, we introduce the professionalism of users, which is not considered in most existing crowdsourcing systems, and define two ways to judge it.

For known expert users who would like to join, we can directly assign them the role of expert user when inserting their information into the database. For unknown users, we prepare ten questions with existing standard answers and ask users to complete when they log in for the first time. The professional level will be comprehensively calculated based on the accuracy of the answers of users and the difficulty of questions. The difficulty of a question dynamically changes based on the initial value obtained by the accuracy of the answers of several volunteers, which can be expressed as the ratio of the number of users with wrong answers to total number of users in the system. The score of a question is proportional to its difficulty, which can be calculated by rounding up ten times the difficulty value of the question. We can obtain the total score by summing the scores of each question and assign expert user role to a user if his or her score is higher than 60% of the total score.

2.2 Result Integration Mechanism

Unlike the existing majority voting method [8] and the method of introducing accuracy [9], we take the degree of professionalism into consideration. Specifically, we assign expert users twice the weight of ordinary users to ensure the accuracy of the results since we believe for classical Chinese corpus annotation tasks that require domain knowledge, users with high professionalism are more likely to make correct annotations.

Improved Majority Voting Strategy. Our system provides the same task to all users and allows users to modify existing annotations. We record the user id and annotation time as well as the annotation content. If multiple users have different annotations for the same entity or entity pair, we will save them respectively rather than overwrite the past annotations. When downloading data, we adopt a weighted majority voting strategy considering the professionalism of users to get the final results if there are multiple records for the same text.

298   Z. Ji et al.

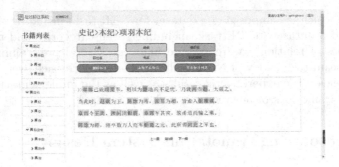

Fig. 2. The annotation page of the crowdsourcing annotation system.

Interference Elimination Strategy. We can delete all annotations made by a user whose accuracy is not satisfactory to further improve the fault tolerance of the system. Besides, we will open data downloads to system users and release new versions of data regularly.

2.3 Crowdsourcing Reward Mechanism

We propose a new reward mechanism based on the existing crowdsourcing systems by considering professionalism, annotation accuracy, and annotation amount synthetically. The reward is settled every fixed period.

Professionalism. We take the professionalism of users as a measurement standard since classical Chinese corpus annotation task requires domain knowledge. Specifically, we give expert users double the rewards of ordinary users.

Annotation Accuracy. We simply treat the final answer after the result integration period as the correct result. If the user makes the same annotation as the correct result, the reward will be granted, otherwise it will not.

Annotation Amount. In order to motivate users to make annotations, we set a threshold for the amount of annotations and the correct rate, and implement multiple rewards for users who exceed this threshold.

A Reward Allocation Example. Suppose that the price of a single annotation is p. If an ordinary user has completed n annotations within a certain reward allocation period, m of which are valid annotations, i.e., they are the same as final answers, and n is higher than the threshold value of the amount a_t, m/n exceeds the threshold value of the accuracy c_t, then the reward is defined as:

$$reward = m * (1 + \frac{m}{n} - c_t) * \frac{n}{a_t} * p \tag{1}$$

The annotation page of the crowdsourcing system is shown in Fig. 2.

3 Experiments

3.1 Task Description

Based on the annotation results of The Twenty-four Histories, we construct a benchmark consisting of two tasks and their corresponding datasets. The fine-grained classical Chinese dataset of NER task consists of text files and label files, including six types of entities: person, location, organization, position, book, and war. The statistics of entities are shown in Table 1.

Furthermore, a dataset for RE task can be built, which is made up of seven categories of relations: organization-organization, location-organization, person-person, person-location, person-organization, person-position, and location-location. The statistics of the relations are shown in Table 2. From original datasets, we can generate a relation classification dataset composed of sentences and relation files, and a sequence labeling dataset similar to the dataset for NER task. Note that, the generated label is not an entity category label, but a label that refers to the subject or object of a relation.

Table 1. Statistics of entities

Entity dataset	Train	Eval	Test	Total
PER	9467	1267	701	11435
LOC	2962	391	167	3520
POS	1750	242	139	2131
ORG	1698	266	100	2064
OTHERS	110	18	9	137

Table 2. Statistics of relations

Dataset	Train	Eval	Test	Total
PER-PER	1139	324	130	1593
PER-ORG	231	60	38	329
PER-LOC	462	129	53	644
PER-POS	1093	319	162	1574
OTHERS	157	40	28	225

3.2 BaseLines

We evaluate the following pre-training models on our benchmark: BERT-base [1], BERT-wwm [2], Roberta-zh [3], and Zhongkeyuan-BERT (ZKY-BERT). For the detailed introduction of the baseline models, please refer to our github project.

3.3 Implementation Details

For fine-tuning, most hyper-parameters are the same as pre-training of BERT, except batch size, learning rate, and number of training epochs. We find the following ranges of possible values work well on fine-tuning, i.e., batch size: 32, learning rate (Adam): $5e^{-5}$, $3e^{-5}$, $2e^{-5}$, and number of epochs ranging from 3 to 10. For the NER task and RE task, we provide a detailed evaluation process in our github project for reference.

3.4   Results and Analysis

The experimental results on the dataset with six categories of entities and the dataset with four categories of entities (without book and war) are as follows.

From the results in Table 3, it can be seen that when handling fine-grained NER, the ZKY-BERT model trained on the classical Chinese corpus performs best, and the BERT-wwm model adapted to Chinese characteristics is the second. From the results in Table 4, we can see that the pre-trained models all achieve relatively better performance due to the fewer number of entity types.

Table 3. Results of NER on 6 types (%)

Model	Precision	Recall	F_1
BERT-base	29.82	35.59	32.12
BERT-wwm	32.98	**43.82**	35.40
Roberta-zh	28.28	34.93	31.09
ZKY-BERT	**33.32**	42.71	**36.16**

Table 4. Results of NER on 4 types (%)

Model	Precision	Recall	F_1
BERT-base	44.33	53.60	48.11
BERT-wwm	**45.42**	**54.33**	**48.95**
Roberta-zh	45.40	53.00	48.61
ZKY-BERT	44.35	53.69	48.09

For the RE task, we split it into two subtasks: relation classification and sequence labeling. Experiments show that the baseline models can achieve an accuracy of 47.61% on the relation classification task.

4   Conclusion

To construct a knowledge graph from the classical Chinese corpus, entities and relations should be extracted efficiently and accurately. Therefore, it is attractive to utilize an elaborately designed crowdsoucing annotation system that considers the professionalism of users during the whole process, aiming to combine swarm intelligence with domain knowledge. Based on annotations obtained from our system, we establish a benchmark with NER and RE tasks on classical Chinese, on which we can evaluate the state-of-the-art methods. Benchmark datasets, implementation details, and evaluation processes are publicly available.

Acknowledgement. This work is supported by the China Universities Industry, Education and Research Innovation Foundation Project (2019ITA03006) and the National Natural Science Foundation of China (61972275).

References

1. Devlin, J., Chang, M.W., Lee, K.: BERT: pre-training of deep bidirectional transformers for language understanding. In: Proceedings of NAACL, pp. 4171–4186 (2019)
2. Cui, Y., Che, W., Liu, T.: Pre-training with whole word masking for Chinese BERT. arXiv preprint arXiv:1906.08101 (2019)

3. Liu, Y., Ott, M., Goyal, N.: RoBERTa: a robustly optimized BERT pretraining approach. arXiv preprint arXiv:1907.11692 (2019)

4. Xu, L., Hu, H., Zhang, X.: CLUE: a Chinese language understanding evaluation benchmark. arXiv preprint arXiv:2004.05986 (2020)

5. Wang, A., et al.: GLUE: a multi-task benchmark and analysis platform for natural language understanding. In: Proceedings of EMNLP, pp. 353–355 (2018)

6. Zhonghua Book Company: The registration channel for the "Gulian Cup" Ancient Books NER Evaluation Competition is now open! http://www.zhbc.com.cn/zhsj/ fg/news/info.html?newsid=402885966e259cb10172605463cf25cf. Accessed 29 May 2020

7. Doan, A., Ramakrishnan, R., Halevy, A.Y.: Crowdsourcing systems on the world-wide web. Commun. ACM **54**(4), 86–96 (2011)

8. Franklin, M.J., et al.: CrowdDB: answering queries with crowdsourcing. In: Proceedings of ACM SIGMOD, pp. 61–72 (2011)

9. Liu, X., Lu, M., Ooi, B.C., et al.: CDAS: a crowdsourcing data analytics system. arXiv preprint arXiv:1207.0143 (2012)

RCWI: A Dataset for Chinese Complex Word Identification

Mengxi Que[ID], Yufei Zhang, and Dong Yu[✉]

Beijing Language and Culture University, Beijing, China

Abstract. Reasonable evaluation of lexical complexity is the premise of multiple downstream NLP tasks such as text simplification. At present, there lacks of reliable Chinese lexical complexity datasets, while most of the existing foreign datasets only focus on the words that cause reading difficulty. This paper constructs a RCWI-Dataset for native Chinese speakers, which contains 40613 examples and three complexity categories. Each example is annotated by at least three annotators. We adopt comparison method to annotate words that are more difficult than average lexical complexity in sentences, so that we can get more information about word complexity and improve the reliability of our dataset. We provide baseline experiments based on feature engineering, the results show the validity of RCWI-Dataset.

Keywords: Chinese lexical complexity · Lexical complexity evaluation · Feature engineering

1 Introduction

Lexical complexity refers to the cognitive load brought by understanding a given word. Evaluating lexical complexity is an indispensable part of many downstream tasks such as text simplification. It can help second language learners, dyslexics and other groups to obtain text information more easily.

At present, there lacks the datasets for lexical complexity evaluation task. So we construct a lexical complexity resource for native Chinese speakers: Relatively Complex Word Identification Dataset (RCWI-Dataset), which contains 40613 sentences. We built it by collecting and annotating Chinese textbooks in the stage of compulsory education. We hypothesize that lexical complexity can be evaluated by comparing word in same sentences, so we annotate the words whose complexity exceed the average word complexity in the sentence and call these words relatively complex words. RCWI-Dataset has three categories: Normal, Complex and Hard. Compared with [5] and [10], our dataset not only focuses on the words causing reading difficulty, but also contains richer information of lexical complexity relationship. Besides, we select multiple features and provide a baseline for lexical complexity evaluation to verify the effectiveness of the corpus. Our data can be found at https://github.com/blcunlp/RCWI-Dataset.

© Springer Nature Singapore Pte Ltd. 2021
B. Qin et al. (Eds.): CCKS 2021, CCIS 1466, pp. 302–307, 2021.
https://doi.org/10.1007/978-981-16-6471-7_25

2   Related Works

Chinese lexical complexity resource for teaching are mainly graded thesaurus, which are generally constructed by expert annotation. [3] comprehensively considers the situation of all kinds of Chinese learners and divides five thousand words into six levels according to the requirements of different period. Similarly, [9] contains more than 14323 words and divides them into four grades according to the learning difficulty of the four period from primary school to junior middle school.

The lexical complexity resources for natural language processing are task-oriented. Usually they are constructed for tasks like lexical simplification. [5] recruited non-native speakers to annotate the words in sentences that cannot be understood independently. [10] required annotators to highlight words or phrases that are difficult to understand in texts. [7] believed that simply dividing words into complex or un-complex can not reflect the complexity of words well. So they proposed Lexical Complexity Prediction task (LCP) and constructed the corresponding dataset CompLex.

[5] and [10] adopted binary judgement to annotate words, which has a strong subjectivity. [7] and [4] used Likert scale, but they lack the object to compare. The result of [2] shows that better results can be obtained by comparison rather than binary judgement.

3   Dataset Construction

3.1   Data Annotation

We use the Chinese sentence readability corpus constructed by [11] as the source of our dataset. The corpus consists of Chinese textbooks. It contains sentences distributed in five difficulty levels, covering popular science, narrative and other genres. Compared with [1], the anchor selection is more rigorous, with higher consistency and better data quality.

We first randomly sample 200 sentences for annotation, and find that the average lexical difficulty in the first level is too simple. So we abandon these sentences. We sample 20000 sentences according to the original ratio. The proportion of sentences in each level is 3:4:2:1. In order to consider the complexity of words in a sentence comprehensively, we define the target to annotate in RCWI-Dataset as: words and phrases in a given sentence that are significantly more difficult than the average lexical difficulty in the sentence. If a word is difficult to understand, annotators should label it with "Hard", otherwise label it as "Complex". According to the results of trial annotation, three common annotating situation are determined, as shown in Table 1.

We ask the annotators to read the whole sentence before annotation. Besides, considering the influence of context, annotators also need to annotate the words that can not be understood without context.

Compared with the process of English complex word annotation, Chinese complex word annotation involves the noise caused by word segmentation.

Table 1. The three common annotating situation, red words belong to "Hard", orange words belong to "Complex"

Complex: Words that can be understood, but whose complexity exceed the average complexity of all words in the sentence
1.万众一心，冒着敌人的炮火，前进！
2. 正因为它不是一般的顽石，当然不能去做墙，做台阶，不能去雕刻，捶布。
Hard: words that cause reading difficulties
1. 见到人们受苦，鲧很着急，就把天上的土偷下来，去堵塞洪水。
2. 于是，伯父家盖房，想以它垒山墙，但苦于它极不规则，没棱角，也没平面儿；用鏨破开吧，又懒得花那么大气力，因为河滩并不甚远，随手去捐一块回来，哪一块也比它强。
Unlabeled: all words in the sentence are simple or there is no significant difference in complexity between words
1. 爸把我从床头打到床尾，外面的雨声混合着我的哭声。
2. 当山间的清泉奔向溪流，当哗啦啦的大雨砸向屋顶，当小水滴清清脆脆地落到盛水的盆里，你总该听到些什么了吧？

To avoid the problem, we stipulate that the target to annotate must be semantically complete words or phrases, and does not contain redundant parts.

We select 10 native college students as the annotators. To ensure the quality, we stipulate that at most five words or phrases should be annotated in a sentence, and each sentence should be annotated by at least three annotators. We use [8] to build the annotation platform. Annotators can select words or phrases by mouse sliding, and determine the corresponding category.

The proportion of annotated sentences in the original corpus is 84.6%, and 41866 annotated words are obtained. "Complex" account for 93% of the total labels, and the number of "Hard" is much lower than that of "Complex". This shows that there are few words in Chinese textbooks that cause reading difficulty for native speakers.

3.2 Data Processing

After obtaining the annotation results, we can not merge them directly due to strong subjectivity, determination of lexical boundary and the phenomenon of overlapping and redundancy in the results.

According to the principle mentioned above, it is necessary to complete the annotations with incomplete semantics and split the annotations with redundant parts.

Because the annotators are college students and their proficiency is higher than others. So it is possible for them to underestimate the lexical complexity. Therefore, if the results of a word contain "Hard", the word should be labeled with "Hard", otherwise should be labeled with "Complex". The process of merging results include three phases: Split, Clean and Combine. The whole process is shown in Fig. 1.

After merging, we remove the words that be labeled with "Complex" only once and the sentence without annotation. Then we construct negative examples. Weselect the unlabeled word with the lowest word frequency in a sentence and label it with "Normal". Finally, the RCWI-Dataset contains 40613 sentences, including 19218 "Complex" labels, 1169 "Hard" labels and 20226

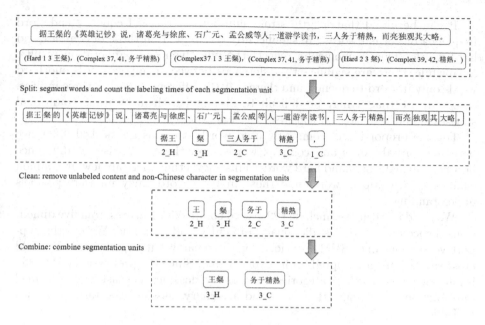

Fig. 1. Example of merging the annotation results

"Normal" labels. The information of each example include sentence, label, word start position, word end position, word and number of annotations.

4 Experiments and Analysis

We analyze the words of three categories in RCWI-Dataset, the result is shown in Table 2.

Table 2. The analysis result of words in different categories in RCWI-Dataset. The lexicology features include average word length (Ave_len) and average stroke number (Ave_stroke). Statistical features include average word frequency (Ave_wfreq) and average character frequency (Ave_cfreq). Dictionary features include proportion of words in dictionary of common words (Common_wrate) and proportion of characters in dictionary of common words (Common_crate)

	Normal	Complex	Hard
Ave_len	2.057	2.597	2.672
Ave_stroke	7.450	8.819	8.803
Ave_cfreq	0.163	0.097	0.101
Ave_wfreq	0.073	0.012	0.006
Common_crate	0.994	0.953	0.824
Common_wrate	0.262	0.094	0.134

It can be found that length, stroke number, character frequency and word frequency of normal words are significantly different from those of complex and hard words. This phenomenon reflects the effectiveness of annotation results. Meanwhile, there are significantly differences between complex words and hard words only in word frequency, and there is little difference in other three indices. This indicate that it may be difficult to distinguish them by using traditional statistical and lexicological features.

For the proportion of commonly used words, there is no marked difference between normal words and complex words, and there is a distinct difference between normal words and hard words. This shows that there are few uncommon characters in complex words and those in hard words may interfere people's understanding.

We build a feature set included 22 lexical complexity features from five dimensions: lexicology, statistics, dictionary, difficulty and semantic. We use the support vector machine (SVM) provided by [6] to carry out experiments. Ten fold cross validation was adopted in the experiments. Since the proportion of "Hard" is much lower than other categories, the ternary classification task is transformed into four binary classification tasks and a ternary classification task, as shown in Table 3.

Table 3. The results of different dimensional features on five tasks. N, C, H denote Normal, Complex, Hard respectively. The dataset involving H consists of all Hard examples and randomly selected examples of other categories.

Feature types	CN	NH	CH	CNH
Lexicology	65.27	75.40	67.38	54.09
Statistic	50.76	53.47	55.61	36.78
Dictionary	74.59	**85.56**	68.69	**59.64**
Difficulty	68.73	77.54	68.69	58.01
Semantic	61.79	71.12	63.10	49.29
All	**74.81**	79.14	**72.73**	57.65

It can be seen that the lexical complexity evaluation model based on feature engineering can effectively model lexical complexity. Comparing the results of NH and CH, it is found that NH can also achieve good results with the small dataset, which shows that feature engineering can capture the difference between them. This also reflects that the difference of lexical complexity between normal words and hard words are significantly. Comparing the features from different dimensions, the effect of difficulty features is second only to that of dictionary feature, while that of lexicology features is weaker than that of difficulty features and that of semantic features is weaker than that of lexicology features. The effect of the statistic features shows that it is almost impossible to classify effectively using statistic features.

5   Conclusion

In this paper, we construct a lexical complexity identification dataset for native Chinese speakers. Annotators annotates the relatively complex words in sentences by comparison, which contains a variety of complexity information. The paper provides experience for the construction of Chinese lexical complexity evaluation dataset. Since the Chinese textbooks are relatively simple, the proportion of "Hard" is small. In future works, we will introduce more difficult corpus to increase the number of Hard words. We also plan to improve our feature engineering based on different dimension features and investigate complex word identification task using deep learning models.

Acknowledgements. This work is funded by the Humanity and Social Science Youth foundation of Ministry of Education (19YJCZH230) and the Fundamental Research Funds for the Central Universities in BLCU (No. 17PT05).

References

1. Dong, Y., Siyuan, W., Zhaoyang, G., Yuling, T.: Assessing sentence difficulty in Chinese textbooks based on crowdsourcing. J. Chin. Inf. Process. **34**(2) (2020). (in Chinese)
2. Gooding, S., Kochmar, E., Blackwell, A., Sakar, A.: Comparative judgements are more consistent than binary classification for labelling word complexity (2019)
3. Hanban: New HSK Syllabus (2009). (in Chinese)
4. Maddela, M., Wei, X.: A word-complexity lexicon and a neural readability ranking model for lexical simplification. In: Proceedings of the 2018 Conference on Empirical Methods in Natural Language Processing (2018)
5. Paetzold, G., Specia, L.: SemEval 2016 task 11: complex word identification. In: Proceedings of the 10th International Workshop on Semantic Evaluation (SemEval-2016) (2016)
6. Pedregosa, F., et al.: Scikit-learn: machine learning in python (2012)
7. Shardlow, M., Cooper, M., Zampieri, M.: Complex – a new corpus for lexical complexity predicition from likert scale data (2020)
8. Stenetorp, P., Pyysalo, S., Topi'C, G., Ohta, T., Ananiadou, S., Tsujii, J.: Brat: a web-based tool for NLP-assisted text annotation. Association for Computational Linguistics (2012)
9. Xinchun, S.: Theory and method in compiling list of common words in compulsory education (draft). Appl. Linguist. **103**(03), 2–11 (2017). (in Chinese)
10. Yimam, S.M., Štajner, S., Riedl, M., Biemann, C.: CWIG3G2 - complex word identification task across three text genres and two user groups. In: Proceedings of the Eighth International Joint Conference on Natural Language Processing (Volume 2: Short Papers), Taipei, Taiwan, pp. 401–407. Asian Federation of Natural Language Processing, November 2017. https://www.aclweb.org/anthology/I17-2068
11. Yuling, T., Dong, Y.: The method of calculating sentence readability combined with deep learning and language difficulty characteristics. In: Proceedings of the 19th Chinese National Conference on Computational Linguistics, Haikou, China, pp. 731–742. Chinese Information Processing Society of China, October 2020. (in Chinese). https://www.aclweb.org/anthology/2020.ccl-1.68

DiaKG: An Annotated Diabetes Dataset for Medical Knowledge Graph Construction

Dejie Chang[1(✉)], Mosha Chen[2], Chaozhen Liu[1], Liping Liu[1], Dongdong Li[1],
Wei Li[1], Fei Kong[1], Bangchang Liu[1], Xiaobin Luo[1], Ji Qi[3], Qiao Jin[3],
and Bin Xu[3]

[1] Miao Health, Singapore, Singapore
{changdejie,liuchaozhen,liuliping,lidongdong,liwei,kongfei,liubangchang,
luoxiaobin}@miao.cn
[2] Alibaba Group, Hangzhou, China
chenmosha.cms@alibaba-inc.com
[3] Tsinghua University, Beijing, China
{jqa14,qj20}@mails.tsinghua.edu.cn, xubin@tsinghua.edu.cn

Abstract. Knowledge Graph has been proven effective in modeling structured information and conceptual knowledge, especially in the medical domain. However, the lack of high-quality annotated corpora remains a crucial problem for advancing the research and applications on this task. In order to accelerate the research for domain-specific knowledge graphs in the medical domain, we introduce DiaKG, a high-quality Chinese dataset for Diabetes knowledge graph, which contains 22,050 entities and 6,890 relations in total. We implement recent typical methods for Named Entity Recognition and Relation Extraction as a benchmark to evaluate the proposed dataset thoroughly. Empirical results show that the DiaKG is challenging for most existing methods and further analysis is conducted to discuss future research direction for improvements. We hope the release of this dataset can assist the construction of diabetes knowledge graphs and facilitate AI-based applications.

Keywords: Diabetes · Dataset · Knowledge graph

1 Introduction

Diabetes is a chronic metabolic disease characterized by high blood glucose level. Untreated or uncontrolled diabetes can cause a range of complications, including acute ones like diabetic ketoacidosis and chronic ones such as cardiovascular diseases and diabetic nephropathy. With the rapid economic developments and changes in lifestyle, China has become the country with the most diabetes patients in the world: the prevalence of diabetes in Chinese adults is about 11.2% and still increasing [1]. The medical expenses from diabetes without complications already account for 8.5% of national health expenditure in China [2]. As a

B. Qin et al. (Eds.): CCKS 2021, CCIS 1466, pp. 308–314, 2021.
https://doi.org/10.1007/978-981-16-6471-7_26

result, diabetes is a serious public health problem in the realization of "Healthy China 2030" that requires interdisciplinary innovations to solve.

Knowledge Graph (KG) has been proven effective in modeling structured information and conceptual knowledge, especially in the medical domain [3]. Medical knowledge graph is attracting attention from both academic and health-care industries due to its power in intelligent healthcare applications, such as clinical decision support systems (CDSSs) for diagnosis and treatment [4,5], self-diagnosis utilities to assist patient evaluating health conditions based on symptoms [6,7]. High-quality entity and relation corpus is crucial for construct-ing knowledge base, however, there is no dataset dedicated to the diabetes dis-ease at the moment. To address this issue, we introduce DiaKG, a high-quality Chinese dataset for Diabetes knowledge graph construction.

The contributions of this work are as follows:

1. To the best of our knowledge, this is the first diabetes dataset for medical knowledge graph construction at home and abroad.
2. In addition to the medical experts, we also introduce AI experts to participate in the annotation process to provide data insight, which improves the usability of DiaKG and finally benefits the end-to-end model performance.

We hope the release of this corpus can help researchers develop knowledge bases for clinical diagnosis, drug recommendation, and auxiliary diagnostics to further explore the mysteries of diabetes. The datasets are publicly available at https:// tianchi.aliyun.com/dataset/dataDetail?dataId=88836

2 DiaKG Construction

2.1 Data Resource

The dataset is derived from 41 diabetes guidelines and consensus, which are from authoritative Chinese journals covering the most extensive fields of research con-tent and hotspot in recent years, including clinical research, drug usage, clini-cal cases, diagnosis and treatment methods, etc. Hence it is a quality-assured resource for constructing a diabetes knowledge base.

2.2 Annotation Guide

Two seasoned endocrinologists designed the annotation guide. The guide focuses on entities and relations since these two types are the fundamental elements of a knowledge graph.

Entity. 18 types of entities are defined (Table 1). Nested entities are allowed; for example, '2型糖尿病' is a 'Disease' entity, and '2型' is a 'Class' one. Enti-ties in DiaKG has two characteristics that stand out: 1. Entities may attribute to different types according to the contextual content; for example, '糖尿病' in sentence '糖尿病患者需控制饮食' is a 'Disease' type, while in the sentence '糖尿病所致肾损伤占1/3' serves as a 'Reason' type; 2. Some entity types are of long spans, like 'Pathogenesis' type is usually consisted of a sentence.

Table 1. List of entities

entity name	example	# num	avg length
疾病(Disease)	运动对1型糖尿病微血管病变的预后无改善作用	5,743	7.3
疾病分期分型(Class)	心功能Ⅲ-Ⅳ级、终末期肾病	1,262	4.3
病因(Reason)	若体重增加，可能加重胰岛素抵抗	175	7.3
发病机制(Pathogenesis)	多数患者的β细胞完全破坏	202	10.3
临床表现(Symptom)	已发生明确的足趾、足掌坏疽创面	479	5.8
检查方法(Test)	进行混合餐耐量试验(MMTT)	489	6.1
检查指标(Test_Items)	测量指血(毛细血管血)血糖	2,718	7.7
检查指标值(Test_Value)	血糖< 3.3mmol/L	1,356	9.5
药物名称(Drug)	包括COX-2抑制剂	4,782	7.8
用药频率(Frequency)	按照0.5mg，1～3次／d	156	4.7
用药剂量(Amount)	可根据0.3～0.5单位/千克体重来估算	301	6.7
用药方法(Method)	短效胰岛素一般在餐前15～30min皮下注射	399	6.1
非药治疗(Treatment)	认知-行为及心理干预是调整患者的生活环境	756	8.0
手术(Operation)	进行胰岛细胞移植手术来改善胰岛情况	133	9.0
不良反应(ADE)	贝特类可使胆结石的发生率升高	874	5.1
部位(Anatomy)	微血管和大血管并发症等方面的证据	1,876	3.1
程度(Level)	对于中到重度肾功能不全患者需减少剂量	280	2.9
持续时间(Duration)	预防治疗维持3～6个月	69	3.7

Relation. Relations are centered on 'Disease' and 'Drug' types, where a total of 15 relations are defined (Table 2). Relations are annotated on the paragraph level, so entities from different sentences may form a relation, which has raised the difficulty for the relation extraction task. Head entity and tail entity existing in the same sentence only account for 43.4% in DiaKG.

2.3   The Annotation Process

The annotated process is shown in Fig. 1. The process can be divided into two steps:

OCR Process. The PDF files are transformed to plain text format via the OCR tool[1], where non-text data like figures and tables are manually removed. Additionally 2 annotators manually check the OCR results character by character to avoid misrecognitions, for example, 'β细胞' may be recognized as 'B细胞'.

Annotation Process. 6 M.D. candidates were employed and were trained thoroughly by our medical experts to have a comprehensive understanding of the annotation task. During the **trial annotation** step, we creatively invited 2 AI experts to label the data simultaneously, based on the assumption that AI experts could provide data insight from the model's perspective. For example, medical experts are inclined to label

---

[1] https://duguang.aliyun.com/.

Table 2. List of relations

relation	example	# num
TestItems_Disease	血浆酮体增加或酮血症倾向低于正常人	1,171
Treatment_Disease	积极进行糖尿病防治知识的宣教，增加运动	354
Class_Disease	分级I-II级的充血性心力衰竭的患者	854
Anatomy_Disease	慢性开发症如各种神经病变、视网膜病变等	195
Drug_Disease	二甲双胍可有效改善糖尿病的IR	1,315
Reason_Disease	慢性梗阻可引起肾积水和肾实质萎缩	164
Symptom_Disease	对糖尿病足溃疡及...更好地体现了创面感染的情况	283
Operation_Disease	接受糖尿病外科手术患者...对接受减重代谢手术的病人	37
Test_Disease	5项检查(...温度觉)等方法半定量评估患者的神经病变程度	271
Pathogenesis_Disease	二甲双胍可改善IR...更全面针对T2DM的生理缺陷的特点	130
ADE_Drug	正确使用磺脲类药物...，轻、中度低血糖发生率为...	693
Amount_Drug	二甲双胍(1000mg/d)起始治疗	195
Method_Drug	短效胰岛素一般在餐前15~30min皮下注射	185
Frequency_Drug	每日1次基础胰岛素或...作为胰岛素起始治疗方案	103
Duration_Drug	持续静脉泵注胰岛素有利于减少血糖波动	61

'成年型糖尿病(maturity-onset diabetes of the young，MODY)' as a whole entity, while AI experts regard '成年型糖尿病', 'maturity-onset diabetes of the young' and 'MODY' as three separate entities are more model-friendly. Feedback from AI experts and the annotators were sent back to the medical expert to refine the annotation guideline iteratively. The **formal annotation** step started by the 6 M.D. candidates and 1 medical experts would give timely help when needed. **The Quility Control (QC)** step was conducted by the medical experts to guarantee the data quality, and common annotation problems were corrected in a batch mode. The final quality is evaluated by the other medical expert via random sampling of 300 records. The accuracy rates of entity and relation are 90.4% and 96.5%, respectively, demonstrating the high-quality of DiaKG. The examined dataset contains 22,050 entities and 6,890 relations, which is empirically adequate for a specified disease.

2.4 Data Statistic

Detailed statistical information for DiaKG is shown in Table 1 and Table 2.

3 Experiments

We conduct Named Entity Recognition (NER) and Relation Extraction (RE) experiments to evaluate DiaKG. The codebase is public on github[2], and the implementation details are also illustrated on the github repository.

---

[2] https://github.com/changdejie/diaKG-code.

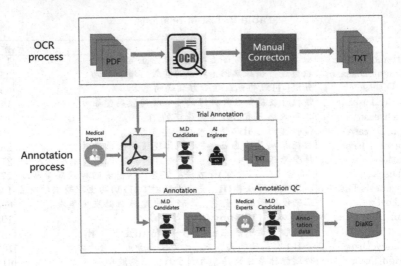

Fig. 1. The annotated process of the diabetes dataset.

3.1 Named Entity Recognition (NER)

We only report results from X Li et al. (2019) [8] since it is the SOTA model for NER with nested settings at the time of this writting.

3.2 Relation Extraction (RE)

The RE task is defined as giving the head entity and the tail entity, to classify the relation type. Due to the simplified setting, we report results from bi-directional GRU-attention [9] in this paper.

4 Analysis

The experimental results are shown in Table 3 and Table 4. We report the total result, plus the top 2 and last 3 types' results for each task to analyze DiaKG.

The **overall** macro-average scores for the two tasks are 83.3% and 83.6%, respectively, which are satisfying considering the multifarious types we define, also demonstrating DiaKG's high quality. For the **NER task**, the results of 'Disease' and 'Drug' types are as expected because these two types exist frequently among the documents, thus leading to a higher score. The average entity length for 'Pathogenesis' type is 10.3, showing that the SOTA MRC-Bert model still can not handle the long spans perfectly; We analyzed errors of the 'Symptom' and 'Reason' types and found that the model is prone to classify entities as other types, mainly contributing to the characteristic that entity may be of different types due to the contextual content. For the **RE task**, the case study shows that entities with long distance are difficult to classify. For example, entities with 'Drug_Diesease' type usually exist in the same sub-sentence, whereas the

Table 3. Selected NER results

Entity	Precision	Recall	F1
Total	0.814	0.853	0.833
Drug	0.881	0.902	0.892
Disease	0.794	0.91	0.848
Pathogenesis	0.595	0.667	0.629
Symptom	0.535	0.535	0.535
Reason	0.333	0.3	0.316

Table 4. Selected RE results

Relation	Precision	Recall	F1
Total	0.839	0.837	0.836
Class_Disease	0.968	0.874	0.918
ADE_Drug	0.892	0.892	0.892
Test_Disease	0.648	0.636	0.642
Pathogenesis_Disease	0.486	0.692	0.571
Operation_Disease	0.6	0.231	0.333

ones with 'Reason_Disease' type are usualy located in different sub-sentences, sometimes even in different sentences. The above experimental results demonstrate that DiaKG is challenging for most current models and it is encouraged to employ more powerful models on this dataset.

5   Conclusion and Future Work

In this paper, we introduce DiaKG, a specified dataset dedicated to the diabetes disease. Through a carefully designed annotation process, we have obtained a high-quality dataset. The experiment results prove the practicability of DiaKG as well as the challenges for the most recent typical methods. We hope the release of this dataset can advance the construction of diabetes knowledge graphs and facilitate AI-based applications. We will further explore the potentials of this corpus and provide more challenging tasks like QA tasks.

Acknowledgments. We want to express gratitude to the anonymous reviewers for their hard work and kind comments. We also thank Tianchi Platform to host DiaKG.

References

1. Li, Y., Teng, D., Shi, X., et al.: Prevalence of diabetes recorded in mainland China using 2018 diagnostic criteria from the American Diabetes Association: national cross sectional study. BMJ **369** (2020)
2. Luo, Z., Fabre, G., Rodwin, V.G.: Meeting the Challenge of Diabetes in China. Int. J. Health Policy Manage. **9**(2) (2020)
3. Nickel, M., et al.: A review of relational machine learning for knowledge graphs. Proc. IEEE **104**(1), 11–33 (2015)
4. Bisson, L.J., Komm, J.T., Bernas, G.A., et al.: Accuracy of a computer-based diagnostic program for ambulatory patients with knee pain. Am. J. Sports Med. **42**(10), 2371–6 (2014)
5. Wang, M., Liu, M., Liu, J., et al.: Safe medicine recommendation via medical knowledge graph embedding. arXiv preprint arXiv:1710.05980.2017
6. Tang, H., Ng, J.H.K.: Googling for a diagnosis use of Google as a diagnostic aid: internet based study. BMJ **333** (2006)
7. Gann, B.: Giving patients choice and control: health informatics on the patient journey. Yearb Med. Inform. **21**(01), 70–73 (2012)

8. Li, X., Feng, J., Meng, Y., et al.: A unified MRC framework for named entity recognition (2019)
9. Peng, Z., Wei, S., Tian, J., et al.: Attention-based bidirectional long short-term memory networks for relation classification. In: Proceedings of the 54th Annual Meeting of the Association for Computational Linguistics (Volume 2: Short Papers) (2016)

Weibo-MEL, Wikidata-MEL and Richpedia-MEL: Multimodal Entity Linking Benchmark Datasets

Xingchen Zhou, Peng Wang$^{(\boxtimes)}$, Guozheng Li, Jiafeng Xie, and Jiangheng Wu

School of Computer Science and Engineering, Southeast University, Nanjing, China
{xczhou_2021,pwang}@seu.edu.cn

Abstract. Multimodal entity linking (MEL) aims to utilize multimodal information to map mentions to corresponding entities defined in knowledge bases. In this paper, we release three MEL datasets: Weibo-MEL, Wikidata-MEL and Richpedia-MEL, containing 25,602, 18,880 and 17,806 samples from social media, encyclopedia and multimodal knowledge graphs respectively. A MEL dataset construction approach is proposed, including five stages: multimodal information extraction, mention extraction, entity extraction, triple construction and dataset construction. Experiment results demonstrate the usability of the datasets and the distinguishability between baseline models. All resources are available at https://github.com/seukgcode/MELBench.

Keywords: Entity linking · Multimodal · Knowledge graph

1 Introduction

Entity linking (EL) is the task of mapping mentions to the corresponding entities in the knowledge bases, which plays a pivotal role in tasks such as semantic retrieval, recommendation system and question answering. Existing approaches mainly address the problem via textual information. However, on the one hand, it is still challenging that linking mentions from short and coarse text. On the other hand, in real-world data, such as social media, encyclopedia, and multimodal knowledge graphs, entities are often described with both textual and visual information. To this end, it is necessary to combine multimodal information to address the EL task, which is called multimodal entity linking (MEL).

To the best of our knowledge, there is still a lack of public MEL datasets. Moon et al. [1] released a dataset for multimodal named entity disambiguation that could be used in MEL, but it is not accessible. Adjali et al. [2] constructed a MEL dataset from Twitter. However, the dataset cannot be reproduced due to a large amount of Twitter contents are expired.

The work is supported by All-Army Common Information System Equipment Pre-Research Project (No. 31514020501, No. 31514020503).

Therefore, to facilitate the MEL research, we release three new MEL datasets: Weibo-MEL, Wikidata-MEL and Richpedia-MEL, for three real-world scenarios: social media, encyclopedia, and multimodal knowledge graphs. Weibo-MEL has more than 25K multimodal samples collected from Weibo and CN-DBpedia [3]. Wikidata-MEL contains more than 18K multimodal samples based on Wikidata and Wikipedia. Richpedia-MEL is constructed based on Richpedia [4] and Wikipedia, contains more than 17K multimodal samples. In addition, we implement six MEL baseline models on the datasets to verify the usability of our datasets and the distinguishability between baseline models.

2　Datasets Construction

2.1　Construction Approach Overview

To construct large-scale MEL datasets, we propose a MEL dataset construction approach, including five stages. In **Multimodal Information Extraction**, we select multimodal data sources and extract textual and visual information. In **Mention Extraction**, we extract mentions from textual information and keep the mentions which corresponding entities may exist. In **Entity Extraction**, we query the knowledge bases with the filtered mentions, gather the entity lists, and save the correct entities. In **Triple Construction**, we merge the corresponding mentions and entities into mention-entity (M-E) pairs, and combine them into triples with textual and visual information. Then, we keep the correct triples as the samples of the MEL dataset. Finally, in **Dataset Construction** stage, we partition the dataset into training set (70%), validation set (10%) and testing set (20%). The overview of the approach is illustrated in Fig. 1.

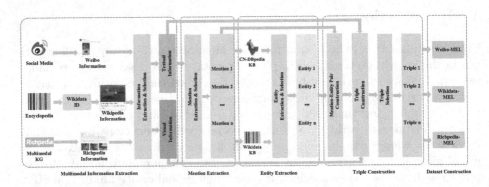

Fig. 1. Overview of the MEL dataset construction approach.

2.2　Weibo-MEL Dataset Construction

Stage 1: Multimodal Information Extraction. We choose Weibo as the data source, randomly select 1500 users and collect their Weibo contents between

January 1, 2021 and April 1, 2021. In this stage, 89,253 original contents are collected and 25,602 contents are preserved.

Stage 2: Mention Extraction. We utilize TexSmart NER model [5] for mention extraction. 64,273 mentions are gathered and 35,716 mentions are saved.

Stage 3: Entity Extraction. we choose CN-DBpedia as the knowledge base and input the mentions to the mention-to-entity API[1] to get the corresponding entities. Finally, 99,290 entities are obtained and 34,204 entities are selected.

Stage 4: Triple Construction. In this stage, 34,204 M-E pairs and 27,532 samples are combined. Finally, we save 31,516 M-E pairs and 25,602 samples of the Weibo-MEL dataset.

Stage 5: Dataset Construction. After division, the training set, validation set and testing set contain 17,921, 2,560 and 5,121 samples, correspondingly.

2.3 Wikidata-MEL Dataset Construction

Stage 1: Multimodal Information Extraction. We choose Wikidata and Wikipedia as data sources. Wikidata is a knowledge base that provides a large amount of structural knowledge. However, Wikidata lacks descriptions of entities. Therefore, we gather entities with Wikidata and multimodal descriptive information using Wikipedia. We randomly generate 30,000 Wikidata IDs and keep the corresponding information through the MediaWiki API[2]. As a consequence, we collect 27,758 items and preserve 25,256 items.

Stage 2: Mention Extraction. To label mentions, we propose a mention-labeling algorithm, including three steps: **Accurate matching** uses the longest common subsequence algorithm and labels the common part of the textual information and the entity names corresponding to the Wikidata ID as mentions; **Fuzzy matching** is to analyze the common prefix and normalized edit distance between textual information and entity names, and the strings which the common prefix lengths are more than zero and the normalized edit distances are less than the threshold are marked as mentions; **Abbreviation matching** is to find the abbreviations of entity names and mark them as mentions. Finally, 26,221 mentions are identified and 23,183 mentions are preserved.

Stage 3: Entity Extraction. We select Wikidata as the knowledge base. In the **Mention Extraction** stage, we preserve the entities corresponding to the identified mentions. After manual selection, we store 23,048 correct entities.

Stage 4: Triple Construction. We combine 23,048 M-E pairs and 19,102 samples, and after manual selection, we save 22,534 M-E pairs and 18,880 samples of the Wikidata-MEL dataset.

Stage 5: Dataset Construction. Finally, the training, validation and testing set of Wikidata-MEL contain 13,216, 1,888 and 3,776 samples, respectively.

---

[1] http://kw.fudan.edu.cn/apis/cndbpedia.
[2] https://www.mediawiki.org/wiki/API.

2.4   Richpedia-MEL Dataset Construction

Stage 1: Multimodal Information Extraction. We choose Richpedia, a multimodal knowledge graph containing a wealth of textual and visual descriptions, as the data source. From Richpedia, We randomly select 20,000 items containing textual and visual information.

Stage 2: Mention Extraction. TexSmart NER model is also used for extracting mentions from Richpedia items. We get 53,246 mentions and, after manual selection, save 23,642 mentions.

Stage 3: Entity Extraction. Wikidata is chosen as the knowledge base. We also use MediaWiki API to gather the entities corresponding to the mentions. Subsequently, we preserve 21,206 entities by manual selection.

Stage 4: Triple Construction. After merging, we obtain 21,206 M-E pairs and 18,153 samples and eventually select 20,752 M-E pairs and 17,806 samples of the Richpedia-MEL dataset.

Stage 5: Dataset Construction. The training, validation and testing sets of Richpedia-MEL contain 12,464, 1,780 and 3,562 samples respectively.

3   Datasets Statistics and Analysis

As described in Sect. 2, we construct three new MEL datasets: Weibo-MEL, Wikidata-MEL and Richpedia-MEL datasets, containing 25,602, 18,880 and 17,806 samples corresponding to 31,516, 22,534 and 20,752 mention-entity (M-E) pairs. The statistics of the datasets are summarized in Table 1, which contain the number of samples, M-E pairs, average text length, and average mention number. In addition, we analyze the text length and mention number distribution of the datasets as Fig. 2 shown.

Table 1. Statistics of the datasets.

Dataset	Samples	M-E Pairs	Text length (avg.)	Mention number (avg.)
Weibo-MEL	25,602	31,516	42.6	1.23
Wikidata-MEL	18,880	22,534	8.4	1.19
Richpedia-MEL	17,806	20,752	13.6	1.17

Text Length. As shown in Table 1, the average text length of Weibo-MEL, Wikidata-MEL and Richpedia-MEL are 42.6, 8.4 and 13.6. Meanwhile, it can be observed from Fig. 2 (a) that the text lengths of Weibo-MEL are composed of 40–60 words. And from Fig. 2 (b), the text lengths of Wikidata-MEL and Richpedia-MEL mainly in 4–16 words. In summary, the text of Weibo-MEL is longer, which provides richer textual information than the text of Wikidata-MEL and Richpedia-MEL.

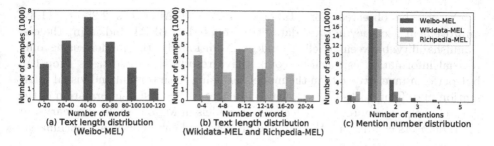

Fig. 2. Text length and mention number distribution of the datasets.

Mention Number. From Table 1, it could be found that the average mention number of Weibo-MEL, Wikidata-MEL and Richpedia-MEL are 1.23, 1.19 and 1.17. Additionally, Fig. 2 (c) depicts the distribution of mention number in the textual information. It can be seen that most samples contain a single mention and few samples contain multiple mentions. Meanwhile, there are some samples that do not contain any mentions.

4 Benchmark Results

In order to verify the usability of the datasets, we evaluate six MEL models on our datasets: **ARNN** [6] utilizes the Attention-RNN to predict associations with candidate entity textual features. **BERT** [7] selects the transformer layers to encode tokens. **JMEL** [8] utilizes fully connected layers to project the textual and visual features into an implicit joint space. **DZMNED** [1] utilizes a concatenated multimodal attention mechanism to fuse textual and visual features. **DZMNED-BERT** replaces the Glove pre-training model with BERT. **HieCoATT-Alter** [9] uses alternating co-attention and three textual levels to calculate co-attention maps. Table 2 shows the Top-1 and Top-10 accuracy results of the models in our datasets.

Table 2. Results of the baseline models at Top-1 and Top-10 accuracies (%). (T: textual modal, V: visual modal).

Modalities	Models	Weibo-MEL		Wikidata-MEL		Richpedia-MEL	
		Top-1	Top-10	Top-1	Top-10	Top-1	Top-10
T	ARNN	41.3	53.4	32.0	56.6	31.2	45.9
T	BERT	42.4	54.0	31.7	57.8	31.6	47.6
T + V	JMEL	41.8	53.9	31.3	57.9	29.6	46.6
T + V	DZMNED	10.6	54.3	30.0	56.0	29.5	45.8
T + V	DZMNED-BERT	46.3	55.5	34.7	58.1	32.4	48.2
T + V	HieCoATT-Alter	**47.2**	**56.2**	**40.5**	**69.6**	**37.2**	**54.2**

As can be observed from Table 2, HieCoATT-Alter has 2.2% to 11.8% improvement accuracy in three datasets compared to BERT, indicating the distinguishability between baseline models. Additionally, the models using only textual information can achieve competitive results in simple tasks (Top-10), but perform much worse than the models utilizing both textual and visual information in difficult tasks (Top-1), which illustrates that visual information of the datasets helps to improve the effect of the models in MEL task. In summary, our MEL datasets can distinct the baseline models, and can be used as benchmarks.

5   Conclusion

To compensate for the lack of public MEL datasets, we release Weibo-MEL, Wikidata-MEL and Richpedia-MEL, three MEL large-scale datasets involving social media, encyclopedia and multimodal knowledge graphs. The datasets can be reproduced using the dataset construction approach proposed in this paper. Moreover, we verify the usability of the datasets and the distinguishability between baseline models, so that the datasets can be used as benchmarks of MEL.

References

1. Moon, S., Neves, L., Carvalho, V.: Multimodal named entity disambiguation for noisy social media posts. In: Proceedings of the 56th ACL, pp. 2000–2008 (2018)
2. Adjali, O., Besançon, R., Ferret, O., et al.: Multimodal entity linking for tweets. Adv. Inf. Retrieval **12035**, 463 (2020)
3. Xu, B., Xu, Y., Liang, J., et al.: Cn-dbpedia: a never-ending Chinese knowledge extraction system. In: Proceedings of the 30th IEA-AIE, pp. 428–438 (2017)
4. Wang, M., Wang, H., Qi, G., et al.: Richpedia: a large-scale, comprehensive multimodal knowledge graph. Big Data Res. 130–145 (2020)
5. Zhang, H., Liu, L., Jiang, H., et al.: Texsmart: a text understanding system for fine-grained ner and enhanced semantic analysis. arXiv:2012.15639 (2020)
6. Eshel, Y., Cohen, N., Radinsky, K., et al.: Named entity disambiguation for noisy text. In: Proceedings of the 21st CoNLL, pp. 58–68 (2017)
7. Devlin, J., Chang, M.W., Lee, K., et al.: Bert: pre-training of deep bidirectional transformers for language understanding. In: Proceedings of the 17th NAACL-HLT, pp. 4171–4186 (2019)
8. Adjali, O., Besançon, R., Ferret, O., et al.: Multimodal entity linking for tweets. In: Proceedings of the 42nd ECIR, pp. 463–478 (2020)
9. Lu, J., Yang, J., Batra, D., et al.: Hierarchical question-image co-attention for visual question answering. In: Proceedings of the 30th NIPS, pp. 289–297 (2016)

MAKG: A Mobile Application Knowledge Graph for the Research of Cybersecurity

Heng Zhou[1,3], Weizhuo Li[2,3,4]($\boxtimes$), Buye Zhang[1,3], Qiu Ji[2], Yiming Tan[1,3], and Chongning Na[5]

[1] School of Cyber Science and Engineering, Southeast University, Nanjing, China
{zhouheng2020,zhangbuye,tt_yymm}@seu.edu.cn
[2] School of Modern Posts, Nanjing University of Posts and Telecommunications, Nanjing, China
qiuji@njupt.edu.cn
[3] Key Laboratory of Computer Network and Information Integration (Southeast University), Ministry of Education, Nanjing, China
[4] State Key Laboratory for Novel Software Technology, Nanjing University, Nanjing, China
liweizhuo@amss.ac.cn
[5] Zhejiang Lab, HangZhou, China
na@zhejianglab.com

Abstract. Large-scale datasets for mobile applications (a.k.a. "app") such as AndroZoo++ and AndroVault have become powerful assets for malware detection and channel monitoring. However, these datasets focus on the scale of apps while most apps in them remain isolated and cannot easily be referenced and linked from other apps. To fill these gaps, in this paper, we present a mobile application knowledge graph, namely MAKG, which aims to collect the apps from various resources. We design a lightweight ontology of apps. It can bring a well-defined schema of collected apps so that these apps could share more linkage with each other. Moreover, we evaluate the algorithms of information extraction and knowledge alignment during the process of construction, and select the competent models to enrich the structured triples in MAKG. Finally, we list three use-cases about MÁKG that are helpful to provide better services for security analysts and users.

Keywords: Mobile applications · Knowledge graph · Knowledge alignment

Resource type: Dataset
OpenKG Repository: http://www.openkg.cn/dataset/makg
Github Repository: https://github.com/Everglow123/MAKG

1 Introduction

With the popularity of smart phones and mobile devices, the number of mobile applications (a.k.a. "app") has been growing rapidly, which provides great convenience to users for online shopping, education, entertainment, financial management etc. [1]. According to a recent report[1], as of January 2021, there were over 2.5 and 3.4 million apps available on Google Play and App Store, respectively, and global downloads of mobile applications have exceeded 218 billion. With large amounts of apps, it also provided the chances for cybercriminals such as thieving private data, propagating false news and pornography, online scam. According to the statistic of National Internet Emergency Center (NIEC) in China, since the first half of 2019, there have been more than 15,000 false loan apps by mobile Internet as a carrier that cause substantial damage to numerous users. Therefore, it is essential to build a mobile application knowledge graph for cybersecurity and provide better services (e.g., semantic retrieval, sensitive detection) for security analysts and users.

There exist several knowledge bases of mobile applications that are constructed in the past decade, and gain a remarkable attraction from researchers of both academia and industry [2]. Most of them are ongoing efforts to gather millions of executable apps from diverse sources, and define dozens of types of apps to share with the research community for relevant research works [3,4]. Meanwhile, other constructed datasets focus on the specific services in cybersecurity such as malware detection [5], channel monitoring [6], vulnerability assessment [7].

Although above knowledge bases contain various apps, they suffer from two limitations. Firstly, most apps in these datasets remain isolated among application markets, which cannot easily be referenced and linked from other apps [2]. Hence, these apps without comprehensive properties and linkages may raise challenges for security analysts and users who need to judge the security of current apps. Secondly, apps in these knowledge bases lack well-defined schema, which may limit resources to share and reuse for above services in cybersecurity [7].

To fill the above gaps, we dedicate to a continuous effort to collect apps and construct **m**obile **a**pplication **k**nowledge **g**raph, namely MAKG, for the research of cybersecurity. Precisely, we design a lightweight ontology that brings a well-defined schema of collected apps including 26 basic classes, 11 relations and 45 properties. It not only can make apps share more linkage with each other, but also bring better services such as resources integration, recommendation. Moreover, we evaluate the algorithms of information extraction and knowledge alignment during the process of construction, and select the competent models to enrich the structured triples for MAKG. Finally, we present three use-cases based on MAKG, which is helpful to provide better services for security analysts and users.

---

[1] https://www.appannie.com/cn/insights/market-data/mobile-2021-new-records-beckon/.

2 The Construction of Mobile Application Knowledge Graph

Figure 1 illustrates the workflow of MAKG construction pipeline, which mainly includes five phases: ontology definition (Sect. 2.1), app crawling (Sect. 2.2), knowledge extraction (Sect. 2.3), knowledge alignment (Sect. 2.4), and knowledge storage (Sect. 2.5). With the help of MAKG, the security analysts and users can employ better services such as semantic retrieval, recommendation, sensitive detection and so on.

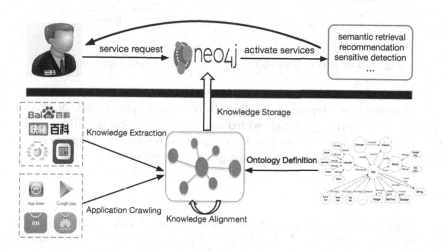

Fig. 1. The workflow of MAKG construction.

2.1 Ontology Definition

To model a well-defined schema of apps for the research of cybersecurity, we discuss with analysts who worked on the China Academy of Industrial Internet, and discover that the vast majority of conceptualizations (e.g., *functionPoint*, *interactionMode*, *availableState*) described for apps (e.g., Facebook, Twitter) are not asserted online. Therefore, we choose appropriate terms based on the survey of existing conceptualizations in view of cybersecurity, and define a set of relations and properties by protégé[2] to cover the related features of mobile applications. On the other hand, we collect the labels from different encyclopedias, and extend the synonyms for relations and properties manually, which is helpful for the models of knowledge extraction and knowledge alignment to enrich structured triples for our knowledge graph.

Figure 2 shows an overview of our lightweight ontology, where red edges and blue ones represent *subclassof* and *rdfs:type* relations, respectively, which are two basic relations. The green ones represent the relations, and the violet ones

---

[2] https://protege.stanford.edu/.

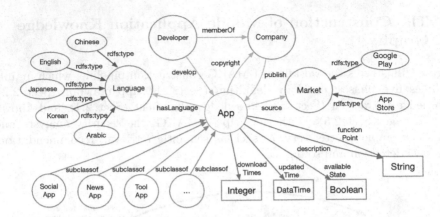

Fig. 2. The overview of lightweight ontology (Color figure online)

represent properties. Overall, we define 26 basic concepts, 11 relations and 45 properties in the ontology. Benefited from a well-defined schema, it not only can make apps present more comprehensive properties to security analysts and users, but also can generate more shared linkages among apps.

2.2 App Crawling

With the help of scrapy framework, we crawl the descriptive information of apps published from Google Play[3], App Store[4], Xiaomi App Store[5] and Huawei App Market[6]. We simply give an introduction for these four markets.

- **Google Play.** Its former is called Android Market, which is the official distribution storefront for Android applications and other digital media. It is available on mobile devices and tablets that run the Android operating system (OS).
- **App Store.** It is a digital distribution platform for mobile applications on iOS and iPadOS operating systems that are developed and maintained by Apple Inc.
- **Xiaomi App Store.** It is an android app recommendation software in the ROM of MIUI Android system of Xiaomi technology, which is one of the largest platforms that can discover entertainment Android apps and games for users.
- **Huawei App Market.** It is the official app distribution platform developed by Chinese technology company Huawei, which is one of the largest platforms that provide Android apps run in EMUI operating system.

---

[3] https://play.google.com/store/apps.
[4] https://www.apple.com/app-store.
[5] https://app.mi.com/game.
[6] https://appgallery.huawei.com.

Table 1 lists the statistics of crawled apps from application markets, in which more than 347 thousand apps are collected and they are transformed into about 7 million triples. Notice that the number of apps in Google Play is larger than others because the language of their names and descriptions is multilingual.

Table 1. The statistics of crawled apps from application markets

Application Market	Google Play	App Store	Xiaomi App Store	Huawei App Market
♯Apps	273802	3137	27694	43287
♯Relations	4	3	3	5
♯Properties	15	5	12	14
♯Entities	382358	5724	47478	70954
♯Triples	5210417	25088	384648	1384854

2.3 Knowledge Extraction

Crawling description information of apps from application markets is the most direct way to build KG. However, existing labels in the application markets are not adequate to cover the value of properties in our designed ontology, which impedes the discovery of the shared linkage among apps. Therefore, we try to collect related web pages from encyclopedias (e.g., Baidu Baike[7], Toutiao Baike[8]), textual descriptions and news related to apps that are published on the website so as to fill the lacked value of these properties.

We mainly consider the following strategies to parse web pages and textual descriptions so as to obtain structured triples of apps and enrich their lacked value.

- **Infobox-Based Ccompletion.** It is a common skill that complements the lacked value of properties. We employ the string matching method (e.g., Levenshtein measure) to calculate the similarity of labels in Infobox and properties in the designed ontology, and further construct their correspondences such as *Market*) in ontology *Platform*.
- **Named Entity Recognition.** For the defined concepts (e.g., *Developer* and *Company*) in our designed ontology, we evaluate several methods named entity recognition and select NcrfPP [8] to capture the related value of apps, which is designed for quick implementation of different neural sequence labeling models with a CRF inference layer and obtain the best performance in our evaluation.
- **Relation Extraction.** For several important relations (e.g., *Headquarter*) and properties (e.g., *publishedTime*) in our designed ontology, we employ promising relation extraction models from two platforms (i.e., OpenNRE[9] and DeepKE[10]), and select the best one based on our dataset to obtain the

---

[7] https://baike.baidu.com.
[8] https://www.baike.com.
[9] https://github.com/thunlp/OpenNRE.
[10] https://github.com/zjunlp/deepke.

structured descriptive information of apps. As several labeled relations in the corpus are not enough, we also try to utilize few-shot relation extraction to achieve our goal.

Finally, we retrieve 2493 and 2503 apps from Baidu Baike and Toutiao Baike, and add nearly 69092 structured triples into MAKG. Relatively, with the help of named entity recognition and relation extraction, we obtain 29651 triples based on our selected basic relations from textual descriptions and encyclopedias for MAKG.

2.4 Knowledge Alignment

We notice that most apps are usually published in various application markets (e.g., Android ones), but their descriptive labels and value from application markets are heterogeneous. Moreover, there exist differences in the perspectives of apps' relations and properties from different application markets. Therefore, it is essential to find the correspondences among apps and entities from different markets, which can share and reuse the information to provide better services for security analysts and users. To achieve this goal, we try to employ the following approaches for knowledge alignment.

- **Rule miner method** [9]. It is a semi-supervised learning algorithm to iteratively refine matching rules and discover new matches of high confidence based on these rules.
- **Knowledge graph embedding-based method** [10]. It encodes entities and relations of knowledge graph in a continuous embedding space and measures entity similarities based on the learned embeddings.

Based on above models of knowledge alignment, we finally obtain the alignments among application markets[11] and then manually verify them. The concrete statistic of correspondences is listed in Table 2.

Table 2. The statistic of correspondences for knowledge alignment

Market	♯ APP	♯ Entities	♯ Triples	♯ Equivalent entities in Huawei App Market	♯ Equivalent entities in Xiaomi App Store	♯ Equivalent entities in Google Play
Huawei App Market	43287	70954	1384854	–	27397	1526
Xiaomi App Store	27694	47478	384648	27397	–	979
Google Play	273802	382358	5210417	1526	979	–

---

[11] As the number of App Store is fewer than other markets, we do not list the results of it.

2.5   Knowledge Storage

After we utilized the crawled data to instantiate the properties based on our designed ontology, and employed knowledge extraction and knowledge alignment to enrich the structured triples to our knowledge graph, we transform them into structured triples $\{(h, r, t)\}$ with specified URL by Jena[12]. For knowledge storage, we employ Neo4j[13] to store the transformed triples, which is one of the efficient graph bases for storing the RDF triples and provide one convenience query language called Cypher. To keep MAKG in sync with the evolving apps, we try to periodically update the descriptive information of apps by crawling above sources and record updated logs.

3   Use-Cases

We list three use-cases based on MAKG (shown in Fig. 1) about cybersecurity. Firstly, it can achieve semantic retrieval. For example, if users query one app, MAKG can present its comprehensive structured information to them. Moreover, we can further employ entity linking techniques [11] to link apps to the news of their appearing textual descriptions. Benefited from above cases, users can fully understand the information of apps and avoid downloading some invalid apps. Secondly, it can help security analysts to detect some sensitive apps, which own more conditions or plausibility than normal apps that become the hotbeds for related cybercriminals. With the help of comprehensive relations and properties, analysts can define some prior rules or employ more promising algorithms to evaluate the sensitivity and rank them. It can lower the risk of some sensitive apps in advance. Thirdly, it can recommend some similar apps for users and security analysts when they request some services. Similarly, if suitable algorithms [12] are utilized for the recommendation, it can reduce the potential risks and maintain the security of the mobile internet.

4   Conclusion

In this paper, we presented a mobile application knowledge graph, namely MAKG. Our work is to collect apps from application markets and external resources. Lots of structured triples are obtained by information extraction techniques according to our designed ontology for enriching MAKG. We further employ knowledge alignment models to generate correspondences among apps from different markets so as to share and reuse the information of apps. The result is a high-quality mobile application dataset, which provides an open data resource to the researchers from The Semantic Web and CyberSecurity. As future work, our plan is to broaden the apps and enrich their structured information, so that MAKG becomes more comprehensive and covers more topics. Besides, we try to explore promising algorithms for the use-cases serviced for security analysts and users.

---

[12] http://jena.apache.org/.
[13] https://neo4j.com/.

Acknowledgements. This work was partially sponsored by the Natural Science Foundation of China grants (No. 62006125, 61602259), the NUPTSF grant (No. NY220171, NY220176).

References

1. Meng, G., Patrick, M., Xue, Y., Liu, Y., Zhang, J.: Securing android app markets via modeling and predicting malware spread between markets. IEEE Trans. Inf. Forensics Secur. **14**(7), 1944–1959 (2019)
2. Geiger, F.X., Malavolta, I.: Datasets of android applications: a literature review. CoRR, abs/1809.10069 (2018)
3. Li, L., et al.: AndroZoo++: collecting millions of android apps and their metadata for the research community. CoRR, abs/1709.05281 (2017)
4. Meng, G., Xue, Y., Siow, J.K., Su, T., Narayanan, A., Liu, Y.: AndroVault: constructing knowledge graph from millions of android apps for automated analysis. CoRR, abs/1711.07451 (2017)
5. Arp, D., Spreitzenbarth, M., Hubner, M., Gascon, H., Rieck, K.: DREBIN: effective and explainable detection of android malware in your pocket. In: NDSS (2014)
6. Yang, X.-Y., Guo-Ai, X.U.: Construction method on mobile application security ecological chain {J}. J. Softw. **28**(11), 3058–3071 (2017)
7. Kiesling, E., Ekelhart, A., Kurniawan, K., Ekaputra, F.: The SEPSES knowledge graph: an integrated resource for cybersecurity. In: Ghidini, C., et al. (eds.) ISWC 2019. LNCS, vol. 11779, pp. 198–214. Springer, Cham (2019). https://doi.org/10.1007/978-3-030-30796-7_13
8. Yang, J, Zhang, Y.: NCRF++: an open-source neural sequence labeling toolkit. In: ACL, pp. 74–79 (2018)
9. Niu, X., Rong, S., Wang, H., Yu, Y.: An effective rule miner for instance matching in a web of data. In: CIKM, pp. 1085–1094 (2012)
10. Sun, Z., et al.: A benchmarking study of embedding-based entity alignment for knowledge graphs. Proc. VLDB Endow. **13**(11), 2326–2340 (2020)
11. Radhakrishnan, P., Talukdar, P., Varma, V.: Elden: improved entity linking using densified knowledge graphs. In: NAACL, pp. 1844–1853 (2018)
12. Cao, D., et al.: Cross-platform app recommendation by jointly modeling ratings and texts. ACM Trans. Inf. Syst. (TOIS) **35**(4), 1–27 (2017)

Author Index

Printed in the United States
by Baker & Taylor Publisher Services